Certificate Mathematics

Volume one

N. Abbott, B.Sc. (Tech.), A.F.I.M.A., Dip. Math.

based on
National Certificate Mathematics Volume 2
by P. A. Abbott, B.A., and H. Marshall, B.A.

HODDER AND STOUGHTON
LONDON SYDNEY AUCKLAND TORONTO

Paperback edition ISBN 0 340 11520 3
Boards edition ISBN 0 340 11521 1

First published (as *National Certificate Mathematics Volume 2*) 1938
Fifteenth impression 1957
Second edition 1960
Fourth impression 1968
Third edition 1972
Second impression 1975

Printed in Great Britain for
Hodder and Stoughton Educational,
a division of Hodder and Stoughton Ltd.,
St Paul's House, Warwick Lane, London EC4P 4AH
by Richard Clay (The Chaucer Press) Ltd, Bungay, Suffolk

General Editor's Foreword

The Technical College Series covers a wide range of technician and craft courses, and includes books designed to cover subjects in National Certificate and Diploma courses, and City and Guilds Technician and Craft syllabuses. This important sector of technical education has been the subject of very considerable changes over the past few years. The more recent of these have been the result of the establishment of the Training Boards, under the Industrial Training Act. Although the Boards have no direct responsibility for education, their activities in ensuring proper training in industry have had a marked influence on the complementary courses which technical colleges must provide. For example, the introduction of the module system of training for craftsmen by the Engineering Industry Training Board led directly to the City and Guilds 500 series of courses.

The Haslegrave Committee on Technician Courses and Examinations reported late in 1969, and made recommendations for far-reaching administrative changes, which will undoubtedly eventually result in new syllabuses and examination requirements.

It should, perhaps, be emphasised that these changes are being made not for their own sake, but to meet the needs of industry and the young men and women who are seeking to equip themselves for a career in industry. And industry and technology are changing at an unprecedented rate, so that technical education must be more concerned with fundamental principles than with techniques.

Many of the books in the Technical College Series are now standard works, having stood the test of time over a long period of years. Such books are reviewed from time to time and new editions published to keep them up to date, in respect of both new technological developments and changing examination requirements. For instance, these books have had to be rewritten in the metric system, using SI units. To keep pace with the rapid changes taking place both in courses and in technology, new works are constantly being added to the list. The Publishers are fully aware of the part that well-written up-to-date textbooks can play in supplementing teaching, and it is their intention that the Technical College Series shall continue to make a substantial contribution to the development of technical education.

E. G. STERLAND

Preface

The rapid development of computation, with its consequent impact on science and technology, has had the effect of increasing the importance of numerical methods in technical mathematics.

The object of *Certificate Mathematics* is to give a full coverage of the new syllabuses reflecting these changes, whilst retaining an adequate mathematical basis. The role of mathematics as a tool in technical subjects, and the needs of the student working alone have been kept in mind throughout. This has led to the introduction of chapters in Volume 1 involving the laws of probability, iterative methods, statistics, desk calculating machines, rounding errors and the slide rule.

Probability is developed with simple experiments for comparison with theoretical models.

Iteration is applied to the determination of square roots, reciprocals, and the solution of 2×2 diagonally dominant simultaneous linear equations. The iteration formulae are developed intuitively, and delayed division is used in some examples, in case desk calculators are not available.

A typical desk calculator and step by step operations are described in the hope that a student without such a resource will gain some idea of the processes, and that a student with access to a calculator will be able to practise. Topics include the operation of a machine, decimal point markers, overflow, back transfer, use of complements, arithmetic operations, iteration, and the solution of linear equations by pivotal condensation with current sum checks.

Relative errors in arithmetic operations are developed algebraically at this stage, rather than by partial differentiation. Relative errors in the circular functions x^n and e^x are developed by ordinary differentiation, using $\Delta y \simeq (\mathrm{d}y/\mathrm{d}x)\, \Delta x$.

The importance of checks is emphasised in the numerical methods and in the sections on the meaning of the solution of an equation, algebraic identities, and binomial expansions.

Statistics is introduced descriptively. Topics include the use and formation of frequency tables and histograms; discrete and continuous variates; uses of the arithmetic mean, median, and mode; and the use of a false mean in the calculation of the arithmetic mean and standard deviation.

On account of its importance as an analogue device, a separate chapter on the use of the slide rule includes the construction of a simple slide rule, the use of log-log scales, and rough approximation. Surds are considered because of the analogy with algebraic complex number operations, with checking by slide rule.

Elimination is treated because of its importance in technical applications, and also in relation to linear simultaneous equations. From the general solution of 2 × 2 linear simultaneous equations comes the idea that linear equations may have a unique solution, no solution, or infinitely many solutions. Here and elsewhere, the fact that division by 0 is undefined is stressed. The determinant is used as a convenient pattern, and Cramer's Rule to provide a computational algorithm. The matrix algebra of 2 × 2 matrices is introduced up to matrix inversion by the adjoint or by row operations, and the solution of linear simultaneous equations with two unknowns.

To anticipate future requirements concerning power series, algebraic long multiplication and division are revised. It seemed necessary to include the transposition of exponential and logarithmic formulae because of their many applications in differential equations, work, energy, entropy, cooling, coil friction, logarithmic mean temperature difference, alternating currents, etc.

Simple flow charts are used informally to illustrate algorithms for the evaluation of square roots, determination of e, and the procedure for multiplication with the slide rule.

Elimination is used to obtain the form $a^{\log_a N} = N$, from which may be derived the laws of logarithms, computation using logarithms to the base 10, and the direct use of hyperbolic logarithm tables.

Also related to numerical work are the graphical solution of simpler polynomial equations, the remainder and factor theorem, equations involving exponential, logarithmic and circular functions, the determination of laws, the arithmetic, geometric, and binomial series.

The usual topics are presented in trigonometry. The solution of triangles is now treated in earlier work, but some revision examples have been retained on this topic. Sin ϕ and cos ϕ are represented by points S and C on orthogonal diameters of a unit circle. The graphs of the circular functions, and the solution of simple trigonometrical equations are developed. Diagrammatic illustrations are given in connection with the addition, sum, and difference rules, and the form $R \sin (\theta + \alpha)$.

Additions in the Calculus section include the functional notation, the notion of a limit, the relative error in $\delta y \simeq (dy/dx)\delta x$, the circular functions, the checking of indefinite integrals, integration from first principles, the definite integral and an introduction to differential equations.

It is not usual to expect the memorisation of proofs. The present tendency to provide lists of formulae for examination use, to shift the emphasis from memory to understanding is welcome. It is at least arguable that manuscript notes could be allowed in examinations as in real life situations.

Acknowledgements are extended to the various examining bodies, colleges and education authorities for permission to print questions from their examinations. The place of origin is given after each question quoted.

I will be pleased to acknowledge suggestions for improvement of *Certificate Mathematics*, care of the Publishers.

N. Abbott

Contents

Chapter One
Equations and Identities

Meaning of the solution of an equation

Consider the equation $9(x + 1) = 4(x + 4) + 3(x - 1)$.

If	$9(x + 1) = 4(x + 4) + 3(x - 1)$
then	$9x + 9 = 4x + 16 + 3x - 3$
then	$9x + 9 = 7x + 13$
then	$9x - 7x = 13 - 9$
then	$2x = 4$
and	$x = 2$

What is meant by the statement that $x = 2$ is the solution of the equation $9(x + 1) = 4(x + 4) + 3(x - 1)$? We claim that, if the variable x is replaced by 2 in the left-hand side expression $9(x + 1)$ and also in the right-hand side expression $4(x + 4) + 3(x - 1)$, then *identical numerical values* will be obtained. The student may note that until this claim is justified (from a strictly logical point of view) the equation as presented above has not been solved. This justification is usually described as a check and may be presented as follows.

When $x = 2$ L.H.S. $= 9(2 + 1) = 9 \cdot 3 = 27$
R.H.S. $= 4(2 + 4) + 3(2 - 1) = 4 \cdot 6 + 3 \cdot 1 = 27$

i.e., when $x = 2$, L.H.S. $= 27 =$ R.H.S.

$\therefore$ $x = 2$ is the solution of the equation

The student should always check in this way, he will then *actively* assimilate the meaning of the solution of an equation.

Examples

1.1 Verify that $x = 11$, $x = -20$ are the solutions of the equation $x^2 + 9x - 220 = 0$.

When $x = 11$ L.H.S. $= 11^2 + 9 \cdot 11 - 220$
$= 121 + 99 - 220 = 0$
R.H.S. $= 0$

i.e., when $x = 11$, L.H.S. $= 0 =$ R.H.S.

$\therefore$ $x = 11$ is a solution of the equation

When $x = -20$ L.H.S. $= (-20)^2 + 9(-20) - 220$
$= 400 - 180 - 220 = 0$
R.H.S. $= 0$

i.e., when $x = -20$, L.H.S. $= 0 =$ R.H.S.

$\therefore$ $x = -20$ is a solution of the equation

1.2 Verify that $\theta = 15°$ is a solution of the equation $10 \sin \theta + 3 \cos \theta = 5\cdot486$.

When $\theta = 15°$ L.H.S. $= 10 \sin 15° + 3 \cos 15°$
$$= 10 \times 0\cdot2588 + 3 \times 0\cdot9659$$
$$= 2\cdot588 + 2\cdot898$$
$$= 5\cdot486$$
R.H.S. $= 5\cdot486$

i.e., when $\theta = 15°$, L.H.S. $= 5\cdot486 =$ R.H.S.

$\therefore$ $\theta = 15°$ is a solution of the equation

1.3 Verify that $x = 1$, $y = -1$ is the solution of the simultaneous equations $3x + 2y = 1$ (i), $2x - 3y = 5$ (ii).

If $x = 1$, $y = -1$ in equation (i)
L.H.S. $= 3(1) + 2(-1) = 3 - 2 = 1$
R.H.S. $= 1$

i.e.   L.H.S. $= 1 =$ R.H.S.

In equation (ii) L.H.S. $= 2(1) - 3(-1) = 2 + 3 = 5$
R.H.S. $= 5$

i.e.   L.H.S. $= 5 =$ R.H.S.

Hence, $x = 1$, $y = -1$ is the solution of the pair of equations (i) and (ii).

1.4 Verify that $x = 1$, $y = -1$, $z = 2$ is the solution of the simultaneous equations $3x + 2y + z = 3$ (i), $2x - 3y - 2z = 1$ (ii), $4x + 2y - z = 0$ (iii).

If $x = 1$, $y = -1$, $z = 2$ in equation (i)
L.H.S. $= 3(1) + 2(-1) + 2 = 3$
R.H.S. $= 3$

i.e.   L.H.S. $= 3 =$ R.H.S.

In equation (ii) L.H.S. $= 2(1) - 3(-1) - 2(2) = 1$
R.H.S. $= 1$

i.e.   L.H.S. $= 1 =$ R.H.S.

In equation (iii) L.H.S. $= 4(1) + 2(-1) - 2 = 0$
R.H.S. $= 0$

i.e.   L.H.S. $= 0 =$ R.H.S.

Hence, $x = 1$, $y = -1$, $z = 2$ is the solution of the trio of equations (i), (ii) and (iii).

EXERCISE 1.1

1 Verify that $x = -21\frac{1}{2}$ is the solution of the equation $2(x - 5) - 3(x + 7) = x + 12$.

2 Verify that $\theta = 60°$ is a solution of the equation $12 - 9\cos\theta - 10\sin^2\theta = 0$.

3 Verify that $x = 2, y = -2$ is the solution of the simultaneous equations $5x + 4y = 2$, $2x - 7y = 18$.

4 Verify that $x = 2, y = 0, z = 1$ is the solution of the simultaneous equations $3x + 7y - 6z = 0$, $x - 3y + 2z = 4$, $4x + 3y - 5z = 3$.

5 Verify that $x = \dfrac{-b + \sqrt{(b^2 - 4ac)}}{2a}$ and also $x = \dfrac{-b - \sqrt{(b^2 - 4ac)}}{2a}$

are the solutions of the quadratic equation $ax^2 + bx + c = 0$ and that the sum of these roots (solutions) is $-b/a$, and their product c/a.

MID-CHESHIRE

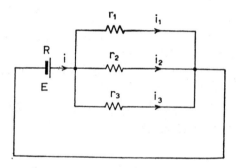

Fig. 1.1

6 A battery of internal resistance R ohms and electromotive force E volts is connected in series with resistors of r_1, r_2 and r_3 ohms respectively arranged in parallel with driving currents of i_1, i_2 and i_3 amperes respectively through them (Fig. 1.1). The equations

$$iR + i_1r_1 = E \quad \text{(i)} \qquad iR + i_2R_2 = E \quad \text{(ii)}$$
$$iR + i_3R_3 = E \quad \text{(iii)} \qquad i = i_1 + i_2 + i_3 \quad \text{(iv)}$$

apply. By eliminating i from equations (i), (ii), (iii) and (iv), the following equations are obtained.

$$i_1(R + r_1) + i_2R + i_3R = E \quad \text{(v)}$$
$$i_1R + i_2(R + r_2) + i_3R = E \quad \text{(vi)}$$
$$i_1R + i_2R + i_3(R + r_3) = E \quad \text{(vii)}$$

Verify that $i_1 = \dfrac{Er_2r_3}{\Delta}$, $i_2 = \dfrac{Er_3r_1}{\Delta}$ and $i_3 = \dfrac{Er_1r_2}{\Delta}$ are the solutions of (v), (vi) and (vii), where Δ is a symbol for (short way of writing) $r_1r_2r_3 + R(r_1r_2 + r_2r_3 + r_3r_1)$. Using (iv) show that

$$i = \frac{E(r_1r_2 + r_2r_3 + r_3r_1)}{\Delta}$$

and hence verify that equations (i), (ii) and (iii) are satisfied also.

MID-CHESHIRE

7 Verify that $x = \frac{1}{2}$, $x = -2$, $x = \frac{1}{3}$ and $x = -3$ are the solutions of the equation $6x^4 + 25x^3 + 12x^2 - 25x + 6 = 0$.

8 Verify that $\phi = 30° 30'$ is a solution of the equation $3 \sin (2\phi + 10°) - 4 \cos (3\phi - 5°) = 2·593$, and that it is also a solution of the equation $\tan 2\phi + 2 \sin 2\phi + \cos 2\phi = 4·308$.

9 Verify that $u = +2$ and $u = -2$ are the solutions of the equation $\frac{3}{5} = \frac{u - 1/u}{u + 1/u}$.

10 Verify that $t = 2 + j$ and $t = 2 - j$ are the solutions of the equation $t^2 - 4t + 5 = 0$, providing that $j^2 = -1$.

Algebraic identities

A sound understanding of the nature of identities is vital in subsequent work in trigonometry, partial fractions, etc.

Consider the expression $(x + 1)^2$, which is called a function of x, since its value depends upon the value given to x.

$$(x + 1)(x + 1) = x^2 + x + x + 1 = x^2 + 2x + 1$$

Geometrical illustration (Fig. 1.2)

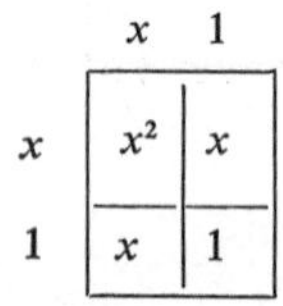

Area number $= (x + 1)(x + 1)$

$= x^2 + x + x + 1$

$= x^2 + 2x + 1$

Fig. 1.2

We can write $(x + 1)^2 \equiv x^2 + 2x + 1$ (i) which reads $(x + 1)^2$ is identically equal to $x^2 + 2x + 1$ and implies that, for *any* particular value we choose to assign to x everywhere it occurs in (i), we shall always find that the functions on the left-hand side and the right-hand side will have *identical numerical values*, since they are *different forms* of the *same function*.

The student is challenged to find a value for x for which the statement (i) is not true. After testing many values of x, and preferably illustrating geometrically, he will become convinced of the truth of statement (i). For example, if in (i) we choose

$x = 0$ L.H.S. $= (0 + 1)^2 = 1$

          R.H.S. $= 0^2 + 2.0 + 1 = 1$

i.e.   L.H.S. $= 1 =$ R.H.S.

$x = -1$ L.H.S. $= (-1 + 1)^2 = 0$
R.H.S. $= (-1)^2 + 2(-1) + 1 = 0$
i.e.   L.H.S. $= 0 = $ R.H.S.
$x = 9$ L.H.S. $= (9 + 1)^2 = 100$
R.H.S. $= 9^2 + 2 . 9 + 1 = 81 + 18 + 1 = 100$
i.e.   L.H.S. $= 100 = $ R.H.S.

Consider now the identity $\dfrac{1}{(x - y)^2} - \dfrac{1}{(x + y)^2} \equiv \dfrac{4xy}{(x^2 - y^2)^2}$.

In order to establish the truth of the identity we may start with the L.H.S. and transform to the R.H.S., or vice versa. Experience will convince the student that the more direct method is to combine the fractions by expressing them over a common denominator, i.e. in this case to start with the L.H.S. and transform to the R.H.S. The use of the sign of identical equality ($\equiv$) is not usually strictly enforced, since confusion with equations is not likely. We will follow this convention.

$$L.H.S. = \frac{1}{(x - y)^2} - \frac{1}{(x + y)^2} = \frac{1}{(x - y)^2}\frac{(x + y)^2}{(x + y)^2} - \frac{1}{(x + y)^2}\frac{(x - y)^2}{(x - y)^2}$$

$$= \frac{(x^2 + 2xy + y^2) - (x^2 - 2xy + y^2)}{\{(x - y)(x + y)\}^2}$$

$$= \frac{x^2 + 2xy + y^2 - x^2 + 2xy - y^2}{(x^2 - y^2)^2}$$

$$= \frac{4xy}{(x^2 - y^2)^2} = R.H.S., \text{ which proves the identity.}$$

i.e. $\dfrac{1}{(x - y)^2} - \dfrac{1}{(x + y)^2} \equiv \dfrac{4xy}{(x^2 - y^2)^2}$ $\cdot$ $\cdot$ $\cdot$ $\cdot$ $\cdot$ $\cdot$ (ii)

Suppose we test the truth of the identity by choosing at random particular values of x and y, say $x = 2, y = 1$.

When $x = 2, y = 1$

$$L.H.S. = \frac{1}{(2 - 1)^2} - \frac{1}{(2 + 1)^2} = 1 - \frac{1}{9} = \frac{8}{9}$$

$$R.H.S. = \frac{4 . 2 . 1}{(2^2 - 1^2)^2} = \frac{8}{9}$$

i.e.

$$L.H.S. = \frac{8}{9} = R.H.S.$$

Suppose we choose $x = 1$, $y = 1$.

When $x = 1$, $y = 1$

L.H.S. $= \dfrac{1}{(1-1)^2} - \dfrac{1}{(1+1)^2} = \dfrac{1}{0} - \dfrac{1}{4}$

R.H.S. $= \dfrac{4 \cdot 1 \cdot 1}{(1^2 - 1^2)} = \dfrac{4}{0}$

But division by zero is a meaningless operation and fractions with zero denominator are said to be undefined. We are thus led to the statement that the identity (ii) is true for all values of x and y for which all its terms are defined. Henceforward in this book it will be understood that zero denominators are excluded.

Example
1.5 Prove the identities

$\qquad$ (a) $x^2 - y^2 \equiv (x - y)(x + y)$
$\qquad$ (b) $x^3 - y^3 \equiv (x - y)(x^2 + xy + y^2)$
$\qquad$ (c) $x^3 + y^3 \equiv (x + y)(x^2 - xy + y^2)$

Use (a) and (b) to evaluate $101^2 - 100^2$, 101^2, $101^3 - 100^3$, 101^3.

MID-CHESHIRE

(a) R.H.S. $= (x - y)(x + y) = x^2 + xy - yx - y^2$
$\qquad\quad = x^2 - y^2 = $ L.H.S.

$\quad \therefore \quad x^2 - y^2 \equiv (x - y)(x + y)$ when x is 101 and y is 100
$\qquad 101^2 - 100^2 = (101 - 100)(101 + 100) = 1 \cdot 201 = 201$
$\quad$ Hence, $101^2 = 100^2 + 201 = 10\,000 + 201 = 10\,201$

(b) R.H.S. $= (x - y)(x^2 + xy + y^2) = x^3 + x^2y + xy^2 - yx^2 - xy^2 - y^3$
$\qquad\quad = x^3 - y^3 = $ L.H.S.

$\quad \therefore \quad x^3 - y^3 \equiv (x - y)(x^2 + xy + y^2)$ when x is 101 and y is 100
$\qquad 101^3 - 100^3 = (101 - 100)(101^2 + 101 \cdot 100 + 100^2)$
$\qquad\qquad\qquad\quad\ = 1(10\,201 + 10\,100 + 10\,000)$
$\qquad\qquad\qquad\quad\ = 30\,301$

$\quad$ Hence, $101^3 = 100^3 + 30\,301 = 1\,030\,301$

(c) R.H.S. $= (x + y)(x^2 - xy + y^2) = x^3 - x^2y + xy^2 + yx^2 - xy^2 + y^3$
$\qquad\quad = x^3 + y^3 = $ L.H.S.

$\quad \therefore \quad x^3 + y^3 \equiv (x + y)(x^2 - xy + y^2)$

EXERCISE 1.2

1 Prove the identity $\dfrac{1}{y^3} - \dfrac{1}{x^3} \equiv \dfrac{(x - y)(x^2 + xy + y^2)}{y^3 x^3}$.

Verify that when $y = 1$ and $x = 2$, then L.H.S. $= \dfrac{7}{8} = $ R.H.S.

2 Prove the following identities using repeated multiplication:

    (a) $(a + b)^2 \equiv a^2 + 2ab + b^2$
    (b) $(a + b)^3 \equiv a^3 + 3a^2b + 3ab^2 + b^3$
    (c) $(a + b)^4 \equiv a^4 + 4a^3b + 6a^2b^2 + 4ab^3 + b^4$

By replacing b by $-c$ above, deduce the identities

    (d) $(a - c)^2 \equiv a^2 - 2ac + c^2$
    (e) $(a - c)^3 \equiv a^3 - 3a^2c + 3ac^2 - c^3$
    (f) $(a - c)^4 \equiv a^4 - 4a^3c + 6a^2c^2 - 4ac^3 + c^4$

Verify that if $a = 3$, $b = 2$, $c = 1$, *each* side of the above identities has the respective value (a) 25, (b) 125, (c) 625, (d) 4, (e) 8, (f) 16.

3 Prove the identity
$$x^3 + y^3 + z^3 - 3xyz \equiv (x + y + z)(x^2 + y^2 + z^2 - yz - zx - xy)$$
Verify that when $x = 2$, $y = 1$, $z = -1$, then L.H.S. $= 14 =$ R.H.S.

4 Prove the identity

$$u^3 - v^3 \equiv (u - v)(u^2 + uv + v^2) \quad . \quad . \quad . \quad . \quad . \quad \text{(i)}$$

and, replacing u by $3p$ and v by $2q$, prove that

$$27p^3 - 8q \equiv (3p - 2q)(9p^2 + 6pq + 4q^2) \quad . \quad . \quad . \text{(ii)}$$

Verify in (i) that when $u = 2$, $v = 1$, then L.H.S. $= 7 =$ R.H.S., and in (ii) that when $p = 1$, $q = -1$, then L.H.S. $= 35 =$ R.H.S.

5 Prove the identity
$$\frac{4}{x + 3} - \frac{3}{x - 2} \equiv \frac{x - 17}{(x + 3)(x - 2)}$$

Verify that when $x = 1$, then L.H.S. $= 4 =$ R.H.S.

6 Prove the identity
$$\frac{x + 1}{x^2 + 1} - \frac{1}{x - 1} \equiv \frac{-2}{(x^2 + 1)(x - 1)}$$

Verify that when $x = 3$, then L.H.S. $= -\frac{1}{10} =$ R.H.S.

The process of long multiplication illustrated geometrically
Suppose that we wish to multiply $x^2 + x + 1$ by $x^2 + x + 2$. We may proceed by the laws of algebra.

$$(x^2 + x + 1)(x^2 + x + 2) = x^4 + x^3 + 2x^2 + x^3 + x^2 + 2x + x^2 + x + 2$$
$$= x^4 + 2x^3 + 4x^2 + 3x + 2$$

i.e. $(x^2 + x + 1)(x^2 + x + 2) \equiv x^4 + 2x^3 + 4x^2 + 3x + 2$

Verifying, when $x = 1$ we find L.H.S. $= (1 + 1 + 1)(1 + 1 + 2)$
$$= 3 \cdot 4 = 12$$
$$\text{R.H.S.} = 1 + 2 + 4 + 3 + 2 = 12$$
i.e.   L.H.S. $= 12 =$ R.H.S.

We may think of a rectangle whose length and breadth are represented by $x^2 + x + 2$ and $x^2 + x + 1$ units respectively, whilst the area is represented by their product (Fig. 1.3).

	x^2	x	2
x^2	x^4	x^3	$2x^2$
x	x^3	x^2	$2x$
1	x^2	x	2

$$\begin{array}{r} x^2 + x + 2 \\ x^2 + x + 1 \\ \hline x^4 + x^3 + 2x^2 \qquad\qquad \\ x^3 + x^2 + 2x \qquad \\ x^2 + x + 2 \\ \hline x^4 + 2x^3 + 4x^2 + 3x + 2 \\ \hline \end{array}$$

FIG. 1.3

The process of long division

The student will find that in more advanced work in connection with Maclaurin expansions of functions the processes of multiplication and long division in algebra are often required. The technique of long division involves the inversion of the above process. Suppose that we wish to divide $x^3 + 3x^2 + 5x + 4$ by $x + 1$.

$$\begin{array}{r|l} & x^2 \\ \hline x+1 & x^3 + 3x^2 + 5x + 4 \end{array}$$

The first step is to divide x into x^3 giving x^2 which is the first term of the quotient.

$$\begin{array}{r|l} & x^2 \\ \hline x+1 & x^3 + 3x^2 + 5x + 4 \\ & x^3 + x^2 \\ \cline{2-2} & \quad\;\; 2x^2 \end{array}$$

The next step is to multiply $x + 1$ by x^2, place the product thus and subtract, giving $2x^2$

$$\begin{array}{r|l} & x^2 \\ \hline x+1 & x^3 + 3x^2 + 5x + 4 \\ & x^3 + x^2 \\ \cline{2-2} & \quad\;\; 2x^2 + 5x \end{array}$$

The term $5x$ is brought down and x divided into $2x^2$, giving $2x$ the second term of the quotient.

$$\begin{array}{r|l} & x^2 + 2x \\ \hline x+1 & x^3 + 3x^2 + 5x + 4 \\ & x^3 + x^2 \\ \cline{2-2} & \quad\;\; 2x^2 + 5x \\ & \quad\;\; 2x^2 + 2x \\ \cline{2-2} & \qquad\qquad 3x \end{array}$$

$x + 1$ is multiplied back by $2x$ and the product placed thus and subtracted, giving $3x$.

$$\begin{array}{r} x^2 + 2x + 3 \\ x+1 \overline{\smash{\big)}\, x^3 + 3x^2 + 5x + 4} \\ \underline{x^3 + x^2} \\ 2x^2 + 5x \\ \underline{2x^2 + 2x} \\ 3x + 4 \\ \underline{3x + 3} \\ 1 \end{array}$$

The term 4 is brought down and x is divided into $3x$ giving 3, the third term of the quotient. $x + 1$ is multiplied by 3 and the product is subtracted, leaving 1.

Thus $\dfrac{x^3 + 3x^2 + 5x + 4}{x + 1} \equiv x^2 + 2x + 3 + \dfrac{1}{x + 1}$

Check

When $x = 1$, L.H.S. $= \dfrac{1 + 3 + 5 + 4}{1 + 1} = \dfrac{13}{2} = 6\frac{1}{2}$

R.H.S. $= 1 + 2 + 3 + \dfrac{1}{1 + 1} = 6\frac{1}{2}$

i.e. L.H.S. $= 6\frac{1}{2} =$ R.H.S.

EXERCISE 1.3

1 Prove by long division the identity

$$\frac{x^4 - x^3 - 3x^2 + 3x + 4}{x - 2} \equiv x^3 + x^2 - x + 1 + \frac{6}{(x - 2)}$$

Verify that when $x = 1$, then L.H.S. $= -4 =$ R.H.S.

2 Prove by long division the identity

$$\frac{x^6 - 1}{x - 1} = x^5 + x^4 + x^3 + x^2 + x + 1$$

Verify that when $x = -1$, then L.H.S. $= 0 =$ R.H.S.

The student is recommended to write $x^6 - 1$ in the form $x^6 + 0x^5 + 0x^4 + 0x^3 + 0x^2 + 0x + 1$ for the long division process.

3 (*a*) Prove by long division the identity

$$\frac{1}{1 + x} \equiv 1 - x + x^2 - x^3 + x^4 - \frac{x^5}{1 + x} \quad \cdots \quad \text{(i)}$$

Verify that when $x = 2$, then L.H.S. $= \frac{1}{3} =$ R.H.S.

(*b*) By replacing x by $-x$ in (i), deduce the identity

$$\frac{1}{1 - x} = 1 + x + x^2 + x^3 + x^4 + \frac{x^5}{1 - x}$$

Verify that when $x = -1$, then L.H.S. $= \frac{1}{2} =$ R.H.S.

(c) By replacing x by x^2 in (i) deduce the identity

$$\frac{1}{1 + x^2} \equiv 1 - x^2 + x^4 - x^6 + x^8 - \frac{x^{10}}{1 + x^2}$$

Verify that when $x = 3$, then L.H.S. $= \frac{1}{10} = $ R.H.S.

4 If x is a sufficiently small number in radians, $\sin x$ is approximately $x - \frac{x^3}{6} + \frac{x^5}{120}$ and $\cos x$ is approximately $1 - \frac{x^2}{2} + \frac{x^4}{24}$.

(a) Show by long division that $\tan x$, which is $\sin x$ divided by $\cos x$, is approximately given by $x + \frac{x^3}{3} + \frac{2x^5}{15}$.

(b) Show by long multiplication that $\sin 2x$, which is $2 \sin x \cos x$, is approximately given by $2x - \frac{(2x)^3}{6} + \frac{(2x)^5}{120}$.

Chapter Two
Surds and Further Equations

Surds

A number which can be expressed in the form p/q, where p and q are positive or negative whole numbers (integers), is called a rational number. A surd cannot be expressed in this form.

$$\sqrt{a} \cdot \sqrt{a} = a \quad \text{defines the square root of } a.$$

$$\sqrt[3]{a} \cdot \sqrt[3]{a} \cdot \sqrt[3]{a} = a \quad \text{defines the cube root of } a.$$

$$\underbrace{\sqrt[n]{a} \cdot \sqrt[n]{a} \cdot \sqrt[n]{a} \ldots \sqrt[n]{a}}_{\text{to } n \text{ factors}} \quad \text{defines the } n\text{th root of } a.$$

A surd is a root or a number which cannot be determined exactly: thus $\sqrt{2}, 4 + \sqrt{3}, \sqrt{5} + \sqrt{3}$ are examples of surds. We can give approximations to their values to any required number of decimal places, but we cannot express their values in the form p/q. They are irrational numbers.

Since $(+2) \cdot (+2) = 4$, it follows that $\sqrt{4} = +2$ but as $(-2) \cdot (-2) = 4$, also, it follows that $\sqrt{4} = -2$, Thus $\sqrt{4} = \pm 2$. In the following work we shall assume that the positive value of a square root is required. The following laws are useful in working with surds.

$$\sqrt[n]{\frac{p}{q}} = \frac{\sqrt[n]{p}}{\sqrt[n]{q}}$$

$$\sqrt[n]{p} \cdot \sqrt[n]{q} = \sqrt[n]{pq}$$

$$\sqrt[r]{(\sqrt[s]{p})} = \sqrt[rs]{p} = \sqrt[s]{(\sqrt[r]{p})}$$

Thus

$$\sqrt[3]{\frac{2}{27}} = \frac{\sqrt[3]{2}}{\sqrt[3]{27}} = \frac{\sqrt[3]{2}}{3}$$

$$\sqrt[3]{2} \cdot \sqrt[3]{6} = \sqrt[3]{12}$$

$$\sqrt[3]{\sqrt[5]{7}} = \sqrt[15]{7} = \sqrt[5]{\sqrt[3]{7}}$$

$\sqrt{2}, \sqrt{3}, \sqrt{5}$ are examples of monomial surds.
$2 - \sqrt{3}, \sqrt{5} + \sqrt{3}$ are examples of binomial surds.
$2 + \sqrt{3} + \sqrt{5}, \sqrt{2} - \sqrt{3} + \sqrt{5}$ are examples of trinomial surds.

Similar surds which have the same surd factor may be combined into a single surd. Thus

$$2\sqrt{3} + 3\sqrt{3} - 4\sqrt{3} = 5\sqrt{3} - 4\sqrt{3} = \sqrt{3}$$

Dissimilar surds cannot be so combined. Thus $2\sqrt{3} + 3\sqrt{5} - 4\sqrt{7}$ cannot be combined into a single surd.

Suppose that we require to evaluate $2/\sqrt{3}$ correct to three decimal places, given $\sqrt{3} = 1{\cdot}7320$. Instead of evaluating $2/1{\cdot}7320$ we may proceed as follows.

$$\frac{2}{\sqrt{3}} = \frac{2}{\sqrt{3}} \cdot \frac{\sqrt{3}}{\sqrt{3}} = \frac{2\sqrt{3}}{3} \simeq \frac{2 \cdot 1{\cdot}7320}{3}$$

$$= \frac{3{\cdot}4640}{3} = 1{\cdot}1546$$

$$= 1{\cdot}155, \text{ correct to three decimal places.}$$

This process involves rationalising the denominator.

Conjugate surds

$2 - \sqrt{3}$ and $2 + \sqrt{3}$, or $\sqrt{3} + \sqrt{2}$ and $\sqrt{3} - \sqrt{2}$ are examples of conjugate surds.

Consider the identity

$$(x - y)(x + y) \equiv x^2 - y^2$$

Let x be $\sqrt{a}$ and y be $\sqrt{b}$, then

$$(\sqrt{a} - \sqrt{b})(\sqrt{a} + \sqrt{b}) = (\sqrt{a})^2 - (\sqrt{b})^2 = a - b$$

Or let x be a and y be $\sqrt{b}$, then

$$(a - \sqrt{b})(a + \sqrt{b}) = a^2 - (\sqrt{b})^2 = a^2 - b$$

i.e., *the product of conjugate surds is rational.*

To express $(\sqrt{c} + \sqrt{d})/(\sqrt{a} + \sqrt{b})$ with rational denominator, we multiply numerator and denominator by the conjugate surd of the denominator.

i.e. $$\frac{\sqrt{c} + \sqrt{d}}{\sqrt{a} + \sqrt{b}} = \frac{(\sqrt{c} + \sqrt{d})(\sqrt{a} - \sqrt{b})}{(\sqrt{a} + \sqrt{b})(\sqrt{a} - \sqrt{b})} = \frac{\sqrt{ac} + \sqrt{ad} - \sqrt{bc} - \sqrt{bd}}{a - b}$$

Examples
2.1 (*a*) Express $\dfrac{1}{\sqrt{2}} + \dfrac{2}{\sqrt{3}} + \dfrac{4}{\sqrt{5}}$ with a common rational denominator.

(*b*) Express $\dfrac{2 - \sqrt{3}}{2 + \sqrt{3}} - \dfrac{2 + \sqrt{3}}{2 - \sqrt{3}}$ with rational denominators and simplify the result.

(*a*) $$\frac{1}{\sqrt{2}} + \frac{2}{\sqrt{3}} + \frac{4}{\sqrt{5}} = \frac{1}{\sqrt{2}} \cdot \frac{\sqrt{2}}{\sqrt{2}} + \frac{2\sqrt{3}}{\sqrt{3} \cdot \sqrt{3}} + \frac{4\sqrt{5}}{\sqrt{5} \cdot \sqrt{5}}$$

$$= \frac{\sqrt{2}}{4} + \frac{2\sqrt{3}}{3} + \frac{4\sqrt{5}}{5} = \frac{15\sqrt{2} + 40\sqrt{3} + 48\sqrt{5}}{60}$$

(b) $\dfrac{2-\sqrt{3}}{2+\sqrt{3}} - \dfrac{2+\sqrt{3}}{2-\sqrt{3}} = \dfrac{(2-\sqrt{3})(2-\sqrt{3})}{(2+\sqrt{3})(2-\sqrt{3})} - \dfrac{(2+\sqrt{3})(2+\sqrt{3})}{(2-\sqrt{3})(2+\sqrt{3})}$

$= \dfrac{4-4\sqrt{3}+3}{4-3} - \dfrac{(4+4\sqrt{3}+3)}{4-3}$

$= (7-4\sqrt{3}) - (7+4\sqrt{3}) = -8\sqrt{3}$

2.2 Given that $\sqrt{2} = 1{\cdot}4142$, evaluate the following expression by rationalising the denominator, expressing the result correct to three decimal places:

$$\frac{\sqrt{18}-\sqrt{12}}{\sqrt{8}-\sqrt{6}} + \frac{\sqrt{18}+\sqrt{12}}{\sqrt{8}+\sqrt{6}}$$

The expression is $\dfrac{3\sqrt{2}-2\sqrt{3}}{2\sqrt{2}-\sqrt{6}} + \dfrac{3\sqrt{2}+2\sqrt{3}}{2\sqrt{2}+\sqrt{6}}$

$= \dfrac{(3\sqrt{2}-2\sqrt{3})(2\sqrt{2}+\sqrt{6})}{(2\sqrt{2}-\sqrt{6})(2\sqrt{2}+\sqrt{6})} + \dfrac{(3\sqrt{2}+2\sqrt{3})(2\sqrt{2}-\sqrt{6})}{(2\sqrt{2}+\sqrt{6})(2\sqrt{2}-\sqrt{6})}$

$= \dfrac{6{\cdot}2-4\sqrt{6}+3\sqrt{12}-2\sqrt{18}}{4{\cdot}2-6} + \dfrac{6{\cdot}2+4\sqrt{6}-3\sqrt{12}-2\sqrt{18}}{4{\cdot}2-6}$

$= \dfrac{12-4\sqrt{6}+6\sqrt{3}-6\sqrt{2}+12+4\sqrt{6}-6\sqrt{3}-6\sqrt{2}}{2}$

$= \dfrac{24-12\sqrt{2}}{2}$

$= 12-6\sqrt{2} = 6(2-\sqrt{2}) \eqsim 6(2-1{\cdot}4142)$

$= 6(0{\cdot}5858) = 3{\cdot}5148 = 3{\cdot}515$, correct to three decimal places.

2.3 (a) Express $\dfrac{x+\sqrt{(x^2+1)}}{x-\sqrt{(x^2+1)}} + \dfrac{x-\sqrt{(x^2+1)}}{x+\sqrt{(x^2+1)}}$ with rational denominators and simplify.

(b) Simplify $\sqrt{2} - \dfrac{1}{\sqrt{2}} + \sqrt{3} - \dfrac{1}{\sqrt{3}} + \dfrac{1}{\sqrt{3}-\sqrt{2}}$ expressing over a common rational denominator, and check the result.

(a) $\dfrac{\{x+\sqrt{(x^2+1)}\}\{x+\sqrt{(x^2+1)}\}}{\{x-\sqrt{(x^2+1)}\}\{x+\sqrt{(x^2+1)}\}} + \dfrac{\{x-\sqrt{(x^2+1)}\}\{x-\sqrt{(x^2+1)}\}}{\{x+\sqrt{(x^2+1)}\}\{x-\sqrt{(x^2+1)}\}}$

$= \dfrac{x^2+2x\sqrt{(x^2+1)}+x^2+1}{x^2-(x^2+1)} + \dfrac{x^2-2x\sqrt{(x^2+1)}+x^2+1}{x^2-(x^2+1)}$

$= \dfrac{x^2+2x\sqrt{(x^2+1)}+x^2+1+x^2-2x\sqrt{(x^2+1)}+x^2+1}{x^2-x^2-1}$

$= \dfrac{4x^2+2}{-1} = \dfrac{2(2x^2+1)}{-1} = -2(2x^2+1)$

(b) $\sqrt{2} - \dfrac{1}{\sqrt{2}} + \sqrt{3} - \dfrac{1}{\sqrt{3}} + \dfrac{1}{\sqrt{3} - \sqrt{2}}$

$= \sqrt{2} - \dfrac{1}{\sqrt{2}} \cdot \dfrac{\sqrt{2}}{\sqrt{2}} + \sqrt{3} - \dfrac{1}{\sqrt{3}} \cdot \dfrac{\sqrt{3}}{\sqrt{3}} + \dfrac{1}{(\sqrt{3} - \sqrt{2})} \cdot \dfrac{(\sqrt{3} + \sqrt{2})}{(\sqrt{3} + \sqrt{2})}$

$= \sqrt{2} - \dfrac{\sqrt{2}}{2} + \sqrt{3} - \dfrac{\sqrt{3}}{3} + \dfrac{\sqrt{3} + \sqrt{2}}{3 - 2}$

$= \dfrac{\sqrt{2}}{2} + \dfrac{2\sqrt{3}}{3} + \sqrt{3} + \sqrt{2} = \dfrac{5\sqrt{3}}{3} + \dfrac{3\sqrt{2}}{2}$

$= \dfrac{10\sqrt{3} + 9\sqrt{2}}{6}$

Taking $\sqrt{2} \simeq 1 \cdot 414$ and $\sqrt{3} \simeq 1 \cdot 732$

$\sqrt{2} - \dfrac{1}{\sqrt{2}} + \sqrt{3} - \dfrac{1}{\sqrt{3}} + \dfrac{1}{\sqrt{3} - \sqrt{2}}$

$= 1 \cdot 414 - \dfrac{1}{1 \cdot 414} + 1 \cdot 732 - \dfrac{1}{1 \cdot 732} + \dfrac{1}{1 \cdot 732 + 1 \cdot 414}$

$= 1 \cdot 414 - 0 \cdot 707 + 1 \cdot 732 - 0 \cdot 577 + 3 \cdot 145 = 5 \cdot 007$

$\dfrac{10\sqrt{3} + 9\sqrt{2}}{6} = \dfrac{10 \cdot 1 \cdot 732 + 9 \cdot 1 \cdot 414}{6} = \dfrac{30 \cdot 046}{6} = 5 \cdot 007$

i.e. the results agree to three decimal places.

EXERCISE 2.1

1 (a) By expressing $\dfrac{3}{\sqrt{2}} + \dfrac{2}{\sqrt{3}} + \dfrac{1}{\sqrt{5}}$ with common rational denominator, prove that

$$\dfrac{3}{\sqrt{2}} + \dfrac{2}{\sqrt{3}} + \dfrac{1}{\sqrt{5}} \equiv \dfrac{45\sqrt{2} + 20\sqrt{3} + 6\sqrt{5}}{30}$$

Test the identity approximately by slide rule, showing that L.H.S. $\simeq 3 \cdot 72 \simeq$ R.H.S.

(b) By rationalising the denominators of $\dfrac{1 - \sqrt{6}}{2 - \sqrt{2}} + \dfrac{1 + \sqrt{6}}{2 - \sqrt{3}}$, prove that

$$\dfrac{1 - \sqrt{6}}{2 - \sqrt{2}} + \dfrac{1 + \sqrt{6}}{2 - \sqrt{3}} \equiv 3 + \sqrt{6} + \dfrac{7}{2}\sqrt{2}$$

Test the identity approximately by slide rule, showing that L.H.S. $\simeq 10 \cdot 4 \simeq$ R.H.S.

2 Prove by rationalising the denominators that

(a) $\dfrac{1}{x - \sqrt{(x^2 - 1)}} + \dfrac{1}{x + \sqrt{(x^2 - 1)}} \equiv 2x$

(b) $\dfrac{1}{\sqrt{10}-3} + \dfrac{1}{\sqrt{5}-2} + \dfrac{1}{\sqrt{26}-5} - (\sqrt{10}+\sqrt{5}+\sqrt{26}+10) \equiv 0$

Test the identity approximately by slide rule.

3 Prove by rationalising the denominators in the denominator that

$$\frac{1}{1/\sqrt{2}+1/\sqrt{3}} \equiv 3\sqrt{2} - 2\sqrt{3}$$

and test the identity approximately by slide rule, showing that L.H.S. $\simeq$ 0·78 $\simeq$ R.H.S.

4 Prove that $\dfrac{a-\sqrt{b}}{a+\sqrt{b}} + \dfrac{a+\sqrt{b}}{a-\sqrt{b}} \equiv \dfrac{2(a^2+b)}{a^2-b}$.

Verify that when $a = 3$ and $b = 4$ L.H.S. $= 5\cdot2 =$ R.H.S.

5 Show by squaring each expression that $\sqrt{(a+b)}$ is not equivalent to $\sqrt{a}+\sqrt{b}$.

Elimination

Suppose that we are given the equations $x = 3 + t$ (i), $y = 2 - t$ (ii) and we are required to show that $x + y = 5$ (iii), we merely have to add the given equations (i) and (ii).

$$\begin{array}{l} x = 3 + t \\ y = 2 - t \end{array} \text{ giving } x + y = 5$$

t has disappeared: we say that t has been eliminated. Suppose that we tabulate the variables in the given equations (i) and (ii) and in the required relation (iii)

| variables in (i) and (ii) | x | y | t |
| variables in required relation | x | y | |

We are forced to conclude that t must be eliminated in order to obtain the required relation (iii).

Examples

2.4 (Continuation of the work on simultaneous equations.) Solve the simultaneous equations $ax + by = c$ (i), $dx + ey = f$ (ii) and verify the solutions, given that a, b, c, d, e and f are constants and $db - ae \neq 0$.

Multiply equation (i) by d: $dax + dby = dc$
Multiply equation (ii) by a: $adx + aey = af$
Subtract to eliminate x: $(db - ae)y = dc - af$
Providing that $(db - ae) \neq 0$: $y = \dfrac{dc - af}{db - ae}$

Multiply equation (i) by e: $eax + eby = ec$
Multiply equation (ii) by b: $bdx + bey = bf$
Subtract to eliminate y: $(ea - bd)x = ec - bf$
Providing that $(ea - bd) \neq 0$: $x = \dfrac{ec - bf}{ea - bd}$

Check
When $x = \dfrac{ec - bf}{ea - bd}$, $y = \dfrac{dc - af}{db - ae}$

equation (i)   L.H.S. $= \dfrac{a(ec - bf)}{ea - bd} + \dfrac{b(dc - af)}{db - ae}$

$$= \frac{aec - abf - b(dc - af)}{ea - bd}$$

$$= \frac{aec - abf - bdc + baf}{ea - bd} = \frac{aec - bdc}{ea - bd} = c\frac{(ae - bd)}{(ea - bd)} = c = \text{R.H.S.}$$

equation (ii)   L.H.S. $= d\left(\dfrac{ec - bf}{ea - bd}\right) + e\dfrac{dc - af}{db - ae}$

$$= \frac{dec - dbf - e(dc - af)}{ea - bd}$$

$$= \frac{dec - dbf - edc + eaf}{ea - bd} = \frac{eaf - dbf}{ea - bd} = f\frac{(ea - db)}{(ea - bd)} = f = \text{R.H.S.}$$

We have thus solved in a general way two linear equations in two unknowns, providing that $ea - bd$ is non-zero. If $ea - bd$ is zero the above solutions involve division by zero, which is nonsense. However, if $ea - bd = 0$ then $ea = bd$; i.e. $a/d = b/e$, the ratios of the coefficients of x and of y in the equations are equal.

e.g.
$$2x + 3y = 5$$
$$8x + 12y = 20$$
here $\dfrac{2}{8} = \dfrac{3}{12}$

Since the second equation is the first equation with each term multiplied by 4, there is no limit to the number of solutions which will satisfy these equations, e.g. $x = 1, y = 1; x = 0, y = \frac{5}{3}; x = 2, y = \frac{1}{3}$. Geometrically the equations represent the same straight line.

2.5 $2x + 3y = 5$
$\quad\;\; 8x + 12y = 15$
here $\dfrac{2}{8} = \dfrac{3}{12}$

If we multiply the first equation by 4 and subtract from the second equation:

$$8x + 12y = 20$$
$$8x + 12y = 15$$

we arrive at a contradiction:

$$0 + 0 = -5.$$

The assumption that a finite solution of these two simultaneous equations exists is false. Geometrically the equations represent the parallel lines

$$y = \frac{-2x}{3} + \frac{5}{3} \text{ and } y = \frac{-2x}{3} + \frac{15}{8}.$$

Note on determinants

We have shown that the solutions of the equations $ax + by = c$, $dx + ey = f$ are $x = \dfrac{ce - bf}{ae - bd}$ and $y = \dfrac{af - cd}{ae - bd}$. These can be written in the form

$$x = \frac{\begin{vmatrix} c & b \\ f & e \end{vmatrix}}{\begin{vmatrix} a & b \\ d & e \end{vmatrix}} \quad \text{and} \quad y = \frac{\begin{vmatrix} a & c \\ d & f \end{vmatrix}}{\begin{vmatrix} a & b \\ d & e \end{vmatrix}} \tag{i}$$

where $\begin{vmatrix} c & b \\ f & e \end{vmatrix}$ means $ce - bf$, $\begin{vmatrix} a & c \\ d & f \end{vmatrix}$ means $af - cd$

and $\begin{vmatrix} a & b \\ d & e \end{vmatrix}$ means $ae - bd$.

Note that a '+' sign is associated with a diagonal (product) which falls from left to right, and a '−' sign is associated with a diagonal (product) which rises from left to right.

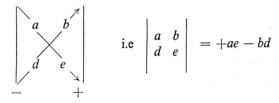

i.e $\begin{vmatrix} a & b \\ d & e \end{vmatrix} = +ae - bd$

The advantage of the determinant form in (i) is that the solutions have a recognisable pattern. The determinant in the denominator is the determinant of the coefficients. The determinants in the numerator are obtained by replacing the coefficients of the respective variable, x or y, by c or f.

Thus $x = \dfrac{\begin{vmatrix} \cancel{a}c & b \\ \cancel{d}f & e \end{vmatrix}}{\begin{vmatrix} a & b \\ d & e \end{vmatrix}} \qquad y = \dfrac{\begin{vmatrix} a & \cancel{b}c \\ d & \cancel{e}f \end{vmatrix}}{\begin{vmatrix} a & b \\ d & e \end{vmatrix}}$

2.6 Solve the equations $2x + 3y = 11, 8x - 10y = 66$.

From above

$$x = \frac{\begin{vmatrix} 11 & 3 \\ 66 & -10 \end{vmatrix}}{\begin{vmatrix} 2 & 3 \\ 8 & -10 \end{vmatrix}} = \frac{11 . -10 - 3 . 66}{2 . -10 - 3 . 8} = \frac{-308}{-44} = 7$$

$$y = \frac{\begin{vmatrix} 2 & 11 \\ 8 & 66 \end{vmatrix}}{\begin{vmatrix} 2 & 3 \\ 8 & -10 \end{vmatrix}} = \frac{2 . 66 - 11 . 8}{-44} = \frac{132 - 88}{-44}$$
$$= \frac{44}{-44} = -1$$

which may be checked by inspection.

2.7 If $x = 2 \cos \theta$, $y = 3 \sin \theta$ and $\cos^2 \theta + \sin^2 \theta = 1$, prove that $\frac{x^2}{4} + \frac{y^2}{9} = 1$.

Variables in the given equations: $\boxed{x}$ $\boxed{y}$ θ
Variables in the required relation: $\boxed{x}$ $\boxed{y}$

∴ θ must be eliminated.

$x = 2 \cos \theta$, ∴ $\cos \theta = \dfrac{x}{2}$

$y = 3 \sin \theta$, ∴ $\sin \theta = \dfrac{y}{3}$

Substitute in $\sin^2 \theta + \cos^2 \theta = 1$, then $\left(\dfrac{x}{2}\right)^2 + \left(\dfrac{y}{3}\right)^2 = 1$

i.e. $\dfrac{x^2}{4} + \dfrac{y^2}{9} = 1$

2.8 The following relations arise in the theory of a balanced alternating current bridge: $RR_3 + \dfrac{L}{C_3} = R_2R_4$ (i) and $\omega LR_3 - \dfrac{R}{\omega C_3} = 0$ (ii).

Prove that $L = \dfrac{R_2R_4C_3}{1 + \omega^2 R_3^2 C_3^2}$.

Variables in the given relations: R $\boxed{R_2}$ $\boxed{R_3}$ $\boxed{R_4}$ $\boxed{L}$ $\boxed{C_3}$ $\boxed{\omega}$
Variables in the required relation: $\boxed{R_2}$ $\boxed{R_3}$ $\boxed{R_4}$ $\boxed{L}$ $\boxed{C_3}$ $\boxed{\omega}$
We see that R must be eliminated.

$\omega LR_3 - \dfrac{R}{\omega C_3} = 0$, ∴ $\omega LR_3 = \dfrac{R}{\omega C_3}$

∴ $\omega^2 LC_3R_3 = R$ (iii)

Substituting (iii) in (i), and so eliminating R,

$$(\omega^2 LC_3 R_3)R_3 + \frac{L}{C_3} = R_2 R_4$$

$$\therefore \quad \frac{\omega^2 LC_3^2 R_3^2 + L}{C_3} = R_2 R_4$$

$$\therefore \quad \frac{L(1 + \omega^2 C_3^2 R_3^2)}{C_3} = R_2 R_4$$

$$\therefore \quad L = \frac{R_2 R_4 C_3}{(1 + \omega^2 C_3^2 R_3^2)}$$

EXERCISE 2.2

1 Given that $x = r \cos \theta$, $y = r \sin \theta$ and $\cos^2 \theta + \sin^2 \theta = 1$, prove that $x^2 + y^2 = r^2$.

2 Prove the identity $\alpha^2 + \beta^2 \equiv (\alpha + \beta)^2 - 2\alpha\beta$. Hence or otherwise, prove that $r = p^2 - 2q$, given that $\alpha + \beta = p$, $\alpha\beta = q$ and $\alpha^2 + \beta^2 = r$.

3 Prove the identity $(l + m)^3 \equiv l^3 + m^3 + 3lm(l + m)$. Hence or otherwise, prove that $s^3 = u + 3ts$, given that $l + m = s$, $lm = t$ and $l^3 + m^3 = u$.

4 Given (a) $x = at^2$, $y = 2at$, prove $y^2 = 4ax$, (b) $x = ct$, $y = \frac{c}{t}$, prove $xy = c^2$, (c) $x = t^2$, $y = t^3$, prove $x^3 = y^2$.

5 Given $\theta = \left(\frac{\omega_1 + \omega_2}{2}\right)t$ and $\omega_2 = \omega_1 + \alpha t$, prove that

(a) $\omega_2^2 = \omega_1^2 + 2\alpha\theta$, (b) $\theta = \omega_1 t + \frac{1}{2}\alpha t^2$.

6 (a) Given $P = \frac{W}{g}f$ and $v^2 = u^2 + 2fs$, prove that $Ps = \frac{1}{2}\frac{W}{g}(v^2 - u^2)$.

(b) Given $P = \frac{W}{g}f$ and $v = u + ft$, prove that $Pt = \frac{W}{g}(v - u)$.

7 Given $T = I\alpha$, $\omega_2^2 = \omega_1^2 + 2\alpha\theta$ and $I = \frac{\omega}{g}k^2$, prove that

$$T\theta = \frac{1}{2}\frac{\omega}{g}k^2(\omega_2^2 - \omega_1^2).$$

8 Given that $\frac{M}{I} = \frac{f}{y}$, $M = \omega(200D)^2$, $I = \frac{D^4}{100}$ and $y = \frac{D}{2}$, prove that

$$f = 2 \times 10^6 \times \frac{\omega}{D}.$$

9 Given that $b^2 = h^2 + (c - x)^2$, $a^2 = h^2 + x^2$ and $x = b \cos B$, prove that $b^2 = a^2 + c^2 - 2ac \cos B$.

10 Given that $Q = \dfrac{\pi d^2}{4} v$, $H = \dfrac{f l v^2}{2 g m}$ and $m = \dfrac{d}{4}$, prove that $Q = K \sqrt{\dfrac{H}{l}} \cdot d^{5/2}$

where $K = \sqrt{\dfrac{g}{2f} \cdot \dfrac{\pi}{4}}$

11 Given that $v = u + ft$ and $s = \left(\dfrac{u + v}{2}\right) t$, prove that $v^2 = u^2 + 2fs$ and

$s = ut + \dfrac{1}{2} f t^2$.

12 (a) If $x = a + 1$ and $y = a^2 + a$, express y in terms of x only. For what values of a are x and y equal?

(b) If $\dfrac{x}{y} = \dfrac{a}{b} = \dfrac{3}{4}$, find the values of $\dfrac{x + a}{y + b}$, $\dfrac{x^2 + a^2}{y^2 + b^2}$, $\dfrac{xb}{ya}$.

ULCI

13 Solve using determinants:

(a) $2x - y = 1$
$\quad\, x + y = 2$

(b) $7x - 10y = 11$
$\quad\, 6x + 8y = 26$

(c) $3x + y = 5a$
$\quad\, x - 2y = 3a$

Formation of linear simultaneous equations in three unknowns

Consider the array $7(\;) - 3(\;) + 2(\;)$
$\qquad\qquad\quad 4(\;) + (\;) - 3(\;)$
$\qquad\qquad\quad (\;) - 2(\;) + 5(\;)$

Suppose we insert $1, -1, 2$, say, in the columns and evaluate the rows:

$$7(1) - 3(-1) + 2(2) = 7 + 3 + 4 = 14$$
$$4(1) + (-1) - 3(2) = 4 - 1 - 6 = -3$$
$$(1) - 2(-1) + 5(2) = 1 + 2 + 10 = 13$$

we have formed the equations

$$7x - 3y + 2z = 14$$
$$4x + y - 3z = -3$$
$$x - 2y + 5z = 13$$

for which the solution is $x = 1$, $y = -1$, $z = 2$.
The student should form two equations in two unknowns, three equations in three unknowns, four equations in four unknowns, etc. for himself and then solve them by elimination.

Solution of linear simultaneous equations in three unknowns

Solution of the equations
$\qquad 7x - 3y + 2z = 14 \qquad$ (i)
$\qquad 4x + y - 3z = -3 \qquad$ (ii)
$\qquad x - 2y + 5z = 13 \qquad$ (iii)

First we eliminate any one of the variables, say y (which has the numerically smallest coefficients), and thus reduce to two equations in two unknowns. Multiply (ii) by 3 and add to (i). Multiply (ii) by 2 and add to (iii).

$$7x - 3y + 2z = 14$$
$$12x + 3y - 9z = -9$$
$$\overline{19x \quad\quad - 7z = \quad 5}$$ (iv)

$$8x + 2y - 6z = \quad 6$$
$$x - 2y + 5z = 13$$
$$\overline{9x \quad\quad - z = \quad 7}$$ (v)

The simplest procedure is now to eliminate z. Multiply (v) by 7 and subtract (iv) from the resulting equation.

$$63x - 7z = 49$$
$$19x - 7z = \quad 5$$
$$\overline{44x \quad\quad = 44} \quad\quad \therefore x = 1$$

Substitute in (v): $9 - z = 7$, $\therefore z = 2$.
Substitute for x and z in (ii): $4 + y - 6 = -3$ $y = -1$.
The solution is $x = 1$, $y = -1$ and $z = 2$, which should now be checked in the original equations (i), (ii) and (iii). Geometrically the equations (i), (ii) and (iii) represent planes which intersect at the point $(1, -1, 2)$. This is a unique (one and only one) solution.

If two or more of the planes are parallel the equations are inconsistent: no finite solution exists. If the planes have a common line of intersection, an infinite number of solutions exists. In industrial applications, six equations in six unknowns occur very frequently from dynamical considerations, and much greater numbers of variables are common. Thus the mathematical study of the behaviour of linear equations is of great importance in technology.

EXERCISE 2.3

1 Given the equations $15x + 30y + z = 46$ (i), $37x - 67y - z = -31$ (ii) and $50x + 25y + 2z = 77$ (iii); show that if z is eliminated by adding (i) and (ii), and adding (iii) and twice (ii), the equations $52x - 37y = 15$ and $124x - 109y = 15$ result. By subtracting these equations, show that $x = y$ and complete and check the solution $x = 1$, $y = 1$, $z = 1$.

2 Solve the simultaneous equations
(a) $a + 2b + 3c = 4$
$\quad a + 3b - 4c = -2$
$\quad a - 4b + 2c = -3$
(b) $x^2 - 9y^2 = 24$
$\quad x - 3y = 8$

NCTEC

3 For a law of the form $y = A + Bx + Cx^2$ it is found that $y = 4.5$ when $x = 1$, $y = 16.5$ when $x = 4$, $y = 29.5$ when $x = 6$. Determine the values of the constants A, B and C.

NCTEC (part question)

4 A parabola passes through the points $(x = 2, y = 13)$, $(x = 5, y = 100)$ and $(x = 1, y = -20)$. If the equation is of the form $y = ax^2 + bx + c$, determine the value of a, b and c.

NCTEC (part question)

5 The relationship between the grid potential V volts and the anode current I milliamperes of a triode value for a given constant anode voltage is $I = a + bV + cV^2$. If corresponding values of V and I are as given in the table, calculate the values of the constants a, b and c.

I	1	0·54	0·24
V	−2	−4	−6

Calculate also the voltage required to give a current of 0·8 milliamperes.

ULCI

6 Solve the following simultaneous equations
 (a) $\log_4(A + B) = 2$, $\log_9(A - B + 1) = 1$
 (b) $A + B - C = 1$, $2A + B - 3C = -7$, $A - B - C = -7$

ULCI

Matrices (2 × 2 only)

Consider the equations $ax + by = c$, $dx + ey = f$, previously studied.

Writing in the form $\begin{pmatrix} a & b \\ d & e \end{pmatrix} \begin{pmatrix} x \\ y \end{pmatrix} = \begin{pmatrix} c \\ f \end{pmatrix}$

$\underbrace{}_{2 \times 2 \text{ matrix}}$ $\underbrace{}_{\substack{\text{column} \\ \text{vector}}}$ $\underbrace{}_{\substack{\text{column} \\ \text{vector}}}$

we have a matrix form of the equations which may be symbolised as $Ax = c$. Note that a matrix is not a number, i.e. has not got a value: it is an array of elements in space.

Addition of matrices

If $ax + by = c$, $px + qy = r$
 $dx + ey = f$, $lx + my = h$

adding horizontally: $(a + p)\, x + (b + q)\, y = c + r$
 $(d + l)\, x + (e + m)\, y = f + h$

By analogy with adding equations,

if $\qquad A = \begin{pmatrix} a & b \\ d & e \end{pmatrix}$ and $B = \begin{pmatrix} p & q \\ l & m \end{pmatrix}$

then $\qquad A + B = \begin{pmatrix} a + p & b + q \\ d + l & e + m \end{pmatrix}$ gives matrix addition

and $\qquad A - B = \begin{pmatrix} a - p & b - q \\ d - l & e - m \end{pmatrix}$ gives matrix subtraction.

It follows that if n is an integer (whole number) then $nA = \begin{pmatrix} na & nb \\ nd & ne \end{pmatrix}$

since n of each element are added together. This leads to the rule

$$\lambda A = \lambda \begin{pmatrix} a & b \\ d & e \end{pmatrix} = \begin{pmatrix} \lambda a & \lambda b \\ \lambda d & \lambda e \end{pmatrix}$$

where λ is a scalar (any number). Note that *all the elements* are multiplied by λ.

Example

2.8 If $C = \begin{pmatrix} 0\cdot1 & 0\cdot2 \\ -0\cdot1 & 0\cdot3 \end{pmatrix}$ and $D = \begin{pmatrix} 0\cdot05 & 0\cdot1 \\ -0\cdot1 & 0\cdot0 \end{pmatrix}$

find $5C + 3D$ and $5C - 3D$.

$$5C = \begin{pmatrix} 0\cdot5 & 1\cdot0 \\ -0\cdot5 & 1\cdot5 \end{pmatrix}, 3D = \begin{pmatrix} 0\cdot15 & 0\cdot3 \\ -0\cdot3 & 0\cdot0 \end{pmatrix}$$

$$\therefore 5C + 3D = \begin{pmatrix} 0\cdot65 & 1\cdot3 \\ -0\cdot8 & 1\cdot5 \end{pmatrix}, 5C - 3D = \begin{pmatrix} 0\cdot35 & 0\cdot7 \\ -0\cdot2 & 1\cdot5 \end{pmatrix}$$

Produce of matrices

If $A = \begin{pmatrix} a & b \\ d & e \end{pmatrix}$ and $B = \begin{pmatrix} p & q \\ l & m \end{pmatrix}$

then $\quad AB = \begin{pmatrix} a & b \\ d & e \end{pmatrix} \begin{pmatrix} p & q \\ l & m \end{pmatrix} = \begin{pmatrix} (a.p + b.l) & (a.q + b.m) \\ (d.p + e.l) & (d.q + e.m) \end{pmatrix}$

$$BA = \begin{pmatrix} p & q \\ l & m \end{pmatrix} \begin{pmatrix} a & b \\ d & e \end{pmatrix} = \begin{pmatrix} (p.a + q.d) & (p.b + q.e) \\ (l.a + m.d) & (l.b + m.e) \end{pmatrix}$$

Note that $AB \neq BA$ in general. Arithmetical products are commutative, i.e. $3 \times 2 = 2 \times 3$, but matrix products in general are not.

Example

2.10 If $A = \begin{pmatrix} 3 & 2 \\ 1 & -1 \end{pmatrix}$, $B = \begin{pmatrix} 4 & 1 \\ -1 & 2 \end{pmatrix}$ and $I = \begin{pmatrix} 1 & 0 \\ 0 & 1 \end{pmatrix}$, find AI, IA, AB and BA.

$$AI = \begin{pmatrix} 3 & 2 \\ 1 & -1 \end{pmatrix} \begin{pmatrix} 1 & 0 \\ 0 & 1 \end{pmatrix} = \begin{pmatrix} (3.1 + 2.0) & (3.0 + 2.1) \\ (1.1 - 1.0) & (1.0 - 1.1) \end{pmatrix} = \begin{pmatrix} 3 & 2 \\ 1 & -1 \end{pmatrix} = A$$

Similarly $\quad IA = \begin{pmatrix} 1 & 0 \\ 0 & 1 \end{pmatrix} \begin{pmatrix} 3 & 2 \\ 1 & -1 \end{pmatrix} = \begin{pmatrix} 3 & 2 \\ 1 & -1 \end{pmatrix} = A$

(Note that I is called the 2×2 *identity matrix*, that $AI = A$ and $IA = A$. I has a role in matrix work similar to 1 in arithmetic. Here multiplication *is* commutative.)

$$AB = \begin{pmatrix} 3 & 2 \\ 1 & -1 \end{pmatrix} \begin{pmatrix} 4 & 1 \\ -1 & 2 \end{pmatrix} = \begin{pmatrix} 10 & 7 \\ 5 & -1 \end{pmatrix}$$

$$BA = \begin{pmatrix} 4 & 1 \\ -1 & 2 \end{pmatrix} \begin{pmatrix} 3 & 2 \\ 1 & -1 \end{pmatrix} = \begin{pmatrix} 13 & 7 \\ -1 & -4 \end{pmatrix}$$

Again note that the matrix multiplication is *not* commutative.

The null 2×2 matrix

$\begin{pmatrix} 0 \\ 0 \end{pmatrix}$ is a null column vector and $O = \begin{pmatrix} 0 & 0 \\ 0 & 0 \end{pmatrix}$ is the null 2×2 matrix.

Thus if $\quad A = \begin{pmatrix} 3 & 1 \\ 2 & 0 \end{pmatrix}$ then $AO = \begin{pmatrix} 3 & 1 \\ 2 & 0 \end{pmatrix} \begin{pmatrix} 0 & 0 \\ 0 & 0 \end{pmatrix} = \begin{pmatrix} 0 & 0 \\ 0 & 0 \end{pmatrix} = O$

Similarly $OA = O$; thus the null matrix has a similar role to zero in arithmetic.

Transpose of a 2×2 matrix

If $A = \begin{pmatrix} a & b \\ d & e \end{pmatrix}$ then A' or $\tilde{A}$ or At is called the transposed matrix of A, or the transpose of A. We shall use A'.

$A' = \begin{pmatrix} a & d \\ b & e \end{pmatrix}$ where the rows of A become the columns of A'. Thus if $B = \begin{pmatrix} 3 & -1 \\ 2 & 4 \end{pmatrix}$ then $B' = \begin{pmatrix} 3 & 2 \\ -1 & 4 \end{pmatrix}$.

Inverse of a 2×2 matrix

Consider the matrix $C = \begin{pmatrix} -1 & 2 \\ 2 & -3 \end{pmatrix}$. Suppose that we pre-multiply C by the matrix $C^{-1} = \begin{pmatrix} 3 & 2 \\ 2 & 1 \end{pmatrix}$

Then $\quad C^{-1}C = \begin{pmatrix} 3 & 2 \\ 2 & 1 \end{pmatrix} \begin{pmatrix} -1 & 2 \\ 2 & -3 \end{pmatrix} = \begin{pmatrix} 1 & 0 \\ 0 & 1 \end{pmatrix} = I$

Similarly, post-multiply C by this matrix.

Then $\quad C C^{-1} = \begin{pmatrix} -1 & 2 \\ 2 & -3 \end{pmatrix} \begin{pmatrix} 3 & 2 \\ 2 & 1 \end{pmatrix} = \begin{pmatrix} 1 & 0 \\ 0 & 1 \end{pmatrix} = I$

Thus C^{-1} is called the inverse matrix of C, since $C^{-1}C = CC^{-1} = I$ (i), just as in arithmetic $3^{-1}.3 = \frac{1}{3}.3 = 1$; but note that in matrix work we do not form a reciprocal—division is not defined. We mean by C^{-1} that matrix *which multiplies with C to give I*, as shown by expression (i).

Rules for the formation of the inverse of a 2×2 matrix

Consider the matrix $A = \begin{pmatrix} a & b \\ d & e \end{pmatrix}$ from $\begin{pmatrix} a & b \\ d & e \end{pmatrix} \begin{pmatrix} x \\ y \end{pmatrix} = \begin{pmatrix} c \\ f \end{pmatrix}$, or $Ax = c$(ii)

We first evaluate the determinant of the elements:

$$|A| = \begin{vmatrix} a & b \\ d & e \end{vmatrix} = ae - bd.$$

If this is zero, the matrix A *is said to be singular and no inverse exists.*

This corresponds to the result obtained earlier in this chapter: that if $ae - bd = 0$, then equations (ii) have no *unique* solution.

Next form the transpose of A:

$$A' = \begin{pmatrix} a & d \\ b & e \end{pmatrix}$$

Then diagonally interchange the elements, insert minus signs on the elements of the diagonal which rises from left to right, and finally divide by $|A|$.

Thus $\quad A^{-1} = \dfrac{1}{|A|} \begin{pmatrix} e & -b \\ -d & a \end{pmatrix} = \dfrac{1}{ae - bd} \begin{pmatrix} e & -b \\ -d & a \end{pmatrix}$

Finally check that $AA^{-1} = I$, or $A^{-1}A = I$.

Thus $\quad AA^{-1} = \begin{pmatrix} a & b \\ d & e \end{pmatrix} \dfrac{1}{ae - bd} \begin{pmatrix} e & -b \\ -d & a \end{pmatrix}$

$$= \dfrac{1}{ae - bd} \begin{pmatrix} ae - bd & -ab + ba \\ de - de & -db + ea \end{pmatrix}$$

$$= \dfrac{1}{ae - bd} \begin{pmatrix} ae - bd & 0 \\ 0 & ae - bd \end{pmatrix} = \begin{pmatrix} 1 & 0 \\ 0 & 1 \end{pmatrix} = I$$

Example
2.11 Find the inverse, where possible, of

(a) $B = \begin{pmatrix} 1 & -3 \\ 4 & -12 \end{pmatrix}$ $\qquad$ (b) $C = \begin{pmatrix} 2 & -1 \\ 3 & -2 \end{pmatrix}$

(a) $|B| = \begin{vmatrix} 1 & -3 \\ 4 & -12 \end{vmatrix} = 1(-12) - 4(-3) = -12 + 12 = 0$

i.e. *no inverse exists*

(b) $|C| = \begin{vmatrix} 2 & -1 \\ 3 & -2 \end{vmatrix} = 2(-2) - 3(-1) = -4 + 3 = -1$

i.e. *an inverse exists*

$$C' = \begin{pmatrix} 2 & 3 \\ -1 & -2 \end{pmatrix}, \therefore C^{-1} = \dfrac{1}{-1} \begin{pmatrix} -2 & 1 \\ -3 & 2 \end{pmatrix} = \begin{pmatrix} 2 & -1 \\ 3 & -2 \end{pmatrix}$$

Check

$$C^{-1}C = \begin{pmatrix} 2 & -1 \\ 3 & -2 \end{pmatrix} \begin{pmatrix} 2 & -1 \\ 3 & -2 \end{pmatrix} = \begin{pmatrix} 1 & 0 \\ 0 & 1 \end{pmatrix} = I$$

Formation of the inverse by row operations

The previous method is open to two objections. Firstly, a rule has been stated rather than derived. This need not worry us unduly, as the inverse has been checked in a general way. To explain why we carry out the interchanges, etc. would require determinant theory. The second objection is more vital: the method rapidly becomes impractically complicated and time consuming as the order of the matrix increases.

Consider the matrix $C = \begin{pmatrix} 2 & -1 \\ 3 & -2 \end{pmatrix}$ and augment this matrix with the unit matrix, thus $\begin{pmatrix} 2 & -1 & | & 1 & 0 \\ 3 & 12 & | & 0 & 1 \end{pmatrix}$

Now multiply or divide the rows by suitable constants, and add or subtract the rows so as to produce the unit matrix to the left of the dotted line and C^{-1} to the right of the dotted line. $R \equiv$ row.

$$\begin{pmatrix} 2 & -1 & | & 1 & 0 \\ 3 & -2 & | & 0 & 1 \end{pmatrix} \xrightarrow{2 \times R_1} \begin{pmatrix} 4 & -2 & | & 2 & 0 \\ 3 & -2 & | & 0 & 1 \end{pmatrix} \xrightarrow{R_1 - R_2} \begin{pmatrix} 1 & 0 & | & 2 & -1 \\ 3 & -2 & | & 0 & 1 \end{pmatrix}$$

$$\xrightarrow{3 \times R_1} \begin{pmatrix} 3 & 0 & | & 6 & -3 \\ 3 & -2 & | & 0 & 1 \end{pmatrix} \xrightarrow{R_2 - R_1} \begin{pmatrix} 3 & 0 & | & 6 & -3 \\ 0 & -2 & | & -6 & 4 \end{pmatrix}$$

$$\xrightarrow{R_1/3 \text{ and } R_2/-2} \begin{pmatrix} 1 & 0 & | & 2 & -1 \\ 0 & 1 & | & 3 & -2 \end{pmatrix}$$

Thus $\qquad C = \begin{pmatrix} 2 & -1 \\ 3 & -2 \end{pmatrix}$, $C^{-1} = \begin{pmatrix} 2 & -1 \\ 3 & -2 \end{pmatrix}$ and $C^{-1}C = I$,

as obtained previously by the first method.

It may be noted that this is essentially the method of elimination, and is used with desk calculating machines in a different (tabular) form in Chapter 18.

Matrix solution of two simultaneous linear equations

Consider the equations $ax + by = c$, $dx + ey = f$

$$\text{i.e.} \begin{pmatrix} a & b \\ d & e \end{pmatrix} \begin{pmatrix} x \\ y \end{pmatrix} = \begin{pmatrix} c \\ f \end{pmatrix} \text{ or } Ax = c$$

where $A = \begin{pmatrix} a & b \\ d & e \end{pmatrix}$ x denotes $\begin{pmatrix} x \\ y \end{pmatrix}$ and c denotes $\begin{pmatrix} c \\ f \end{pmatrix}$.

Notice that we use A, x and c as symbols for more involved expressions, to save writing.

If $Ax = c$, pre-multiply by A^{-1}, then $A^{-1}Ax = A^{-1}c$.

But $A^{-1}A = I$, $\therefore Ix = A^{-1}c$
    i.e. $x = A^{-1}c$.

Previously $\qquad A^{-1} = \dfrac{1}{ae - bd} \begin{pmatrix} e & -b \\ -d & a \end{pmatrix}$

$$\therefore \begin{pmatrix} x \\ y \end{pmatrix} = \dfrac{1}{ae - bd} \begin{pmatrix} e & -b \\ -d & a \end{pmatrix} \begin{pmatrix} c \\ f \end{pmatrix}$$

gives the solution of the equations, after multiplication and division.

Example

2.12 Solve the equations $3x - y = 5$(i), $2x + 4y = 8$(ii) by matrix methods, and check.

$$\begin{pmatrix} 3 & -1 \\ 2 & 4 \end{pmatrix} \begin{pmatrix} x \\ y \end{pmatrix} = \begin{pmatrix} 5 \\ 8 \end{pmatrix}$$

The inverse matrix is obtained by one of the previous methods, and is used to pre-multiply each side.

Thus
$$\frac{1}{14} \begin{pmatrix} 4 & 1 \\ -2 & 3 \end{pmatrix} \begin{pmatrix} 3 & -1 \\ 2 & 4 \end{pmatrix} \begin{pmatrix} x \\ y \end{pmatrix} = \frac{1}{14} \begin{pmatrix} 4 & 1 \\ -2 & 3 \end{pmatrix} \begin{pmatrix} 5 \\ 8 \end{pmatrix}$$

$$\therefore \frac{1}{14} \begin{pmatrix} 14 & 0 \\ 0 & 14 \end{pmatrix} \begin{pmatrix} x \\ y \end{pmatrix} = \frac{1}{14} \begin{pmatrix} 28 \\ 14 \end{pmatrix}$$

$$\therefore \begin{pmatrix} 1 & 0 \\ 0 & 1 \end{pmatrix} \begin{pmatrix} x \\ y \end{pmatrix} = \begin{pmatrix} 2 \\ 1 \end{pmatrix}$$

$$\therefore \begin{pmatrix} x \\ y \end{pmatrix} = \begin{pmatrix} 2 \\ 1 \end{pmatrix} \text{ and } x = 2, y = 1 \text{ is the solution.}$$

Check

When $x = 2, y = 1$ in equation (i) : L.H.S. $= 3(2) - 1 = 5 =$ R.H.S.

in equation (ii): L.H.S. $= 2(2) + 4(1) = 8 =$ R.H.S.

We have seen how the inverse of a 2×2 matrix may be used to solve the corresponding equations; the same applies to 3×3 and 4×4 matrices applied to linear equations with 3 and 4 unknowns respectively. It should not be imagined that the same is true for equations with a large number of unknowns, which are solved using a computer. Finding the inverse is a lengthy process not necessary for the solution of such equations. Iterative methods (see Chapter 16) are sometimes applicable, or methods which involve factorising the original matrix into a number of subsidiary matrices with special properties are used. Matrix algebra provides a method of organising the way we think about such linear simultaneous equations of high order.

EXERCISE 2.4

1 If $A = \begin{pmatrix} 2 & -2 \\ 1 & 3 \end{pmatrix}$ and $B = \begin{pmatrix} 3 & -1 \\ 1 & 3 \end{pmatrix}$ and $C = 6A + 5B$ and $D =$

$6A - 5B$, find C and D. Check your results by verifying that $C + D = 12A$.

2 Invent a number of problems similar to question 1 using different elements (numbers) in A and B and different multipliers instead of 6 and 5, and carry out a check.

3 If $A = \begin{pmatrix} -2 & -9 \\ 1 & 4 \end{pmatrix}$, prove that $A.A$ or $A^2 = \begin{pmatrix} -5 & -18 \\ 2 & 7 \end{pmatrix}$,

$A^3 = \begin{pmatrix} -8 & -27 \\ 3 & 10 \end{pmatrix}$ $A^4 = \begin{pmatrix} -11 & -36 \\ 4 & 13 \end{pmatrix}$ and $A^5 = \begin{pmatrix} -14 & -45 \\ 5 & 16 \end{pmatrix}$.

4 If $B = \begin{pmatrix} 11 & -25 \\ 4 & 9 \end{pmatrix}$ prove that $B^5 = \begin{pmatrix} 51 & -125 \\ 20 & -49 \end{pmatrix}$

5 If $A = \begin{pmatrix} 3 & -4 \\ 5 & 2 \end{pmatrix}$, $B = \begin{pmatrix} 0 & 1 \\ 1 & 0 \end{pmatrix}$, $C = \begin{pmatrix} 3 & 0 \\ 0 & 1 \end{pmatrix}$, $D = \begin{pmatrix} 1 & 0 \\ 0 & 5 \end{pmatrix}$,

$E = \begin{pmatrix} 1 & 7 \\ 0 & 1 \end{pmatrix}$, $F = \begin{pmatrix} 1 & 0 \\ 8 & 1 \end{pmatrix}$, $I = \begin{pmatrix} 1 & 0 \\ 0 & 1 \end{pmatrix}$ and $O = \begin{pmatrix} 0 & 0 \\ 0 & 0 \end{pmatrix}$

Show that (a) $OA = O$,
       (b) $IA = I$,
       (c) BA interchanges the rows of A,
       (d) CA multiplies the top row of A by 3,
       (e) DA multiplies the bottom row of A by 5,
       (f) EA adds 7 times the bottom row of A on to the top,
       (g) FA adds 8 times the top row of A on to the bottom.
Show also that AB, AC, AD, AE, AF have a similar effect on the columns of A.

6 If $A = \begin{pmatrix} 1 & -1 \\ 1 & 0 \end{pmatrix}$ and $B = \begin{pmatrix} 1 & 0 \\ 1 & 0 \end{pmatrix}$, find matrices A^2, B^2, AB, A^2B, AB^2

and $AB - B^2$. Check your results by showing that $A^2B - AB^2 = A(AB - B^2)$.

7 If $C = \begin{pmatrix} 1 & 2 \\ 2 & 1 \end{pmatrix}$ and $D = \begin{pmatrix} 1 & 1 \\ 1 & 1 \end{pmatrix}$, find the matrix represented by

$C^2 - DC + CD - D^2$. Check this result by showing that it is equal to the product of the matrices $C - D$ and $C + D$.

8 Show by either one or two methods that if $A = \begin{pmatrix} 3 & 2 \\ 2 & 4 \end{pmatrix}$ then

$A^{-1} = \frac{1}{8} \begin{pmatrix} 4 & -2 \\ -2 & 3 \end{pmatrix}$. Check that $A^{-1}A = I$. Hence solve the equa-

tions $3x + 2y = 2$, $2x + 4y = -12$ by matrix methods. Show that $x = 4$ and $y = -5$, and check these solutions.

9 Solve question 13, Exercise 2.2. by matrix methods.
Apply the method of question 8 to the equations given in questions 10, 11, 12, 13, 14 and 15. Check your inverse matrix and the solutions.

10 $4x + 5y = -1$ **11** $7x + 4y = 2$ **12** $5x - 6y = 2$
    $3x - 2y = 5$ $3x - 2y = 12$ $3x + 4y = 24$
13 $bx + y = 2$ **14** $3i_1 + 4i_2 = 2{\cdot}5$ **15** $Ri_1 + ri_2 = E$
    $x - by = c$ $5i_1 - 6i_2 = 1{\cdot}0$ $ri_1 + Ri_2 = e$

Chapter Three
Indices and Logarithms

In a previous course the student will have learnt how to use indices and logarithms in elementary applications. A general treatment will now be given, so that it may be possible to proceed to more difficult examples.

The index notation
If a is any number, and n is a positive integer, then a^n means $a \times a \times a \times$ ... to n factors, or, by a^n is meant the product of n factors each of which is a; a^n is termed the nth power of a.

Laws of indices
(1) Multiplication
To prove $a^m \times a^n = a^{m+n}$, where m and n are positive integers.

By definition above,

$$a^m = a \times a \times a \times \ldots \text{ to } m \text{ factors}$$
and
$$a^n = a \times a \times a \times \ldots \text{ to } n \text{ factors}$$
$$\therefore \quad a^m \times a^n = (a \times a \times a \times \ldots \text{ to } m \text{ factors})$$
$$\times (a \times a \times a \times \ldots \text{ to } n \text{ factors}).$$

Thus there are $(m + n)$ factors, each of which is a, on the right-hand side.
$$\therefore \quad a^m \times a^n = a \times a \times a \times \ldots \text{ to } (m + n) \text{ factors,}$$
or
$$a^m \times a^n = a^{m+n} \text{ (by definition).}$$

(2) Division
To prove $a^m \div a^n = a^{m-n}$, where m and n are positive integers, and m is greater than n.

By definition
$$a^m = a \times a \times a \times \ldots \text{ to } m \text{ factors}$$
and
$$a^n = a \times a \times a \times \ldots \text{ to } n \text{ factors}$$
$$\therefore \quad a^m \div a^n = \frac{a \times a \times a \times \ldots \text{ to } m \text{ factors}}{a \times a \times a \times \ldots \text{ to } n \text{ factors}}$$

Cancelling n of the m factors in the numerator by a corresponding number of factors in the denominator, we are left with $(m - n)$ factors in the numerator.

$$\therefore \quad a^m \div a^n = a \times a \times a \times \ldots \text{ to } (m - n) \text{ factors}$$
or
$$a^m \div a^n = a^{m-n} \text{ (by definition).}$$

The case in which n is greater than m will be dealt with later.

(3) Power of a power

To prove $(a^m)^n = a^{mn}$.

By definition,

$$(a^m)^n = a^m \times a^m \times a^m \times \ldots \text{ to } n \text{ factors}$$
$$= a^{m+m+m \ldots \text{ to } n \text{ terms}} \quad \text{(first law of indices)}$$
$$= a^{mn}$$

Extension of the meaning of an index

The laws of indices which we have considered are based on the definition of a power in which the index is a positive integer. If the index is a fraction, or a negative number, or zero, the definition ceases to have any literal meaning. But algebra aims at generalisation, and if indices are to have a really practical value, we must be able to use them in *all* cases, and not merely subject to the restriction stated above.

It becomes necessary, therefore, to consider what meanings can be attached to powers in which the indices are no longer positive integers.

It is important that we should be clear as to what principle must guide us in thus extending the meaning of a power. The principle, clearly, must be as follows.

If the new quantities are to be regarded as powers, *they must obey the laws of indices which have already been formulated.* They must be governed by the same laws as when the indices are positive integers.

1. Consider the meaning of $a^{\frac{1}{n}}$.

First let us consider a simple case, when $n = 2$.

By the principle stated above, $a^{\frac{1}{2}}$ must obey the fundamental laws of indices.

$$\therefore \quad a^{\frac{1}{2}} \times a^{\frac{1}{2}} = a^{\frac{1}{2}+\frac{1}{2}}$$
$$= a$$

Thus $a^{\frac{1}{2}}$ must be such a quantity that, on being multiplied by itself, the result is a. But, by definition, such a quantity is the square root of a.

$$\therefore \quad \text{by } a^{\frac{1}{2}} \text{ we mean } \sqrt{a}$$

Similarly, $$a^{\frac{1}{3}} = \sqrt[3]{a}$$

We may now proceed to the general case.

Since the first law of indices must be obeyed, then

$$a^{\frac{1}{n}} \times a^{\frac{1}{n}} \times a^{\frac{1}{n}} \times \ldots \text{ to } n \text{ factors}$$
$$= a^{\frac{1}{n}+\frac{1}{n}+\frac{1}{n} \ldots \text{ to } n \text{ terms}}$$
$$= a^1$$
$$= a$$

Hence $$a^{\frac{1}{n}} = \sqrt[n]{a}$$

2. Consider now the meaning of $a^{\frac{m}{n}}$.

In a simple case,

$$a^{\frac{2}{3}} \times a^{\frac{2}{3}} \times a^{\frac{2}{3}} = a^{\frac{2}{3}+\frac{2}{3}+\frac{2}{3}} \text{ (by first law of indices)}$$
$$= a^2$$

Hence $a^{\frac{2}{3}}$ must be the cube root of a^2,

or $$a^{\frac{2}{3}} = \sqrt[3]{a^2}$$

In general,

$$a^{\frac{m}{n}} \times a^{\frac{m}{n}} \times a^{\frac{m}{n}} \ldots \text{ to } n \text{ factors}$$

$$= a^{\frac{m}{n}+\frac{m}{n}+\frac{m}{n} \ldots \text{ to } n \text{ terms}} \quad \text{(by first law of indices)}$$

$$= a^{\frac{m}{n} \times n}$$

$$= a^m$$

Hence $a^{\frac{m}{n}}$ must be the nth root of a^m,

or $$a^{\frac{m}{n}} = \sqrt[n]{a^m}$$

3. If the first law of indices is to hold for a^0, i.e. when the index is zero, then

$$a^0 \times a^n = a^{0+n}$$
$$= a^n$$
$$\therefore \quad a^0 = a^n \div a^n$$
$$= 1$$

This result is independent of the value of a,

$\therefore$ *for all values of a we define a^0 as* 1.

4. To find a meaning for a^{-n}.

Assuming as before that the laws of indices hold for negative indices, then by the first law

$$a^n \times a^{-n} = a^{n-n}$$
$$= a^0$$
$$= 1$$
$$\therefore \quad a^{-n} = 1 \div a^n = \frac{1}{a^n}$$

Thus a^{-n} is defined as the reciprocal of a^n.

For example,

$$a^{-2} = \frac{1}{a^2}$$

$$a^{-1} = \frac{1}{a}$$

$$a^{-\frac{1}{2}} = \frac{1}{a^{\frac{1}{2}}} = \frac{1}{\sqrt{a}}$$

Similarly

$$\frac{1}{a^{-2}} = a^2$$

or generally

$$\frac{1}{a^{-n}} = a^n$$

Examples

3.1 Find the value of $2^{\frac{3}{2}}$, given $\sqrt{2} = 1\cdot414$.

$$2^{\frac{3}{2}} = 2^{1+\frac{1}{2}}$$
$$= 2^1 \times 2^{\frac{1}{2}}$$
$$= 2 \times \sqrt{2} = 2\cdot828$$

3.2 Find the value of $2^{-\frac{1}{2}}$.

$$2^{-\frac{1}{2}} = \frac{1}{2^{\frac{1}{2}}} = \frac{1}{\sqrt{2}}$$

$$= \frac{\sqrt{2}}{2} \text{ (rationalising)}$$

$$= \frac{1\cdot414}{2} = 0\cdot707$$

3.3 Find the value of $(16^{\frac{1}{4}})^3$.

$$(16^{\frac{1}{4}})^3 = (\sqrt[4]{16})^3$$
$$= 2^3 = 8$$

3.4 Find the value of $(\frac{2}{3})^{-\frac{3}{2}}$.

$$(\tfrac{2}{3})^{-\frac{3}{2}} = \frac{1}{(\frac{2}{3})^{\frac{3}{2}}} = \frac{1}{\frac{(2)^{\frac{3}{2}}}{(3)^{\frac{3}{2}}}}$$

$$= \frac{(3)^{\frac{3}{2}}}{(2)^{\frac{3}{2}}} = (\tfrac{3}{2})^{\frac{3}{2}}$$
$$= \sqrt{(\tfrac{3}{2})^3} = \sqrt{\tfrac{27}{8}}$$
$$= \sqrt{(\tfrac{9}{4} \times \tfrac{3}{2})} = \tfrac{3}{2}\sqrt{\tfrac{3}{2}}$$
$$= 1\cdot5\sqrt{1\cdot5} = 1\cdot5 \times 1\cdot2247$$
$$= 1\cdot837$$

3.5 Evaluate $(25^{0\cdot125})^{-4}$.

$$(25^{0\cdot125})^{-4} = \frac{1}{(25^{0\cdot125})^4} = \frac{1}{25^{0\cdot5}}$$

$$= \frac{1}{25^{\frac{1}{2}}} = \frac{1}{\sqrt{25}}$$

$$= \frac{1}{5}$$

EXERCISE 3.1

1 Write down the values of $8^{\frac{2}{3}}$, $25^{\frac{3}{2}}$, $(16^{\frac{1}{4}})^{\frac{3}{2}}$.

2 Write down the values of 2^{-2}, $(5^{-1})^2$, $\dfrac{1}{10^{-2}}$, $\dfrac{2}{2^{-3}}$.

3 Write down the values of $(\frac{1}{2})^{-2}$, $(\frac{2}{3})^{-3}$, $(16)^{0.5}$.

4 Write down the values of $(36)^{-0.5}$, $(4)^{1.5}$, $(\frac{1}{4})^{2.5}$.

5 Find the values of $3^{\frac{2}{3}}$, $3^{\frac{3}{8}}$, $4^{\frac{3}{4}}$ (to three places of decimals).

6 Find the values of $10^{\frac{3}{2}}$, 10^{-2}, 10^{-1}, $10^{-0.5}$.

7 Find the values of $2^{\frac{1}{4}}$, $2^{\frac{3}{4}}$ (to two places of decimals).

8 Find the values of $(16^{-0.25})^3$, $(8^{0.5})^{-3}$.

9 Find the simplest form of $a^4 \times a^{-2} \times a^{\frac{1}{2}}$.

10 Find the simplest form of $a^{\frac{1}{2}} \times a^{\frac{1}{3}} \times a^{-1}$.

11 If $10^{\frac{1}{3}} = 2\cdot154$, find the value of $10^{\frac{2}{3}}$ (to two places of decimals).

12 Write down the square root of $81a^4b^2$ and the cube root of $8a^6b^3$.

13 Given that $p_1v_1{}^n = p_2v_2{}^n$ and $\dfrac{p_1v_1}{T_1} = \dfrac{p_2v_2}{T_2}$ prove that $\dfrac{T_1}{T_2} = \left(\dfrac{v_2}{v_1}\right)^{n-1}$ and

$\dfrac{T_1}{T_2} = \left(\dfrac{p_1}{p_2}\right)^{\frac{n-1}{n}}$.

14 By first replacing $(a^2 + 4)^{\frac{1}{2}}$ by u and $(a^2 - 4)^{\frac{1}{2}}$ by v, or otherwise, prove that

$$\frac{(a^2 + 4)^{\frac{1}{2}} + (a^2 - 4)^{\frac{1}{2}}}{(a^2 + 4)^{\frac{1}{2}} - (a^2 - 4)^{\frac{1}{2}}} + \frac{(a^2 + 4)^{\frac{1}{2}} - (a^2 - 4)^{\frac{1}{2}}}{(a^2 + 4)^{\frac{1}{2}} + (a^2 - 4)^{\frac{1}{2}}} \equiv \frac{a^2}{2}$$

Verify that, when $a^2 = 5$, then L.H.S. $= \frac{5}{2} =$ R.H.S.

15 If $x^{p^q} = (x^p)^q$ prove that $p = q^{\frac{1}{q-1}}$.

16 Prove that $(9^{\frac{1}{4}} . 9^{\frac{3}{8}} . 9^{\frac{3}{8}}) \div 9^{\frac{13}{16}}$ on simplification has the value 3.

17 Prove that $[(4^{-2})^{-2}]^{-2} \div [(256)^{\frac{1}{4}}]^{-8}$ on simplification has the value 1.

18 Prove that $[\frac{1}{2}(z + z^{-1})] \times [\frac{1}{2}(z - z^{-1})] \equiv \frac{1}{4}(z^2 - z^{-2})$.

19 Prove that $[\frac{1}{2}(z + z^{-1})]^2 - [\frac{1}{2}(z - z^{-1})]^2 \equiv 1$.

20 Prove that

$$\frac{2\left(\dfrac{z - z^{-1}}{z + z^{-1}}\right)}{1 + \left(\dfrac{z - z^{-1}}{z + z^{-1}}\right)^2} \equiv \frac{z^2 - z^{-2}}{z^2 + z^{-2}}$$

Verify that, when $z = 4$, then L.H.S. $= \frac{15}{17} =$ R.H.S.

21 (a) Solve $\sqrt{(x + 7)} - 2 - \sqrt{(x - 5)} = 0$.

(b) Find A, B and C if $A + B - C = 0$, $3A + B + C = 6$, $A + 2B - 2C = -1$.

(c) Solve $3^{x^2} = 81^{-(x+1)}$.

ULCI

22 (a) What do you understand by the following?

(i) a^0, (ii) a^{-b}.

Justify your statement.

(b) Given that $p = \dfrac{m^x - m^{-x}}{m^x + m^{-x}}$, express $\dfrac{1+p}{1-p}$ as a power of m. Hence,

or otherwise, find the value of x when $\dfrac{1+p}{1-p} = m^{x^2-5}$.

NCTEC

Logarithms

The student has learnt in his previous work that it is possible to express any number as a power of any other number. In particular, he has learnt that any number can be expressed as a power of 10. By the extension of the meaning of an index to include fractions, negative numbers and zero, it is now possible to give as powers of 10 numbers not previously considered. The following table shows a few such cases, all of which can be determined by the student himself by using the previous exercises.

Power	10^{-2}	10^{-1}	$10^{-0.5}$	10	$10^{0.5}$	10^1	$10^{1.5}$	10^2
Number	0·01	0·1	0·3162	1	3·162	10	31·62	100
Index	−2	−1	−0·5	0	0·5	1	1·5	2

This table could be extended indefinitely, both for positive and negative indices. It will be seen that it is possible to obtain a series of indices extending from $-\infty$, through zero, to $+\infty$, which will correspond to a series of numbers, extending from zero to infinity, expressed as powers of 10.

It is also possible to take any other number, say a, and in a similar way to obtain a system of indices corresponding to numbers expressed as powers of a.

Such a system of indices, by means of which a series of numbers is expressed as powers of another number, called the *base* of the system, is called *a system of logarithms*.

Definition of a logarithm

A logarithm of a number to a given base is the index which indicates what power the number is of the base.

For example, since $1000 = 10^3$, the above definition enables us to say that

3 is the logarithm of 1000 to base 10

which may be written

$$3 = \log_{10} 1000$$

Similarly, since $0.01 = 10^{-2}$, -2 is the logarithm of 0.01 to base 10, which may be written

$$-2 = \log_{10} 0.01$$

In general, if $a^x = N$ (where a may be any number greater than 1), then x is the index which indicates what power N is of a: so, by definition,

$$x \text{ is the logarithm of } N \text{ to base } a$$

or
$$x = \log_a N$$

It should be noted that the two equations

$$a^x = N$$

and
$$x = \log_a N$$

both express the relations which, as shown above, exist between the three numbers, x, y and a. They are different forms expressing the same thing, and the student should be able readily to change from one to the other.

Examples
3.6 Convert the following indicial statements to logarithmic statements.

$$10^{-3} = 0\cdot001, \ 10^{-2} = 0\cdot01, \ 10^{-1} = 0\cdot1, \ 10^0 = 1, \ 10^1 = 10,$$
$$10^2 = 100, \ 10^3 = 1000$$

From above,

$$-3 = \log_{10} 0\cdot001, \ -2 = \log_{10} 0\cdot01, \ -1 = \log_{10} 0\cdot1, \ 0 = \log_{10} 1$$
$$1 = \log_{10} 10, \ 2 = \log_{10} 100, \ 3 = \log_{10} 1000$$

(Observe that the index number from the indicial statement appears alone in the logarithmic statement separated from the 'log' symbol by the sign of equality.)

3.7 Convert the following logarithmic statements into indicial statements.

$$\log_2 8 = 3, \ \log_5 25 = 2, \ \log_3 81 = 4, \ \log_4 256 = 4$$

From above,

$$2^3 = 8, \ 5^2 = 25, \ 3^4 = 81, \ 4^4 = 256$$

EXERCISE 3.2

1 Convert the following indicial statements into logarithmic statements.

$$p^q = r, \ u^v = w, \ 8^3 = 512, \ 100^3 = 1\,000\,000$$

$$e^x = y, \ \left(\frac{T_1}{T_2}\right) = e^{\mu\theta}, \ \left(\frac{iR}{V}\right) = e^{\frac{-Rt}{L}}$$

2 Convert the following logarithmic statements into indicial statements.

$$\log_2 16 = 4, \ \log_3 243 = 5, \ \log_4 1024 = 5, \ \log_a x = N,$$
$$\log_b y = M, \ \log_q p = R$$

3 (*a*) Eliminate x from the relations $a^x = N$ and $x = \log_a N$.
  (*b*) Eliminate the number 3 from the relations $10^3 = 1000$ and $3 = \log_{10} 1000$.
  (*c*) Eliminate the number 5 from the relations $2^5 = 32$ and $5 = \log_2 32$.

Laws of logarithms

The student has previously learnt that logarithms are of great practical value in enabling us to carry out operations with numbers. He has also learnt rules for these operations. We will now proceed to examine these rules again, and to consider proofs of them.

The fundamental principle which will guide our work is that a logarithm is an index, and therefore the laws of logarithms must correspond to the laws of indices. The student is advised to revise exercise 3.2 before commencing this work.

To express a logarithm as an index

If $a^x = N$ then $x = \log_a N$. Eliminating x,

$$a^{\log_a N} = N.$$

This form exhibits clearly the fact that a logarithm is an index.

1. Logarithm of a product

Since $a^{\log_a N} = N$ (i) and similarly $a^{\log_a M} = M$ (ii) and $a^{\log_a (MN)} = MN$ (iii), replacing N in (i) by M and MN respectively, substituting (i) and (ii) in (iii),

$$a^{\log_a M} \times a^{\log_a N} = a^{\log_a MN}$$

then $a^{(\log_a M + \log_a N)} = a^{\log_a MN}$ (first law of indices).

Equating indices of the same base,

$$\log_a M + \log_a N = \log_a MN \text{ (first law of logarithms)}$$

∴ *The logarithm of the product of two numbers is the sum of their logarithms.*

The student should compare this rule with the first law of indices on p. 29 (remembering that logarithms are indices). This may be extended to any number of factors. Thus

$$\log_a M + \log_a N + \log_a P = \log_a MNP$$

2. Logarithm of a quotient

Since $a^{\log_a M} = M$ (i), $a^{\log_a N} = N$ (ii) and $a^{\log_a \left(\frac{M}{N}\right)} = \dfrac{M}{N}$ (iii).

Substituting from (i) and (ii) in (iii),

$$a^{\log_a \left(\frac{M}{N}\right)} = \frac{a^{\log_a M}}{a^{\log_a N}}$$

∴ $a^{\log_a \left(\frac{M}{N}\right)} = a^{\log_a M - \log_a N}$ (second law of indices)

Equating indices of the same base,

$$\log_a M - \log_a N = \log_a \left(\frac{M}{N}\right) \quad \text{(second law of logarithms)}$$

Examples
3.8 Simplify $\log_a 20 + \log_a 3 - \log_a 30$.

$$\log_a 20 + \log_a 3 - \log_a 30 = \log_a (20 \times 3) - \log_a 30$$
$$\text{(first law of logarithms)}$$
$$= \log_a 60 - \log_a 30$$
$$= \log_a \left(\frac{60}{30}\right)$$
$$\text{(second law of logarithms)}$$
$$= \log_a 2$$

3.9 Simplify $\log_b 2 - \log_a 8 + \log_b 60 + \log_a 16 - \log_a \frac{1}{2} - \log_b 30$.

We first collect together logarithms of the same base.

$$\begin{aligned}
\text{Expression} &= (\log_b 2 + \log_b 60 - \log_b 30) + (\log_a 16 - \log_a 8 - \log_a \tfrac{1}{2}) \\
&= (\log_b [2 \times 60] - \log_b 30) + (\log_a [\tfrac{16}{8}] - \log_a \tfrac{1}{2}) \\
&= (\log_b 120 - \log_b 30) + (\log_a 2 - \log_a \tfrac{1}{2}) \\
&= \log_b \left(\frac{120}{30}\right) + \log_a \left(\frac{2}{\frac{1}{2}}\right) \\
&= \log_b 4 + \log_a 4
\end{aligned}$$

3. Logarithm of a power
Since $a^{\log M} = M$ (i) and $a^{\log (MK)} = (M^K)$ (ii)
Substituting M from (i) in (ii)
$$a^{\log (MK)} = (a^{\log_a M})^K = a^{K \log_a M} \quad \text{(third law of indices)}$$
Since $a^{\log_a (MK)} = a^{K \log_a M}$, equating indices of the same base
$$\log_a M^K = K \log_a M \quad \text{(third law of logarithms)}$$

Use of the laws of logarithms
Examples
3.10 Express in terms of $\log_a 2$

$$\text{(a) } \log_a 256 \qquad \text{(b) } \log_a \left(\tfrac{1}{64}\right)$$

(a) Since $256 = 2 \times 2 \times 2 \times 2 \times 2 \times 2 \times 2 \times 2 = 2^8$

$$\log_a 256 = \log_a 2^8 = 8 \log_a 2$$

(b) $\log_a \left(\tfrac{1}{64}\right) = \log_a 64^{-1} = -\log_a 64 \quad (64 = 2 \times 2 \times 2 \times 2 \times 2 \times 2 = 2^6)$
$$= -\log_a 2^6 = -6 \log_a 2$$

3.11 If $\log_a 2 = p$ and $\log_a 3 = 9$, express $\log_a 36 + \log_a 16 - \log_a \left(\frac{1}{216}\right)$ in terms of p and q.

$$\begin{aligned}
\text{Expression} &= \log_a (2^2 \times 3^2) + \log_a 2^4 - \log_a (216)^{-1} \\
&= \log_a 2^2 + \log_a 3^2 + \log_a 2^4 + \log_a 216 \\
&= 2\log_a 2 + 2\log_a 3 + 4\log_a 2 + \log_a (2^3 . 3^3) \\
&= 6\log_a 2 + 2\log_a 3 + \log_a 2^3 + \log_a 3^3 \\
&= 6\log_a 2 + 2\log_a 3 + 3\log_a 2 + 3\log_a 3 \\
&= 9\log_a 2 + 5\log_a 3
\end{aligned}$$

replacing $\log_a 2$ by p and $\log_a 3$ by q
$$= 9p + 5q,$$

The previous working is rather tedious but gives good practice in using the laws of logarithms. Alternatively,

$$\begin{aligned}
&\log_a 36 + \log_a 16 - \log_a \left(\tfrac{1}{216}\right) \\
&= \log_a (36 \times 16) - \log_a \left(\tfrac{1}{216}\right) \\
&= \log_a \left(\frac{36 \times 16}{1/216}\right) = \log_a (36 \times 16 \times 216) \\
&= \log_a [(3^2 \times 2^2)(2^4)(2^3 \times 3^3)] \\
&= \log_a (2^9 \times 3^5) \\
&= \log_a 2^9 + \log_a 3^5 \\
&= 9\log_a 2 + 5\log_a 3 \\
&= 9p + 5q
\end{aligned}$$

$$\left.\begin{aligned}
\text{N.B. } 216 &= 2 \times 108 \\
&= 2 \times 3 \times 36 \\
&= 2 \times 3 \times 3 \times 2 \times 3 \times 2 \\
&= 2^3 \times 3^3
\end{aligned}\right.$$

EXERCISE 3.3

1 Show by applying the laws of logarithms that $\log_a 25 + \log_a 9 - \log_a 45$ simplifies to $\log_a 5$.

2 Show by applying the laws of logarithms that $\log_c 441 + \log_d 10 - \log_c 21 + \log_d 25 - \log_c 7 - \log_d 750$ simplifies to $\log_c 3 - \log_d 3$.

3 Prove that $\log_x 2187 = 7\log_x 3$.

4 If $\log_y 3 = a$ and $\log_y 5 = b$, prove that
$\log_y 81 + \log_y 1125 - \log_y (1/15) = 7a + 4b$.

5 If $3\log_a x - 4\log_a y = \log_a 10$, prove that $x^3 = 10y^4$.

6 Prove that $\log_a (625/5) = 3\log_a 5$ but that $\dfrac{\log_a 625}{\log_a 5} = 4$.

To prove that $\log_a 1 = 0$

Since $1 = \dfrac{a^1}{a^1} = a^{1-1} = a^0$

i.e. $a^0 = 1$
then $0 = \log_a 1$ (by the definition of a logarithm).

To prove that $\log_a a = 1$
Since $a^1 = a$
then $1 = \log_a a$ (by the definition of a logarithm).

To prove that $a^{\log_a x} = x$
Let $a^{\log_a x} = z$ (i)
Then $\log_a x = \log_a z$ (by the definition of a logarithm)
$\therefore \quad x = z$
Substituting for z in (i), $a^{\log_a x} = x$

To prove that $1/\log_a b = \log_b a$
Let $\log_b a = z$ (i)
Then $b^z = a$
$\therefore \quad b = a^{1/z}$
$\therefore \quad \frac{1}{z} = \log_a b$
Substituting from (i),

$$1/\log_b a = \log_a b$$
$$\text{or} \quad 1/\log_a b = \log_b a.$$

The following exercise may seem rather repetitive and trivial to the critical student, but he should remember that mathematical progress proceeds by practice not by sheer memorising: results and processes are more easily remembered when familiar through frequent use.

EXERCISE 3.4
1 Prove from first principles that $\log_{10} 1 = 0$, $\log_e 1 = 0$.
2 Prove from first principles that $\log_{10} 10 = 1$, $\log_e e = 1$.
3 Prove from first principles that $10^{\log_{10} x} = x$, $e^{\log_e x} = x$.
4 Prove from first principles that $1/\log_{10} x = \log_x 10$, $1/\log_{10} e = \log_e 10$.

Application of laws of logarithms to calculations
We will now proceed to apply these rules to a few calculations noting that at this stage the base 10 may be omitted as understood. Thus $\log x$ means $\log_{10} x$.

Examples

3.12 Evaluate $\qquad \sqrt{\left(\dfrac{0 \cdot 0972 \times 19 \cdot 98}{6 \cdot 38 \times 0 \cdot 009\,39} \right)}$

Let $\qquad\qquad x = \sqrt{\left(\dfrac{0 \cdot 0972 \times 19 \cdot 98}{6 \cdot 38 \times 0 \cdot 009\,39} \right)}$

Taking logs,

$\log x = \frac{1}{2}\{\log 0 \cdot 0972 + \log 19 \cdot 98 - (\log 6 \cdot 38 + \log 0 \cdot 009\,39)\}$
$\qquad = \frac{1}{2}\{\bar{2} \cdot 9877 + 1 \cdot 3007 - (0 \cdot 8048 + \bar{3} \cdot 9727)\}$
$\qquad = \frac{1}{2}\{0 \cdot 2884 - 2 \cdot 7775\}$
$\qquad = \frac{1}{2}\{\bar{1} \cdot 5109\}$
$\qquad = 0 \cdot 7555$
$\qquad = \log (5 \cdot 696)$
$\therefore \quad x = 5 \cdot 696$

In performing the step $0\cdot2884 - \bar{2}\cdot7775$, first write $0\cdot2884$ as $\bar{1} + 1\cdot2884$.

Then from
take

$$\begin{array}{r} \bar{1} + 1\cdot2884 \\ \bar{2} + 0\cdot7775 \\ \hline 1 + 0\cdot5109 = 1\cdot5109 \end{array}$$

3.13 Evaluate $\sqrt[3]{(0\cdot0838)}$

Let $\qquad\qquad x = \sqrt[3]{(0\cdot0838)}$

Taking logs,

$$\begin{aligned} \log x &= \tfrac{1}{3} \log 0\cdot0838 \\ &= \tfrac{1}{3}\{\bar{2}\cdot9232\} \\ &= \tfrac{1}{3}\{\bar{3} + 1\cdot9232\} \\ &= \bar{1} + 0\cdot6411 \\ &= \bar{1}\cdot6411 \\ &= \log (0\cdot4376) \\ \therefore\quad x &= 0\cdot4376 \end{aligned}$$

It should be remembered that the mantissa (the decimal part) of a logarithm must always be positive, or the log tables cannot be used for it. We must, therefore, adjust $\bar{2}\cdot9232$, so that we can divide it by 3, and still keep a positive mantissa. This is done as shown. By writing $\bar{2}$ as $\bar{3} + 1$, we can get an *exact* division of the negative characteristic, leaving $+1$, which can be carried on to the mantissa, as it also is positive. Study the following which might arise in the process of taking a root.

$$\frac{\bar{1}\cdot6235}{2} \text{ would be written } \frac{\bar{2} + 1\cdot6235}{2} = \bar{1}\cdot8118$$

$$\frac{\bar{3}\cdot6235}{2} \quad ,, \quad ,, \quad ,, \quad \frac{\bar{4} + 1\cdot6235}{2} = \bar{2}\cdot8118$$

$$\frac{\bar{4}\cdot6235}{3} \quad ,, \quad ,, \quad ,, \quad \frac{\bar{6} + 2\cdot6235}{3} = \bar{2}\cdot8745$$

3.14 Evaluate $(0\cdot0273)^{\frac{2}{3}}$

Let $\qquad\qquad x = (0\cdot0273)^{\frac{2}{3}}$

Taking logs,

$$\begin{aligned} \log x &= \tfrac{2}{3} \log (0\cdot0273) \\ &= \tfrac{2}{3} \times (\bar{2}\cdot4362) \\ &= \frac{\bar{4}\cdot8724}{3} \\ &= \frac{\bar{6} + 2\cdot8724}{3} \\ &= \bar{2}\cdot9575 \\ &= \log 0\cdot090\,67 \\ \therefore\quad x &= 0\cdot090\,67 \end{aligned}$$

3.15 Find the value of $(6\cdot023)^{-2\cdot5}$

Let $\qquad x = (6\cdot023)^{-2\cdot5}$
Taking logs,

$$\log x = -2\cdot5 \times \log 6\cdot023$$
$$= -2\cdot5 \times 0\cdot7798$$
$$= -1\cdot9495$$
$$= \bar{2} + 0\cdot0505$$
$$= \bar{2}\cdot0505$$
$$= \log 0\cdot011\,23$$
$$\therefore \quad x = 0\cdot011\,23$$

In this example, it is seen that in multiplying $0\cdot7798$ by $-2\cdot5$, we obtain $-1\cdot9495$, which is wholly negative. Before using the log tables, we adjust this as $-2 + 0\cdot0505$, thus separating it into negative characteristic and positive mantissa, which can then be used in the usual way.

(Alternatively, we could write $(6\cdot023)^{-2\cdot5}$ as $\dfrac{1}{(6\cdot023)^{2\cdot5}}$, etc.)

3.16 Evaluate $(0\cdot1276)^{-1\cdot7}$

Let $\qquad x = (0\cdot1276)^{-1\cdot7}$
Taking logs,

$$\log x = -1\cdot7 \times \log 0\cdot1276$$
$$= -1\cdot7 \times \bar{1}\cdot1059$$
$$= (-1\cdot7 \times -1) + (0\cdot1059 \times -1\cdot7)$$
$$= 1\cdot7 - 0\cdot180\,03$$
$$= 1\cdot519\,97$$
$$= \log 33\cdot11$$
$$\therefore \quad x = 33\cdot11$$

3.17 Evaluate $(0\cdot0719)^{2\cdot4}$

Let $\qquad x = (0\cdot0719)^{2\cdot4}$
Taking logs,

$$\log x = 2\cdot4 \log 0\cdot0719$$
$$= 2\cdot4 \times \bar{2}\cdot8567$$
$$= (2\cdot4 \times -2) + (2\cdot4 \times 0\cdot8567)$$
$$= -4\cdot8 + 2\cdot056\,08$$
$$= 2\cdot056\,08 - 4\cdot8$$
$$= \bar{3}\cdot256\,08$$
$$= \log 0\cdot001\,803$$
$$\therefore \quad x = 0\cdot001\,803$$

The subtraction $(2\cdot056\,08 - 4\cdot8)$ is done as it is usually done in logarithms:

from	2·056 08
take	4·8
	$\overline{3}$·256 08

thus leaving only the characteristic negative.

Evaluations using logarithms

In the evaluations 3.12–3.17 it was possible to put the expression to be evaluated equal to x, and to commence a statement thus:

$$\log x = \ldots$$

This, however, was because only the processes of multiplication, division, the formation of powers, and the extraction of roots were involved in the computations. Logarithms cannot give assistance in addition or subtraction. To evaluate an expression containing the sum or difference of similar expressions logarithms can be used to find numbers for the simpler expressions, which must then be added or subtracted by plain arithmetic. Thus the working of an exercise cannot *as a general rule* commence with '$\log x = \ldots$' Students would be unwise to form a habit of commencing examples in this way.

Every student of mathematics must be able to follow, and if necessary originate, an exact statement in terms of logarithms such as those given in the examples of the last section. But an engineer or other technologist who needs to use logarithms will not write out his work in this form. He will manipulate his logarithms setting out a series of additions and subtractions as in elementary arithmetic, using the briefest possible connecting notes, often according to some personal shorthand. One such lay-out is given to take the place of the previous statement of example 3.12. It should be self-explanatory.

Evaluate $\sqrt{\left(\dfrac{0·0972 \times 19·98}{6·38 \times 0·009\ 39}\right)}$

	Number		*Logarithm*	
	6·38		0·8048	
	0·009 39		$\overline{3}$·9727	
	denominator	←	$\overline{2}$·7775	by addition
	0·0972		$\overline{2}$·9877	
	19·98		1·3007	
	numerator	←	0·2884	by addition
			$\overline{2}$·7775	
Using antilog tables			2)1·5109	by subtraction
5·696		←	0·7555	dividing by 2

Answer = 5·696

Exponential equations

Solution of the equation $a^x = b$

Since $a^x = b$ taking logarithms of each side to the base 10.

$$\log_{10} a^x = \log_{10} b$$
$$\therefore \quad x \log_{10} a = \log_{10} b$$
$$\therefore \quad x = \frac{\log_{10} b}{\log_{10} a}$$

Examples

3.18 Solve the equation $(8\cdot143)^x = 0\cdot0123$

$$(8\cdot143)^x = 0\cdot0123$$
$$\therefore \quad \log(8\cdot143)^x = \log(0\cdot0123)$$
$$\therefore \quad x \log(8\cdot143) = \log(0\cdot0123)$$
$$\therefore \quad x \times 0\cdot9108 = \bar{2}\cdot0899$$
$$\therefore \quad x = \frac{-2 + 0\cdot0899}{0\cdot9108}$$
$$= \frac{-1\cdot9101}{0\cdot9108}$$
$$= -2\cdot079$$

No.	Log.
1·9101	0·2810
0·9108	1·9594
2·079	0·3216

3.19 (*a*) Solve for x: $4^x = (9\cdot1)^{x-2}$.

Taking logs of both sides,

$$x \log 4 = (x - 2)(\log 9\cdot1)$$
$$\therefore \quad 0\cdot6021x = 0\cdot9590(x - 2)$$
$$\therefore \quad 0\cdot6021x = 0\cdot9590x - 1\cdot918$$
$$\therefore \quad 0\cdot3569x = 1\cdot918$$
$$\therefore \quad x = \frac{1\cdot918}{0\cdot3569}$$
$$\therefore \quad x = 5\cdot374$$

The student must be careful *to divide* $1\cdot918$ *by* $0\cdot3569$, either by arithmetic, or by subtracting the log of $0\cdot3569$ from the log of $1\cdot918$ in the usual way: it is incorrect to subtract $0\cdot3569$ from $1\cdot918$ on the assumption that these numbers are logarithms.

(*b*) Solve for x: $2^x \cdot 5^{x-3} = 7^{1-x}$.

Taking logs of both sides,

$$x \log 2 + (x - 3) \log 5 = (1 - x) \log 7$$
$$\therefore \quad 0\cdot3010x + 0\cdot6990(x - 3) = 0\cdot8451(1 - x)$$
$$\therefore \quad 0\cdot3010 + 0\cdot6990x - 2\cdot097 = 0\cdot8451 - 0\cdot8451x$$
$$\therefore \quad 1\cdot8451x = 2\cdot9421$$
$$\therefore \quad x = 1\cdot594$$

(c) Solve for x: $3^{2x} = 5 \times 2^x$.

Taking logs to the base 10 of both sides,
$$\log 3^{2x} = \log 5 + \log 2^x$$

		No.	Log.
$\therefore$	$2x \log 3 = \log 5 + x \log 2$		
$\therefore$	$0{\cdot}4771 \times 2x = 0{\cdot}6990 + 0{\cdot}3010x$		
$\therefore$	$0{\cdot}9542x - 0{\cdot}3010x = 0{\cdot}6990$	0·6990	$\bar{1}{\cdot}8445$
$\therefore$	$0{\cdot}6532x = 0{\cdot}6990$	0·6532	$\bar{1}{\cdot}8150$
$\therefore$	$x = \dfrac{0{\cdot}6990}{0{\cdot}6532}$		
$\therefore$	$x = 1{\cdot}070$	1·070	0·0295

Solution of the equation $a^{2x} + ba^x + c = 0$

Let $z = a^x$, so that $z^2 = a^{2x}$. Substitute in $a^{2x} + ba^x + c = 0$

Then $z^2 + bz + c = 0$, a quadratic equation in z.

This quadratic equation is then solved. Suppose that the solutions or roots are z_1 and z_2. Then $z_1 = a^x$ and $z_2 = a^x$ are solved for x as previously, i.e. $x = \log_a z_1$ or $x = \log_a z_2$.

Examples

3.20 Solve the equation $3^{2x} - 6{\cdot}3^x + 8 = 0$.

Let $z = 3^x$ so that $z^2 = (3^x)^2 = 3^{2x}$.

		No.	Log.
Then	$z^2 - 6z + 8 = 0$		
$\therefore$	$(z - 4)(z - 2) = 0$	0·6021	$\bar{1}{\cdot}7797$
$\therefore$	$z - 4 = 0$ or $z - 2 = 0$	0·4771	$\bar{1}{\cdot}6786$
$\therefore$	$z = 4$ or $z = 2$		
$\therefore$	$3^x = 4$ or $3^x = 2$	1·262	0·1011
$\therefore$	$x \log 3 = \log 4$ or $x \log 3 = \log 3$		

		No.	Log.
$\therefore$	$x = \dfrac{\log 4}{\log 3}$ or $x = \dfrac{\log 2}{\log 3}$	0·3010	$\bar{1}{\cdot}4786$
$\therefore$	$x = \dfrac{0{\cdot}6021}{0{\cdot}4771}$ or $x = \dfrac{0{\cdot}3010}{0{\cdot}4771}$	0·4771	$\bar{1}{\cdot}6786$
$\therefore$	$x = 1{\cdot}262$ or $x = 0{\cdot}6310$	0·6310	$\bar{1}{\cdot}8000$

EXERCISE 3.5 (miscellaneous)

Evaluate the following using four-figure logarithms.

1 $\sqrt[5]{(0{\cdot}0263)}$

2 $(92{\cdot}01)^{-1{\cdot}3}$

3 $(16{\cdot}7)^{-0{\cdot}9}$

4 $2 \times 19{\cdot}02^{-0{\cdot}3}$

5 $(0{\cdot}008\,82)^{-1{\cdot}7}$

6 $(0{\cdot}002\,59)^{0{\cdot}18}$

7 $(0{\cdot}017)^{0{\cdot}017}$

8 $(0{\cdot}25)^{0{\cdot}25}$

9 $\dfrac{2}{(0{\cdot}0619)^{0{\cdot}03}}$

10 $(0{\cdot}626)^{0{\cdot}5} \times (9{\cdot}002)^{-0{\cdot}62}$

11 $(13{\cdot}27)^{0{\cdot}6} \times \log 2{\cdot}718$

12 $(0{\cdot}9929)^{\frac{1}{3}}$

13 $\dfrac{92{\cdot}26^{0{\cdot}7}}{\log 2{\cdot}718}$

14 $(0{\cdot}213)^{\frac{1}{2}} \times (7{\cdot}16)^{-\frac{1}{3}}$

15 $\dfrac{1 \cdot 521^{-0 \cdot 2}}{\sqrt[6]{(1 \cdot 923)}}$

16 ae^{-kt}, when $a = 6$, $e = 2 \cdot 718$, $k = 45$ and $t = 0 \cdot 0037$.

Solve for x in numbers 17 and 18.

17 $x^{3 \cdot 6} = 200$ **18** $500x^{-2 \cdot 8} = 5 \cdot 16$

19 Solve for y: $7 \cdot 16^{y} = 1 \cdot 92^{y+2}$

20 Solve for p: $3^{p} \cdot 5^{p} = 8^{2p} \cdot 7^{1-p}$

21 In the equation $n = K \cdot H^{1 \cdot 25} \cdot P^{-0 \cdot 25}$, find K, when $H = 12$, $n = 200$ and $P = 80$.

22 A gas is expanding according to the law $pv^{n} = C$.
  (a) Find C when $p = 92$, $v = 2 \cdot 6$ and $n = 1 \cdot 3$
  (b) Find v when $C = 310$, $p = 96$ and $n = 1 \cdot 3$
  (c) Find n when $C = 330$, $p = 91$ and $v = 2 \cdot 5$

23 (a) Evaluate (i) $\log_3 27$, (ii) $\log_4 \left(\frac{1}{64}\right)$.

  (b) If $T = 2\pi \sqrt{\dfrac{l}{g}}$, find T when $l = 37 \cdot 25$, $g = 32 \cdot 2$.   RUGBY

24 (a) Express as one power of x $\dfrac{x^{2n+1}\sqrt{(x^{n-1})}}{\sqrt{(x^{3n})} \cdot x^{n-\frac{1}{2}}}$

  (b) Express as one logarithm $\dfrac{2 \log A + \log B}{3} - \log C$

  (c) Evaluate $27 \cdot 5\, e^{-0 \cdot 475}$ COVENTRY

25 Given that $d = \frac{1}{3} \sqrt[5]{\dfrac{G^{2}l}{h}}$, find G when $d = 1 \cdot 6$, $l = 3500$ and $h = 52$.

26 (a) Simplify the following expressions

  (i) $\dfrac{(2a^{3}b^{2})^{2} + (2a^{2}b^{3})^{2}}{4a^{2} + 4b^{2}}$ (ii) $\dfrac{8a^{-1}b^{-1}}{4a^{\frac{1}{2}}b^{\frac{3}{4}}} \times \dfrac{ab^{\frac{5}{8}}}{a^{\frac{3}{4}}b^{\frac{1}{4}}}$

  (b) Rationalise

  (i) $\dfrac{1}{\sqrt{3} - \sqrt{2}}$ (ii) $\dfrac{3 + \sqrt{2}}{\sqrt{2} - 1}$

  and evaluate the expressions given that $\sqrt{(2} = 1 \cdot 414)$ and $\sqrt{(3} = 1 \cdot 732)$.   WORCESTER

27 (a) Evaluate $\sqrt[3]{\dfrac{4 \cdot 156 \times (0 \cdot 006\ 12)^{2}}{0 \cdot 891}}$

  (b) From the formula $T_1 = T_2 e^{\mu\theta}$, find μ when $T_1 = 2$, $T_2 = 1$, $e = 2 \cdot 718$, $\theta = 3 \cdot 142$.

  (c) The distance of the horizon at sea-level varies as the square root of the height of the eye above sea-level. When the eye is at height 16 m, the horizon is 14·5 km away. Find the distance when the eye is 2 m above sea-level.   SUNDERLAND

28 (*a*) Solve the equation $8.41^{x-2} = 24.39$.

(*b*) Simplify

$$\left(\frac{a^8 \cdot b^{-4}}{a^{-1} \cdot b}\right)^{-\frac{1}{3}} + \frac{a^{-\frac{5}{3}} \cdot b}{a^2 \cdot b^{\frac{1}{3}}}$$

WORCESTER

29 (*a*) Evaluate, without using tables,

$$\left[\frac{3^{n+\frac{1}{2}}}{9^{n-\frac{1}{2}}} \div \frac{\sqrt{6}}{8^{1/n}}\right]^{-n/2}$$

when $n = 2$.

(*b*) Solve the equation $e^{2x} = 0.125$, given that $e = 2.718$.

COVENTRY

30 (*a*) Using logarithms, evaluate

$$\frac{(15.83)^{\frac{3}{4}} - 7.259}{(8.297 \times 0.264)^{0.2}}$$

(*b*) The speed v m/s gained by a stone in falling through a distance h m is proportional to the square root of h. If the speed of the stone after falling 10 m is 14 m/s, how far must it fall before its speed becomes 56 m/s?   SUNDERLAND

31 (*a*) Evaluate, using logarithms, (i) $0.062\,37^{-1\frac{1}{2}}$, (ii) $4.6^{-0.361}$.

(*b*) Solve the equation $x^{2\frac{1}{2}} = 10$.

(*c*) Without using logarithm tables show that

$$\frac{\log 8 + \log 4}{\log 8 - \log 2} = \frac{5}{2}$$

SURREY

32 (*a*) Find, using logarithms, the value of

$$\sqrt{\frac{(1.878)^2 + (2.142)^2}{14.4}}$$

(*b*) Given that $p\left(1 + \frac{cl^2}{k^2}\right) = f,$

find an expression for c in terms of the other quantities.   ULCI

33 (*a*) Simplify $16^{\frac{3}{4}} \times 27^{\frac{1}{3}} \times 10^{-2}$.

(*b*) Evaluate p, given that

$$p = \frac{f}{2}\left\{1 + \sqrt{\left(1 + \frac{4s^2}{f^2}\right)}\right\}$$

where $f = 2900$ and $s = 1746$.   WEST RIDING

34 (*a*) Express each of the numbers 0.3802 and $\dfrac{1}{\sqrt[3]{0.3802}}$ as a power of 10.

(*b*) For what value of x is $50^x = 100$?

(*c*) Calculate the value of $\dfrac{4.7(P + Q)}{P\sqrt{P^2 - Q^2}}$ when $P = 18.35$, $Q = 14.85$.

NCTEC

35 In a belt drive, the ratio of the tension T_1 on the tight side of the belt to the tension T_2 on the slack side is given by the equation

$$\frac{T_1}{T_2} = e^{\mu\alpha},$$

where $e = 2 \cdot 718$, μ is the coefficient of friction between the belt and the pulley, and α is the angle of lap of the belt in radians, whilst the effective pull of the belt is $T_1 - T_2$. If $\mu = 0 \cdot 27$, $\alpha = 165$ degrees and the effective pull $= 160$ N, find T_1 and T_2. Without actually working out, state the effect on the ratio of the tensions of doubling the value of α.

36 Solve the equation $10^{2x} - 3.10^x + 2 = 0$, using the substitution $z = 10^x$, or otherwise.

37 Solve the equation $\frac{5}{4} = \frac{1}{2}(2^x + 2^{-x})$, using the substitution $z = 2^x$ or otherwise.

38 Solve the equations (a) $3^{x2} = 9^{x-\frac{1}{2}}$, (b) $3^{2x+1} - 4 \cdot 3^{x+1} + 2 = 0$.

MID-CHESHIRE

39 (a) If $\log_{10} 2 = 0 \cdot 301\ 03$ and $\log_{10} 3 = 0 \cdot 477\ 12$, calculate, without use of tables, the value of (i) $\log_{10} \sqrt{6}$, (ii) $\log_{10} 1 \cdot 5$ and (iii) $\log_{10} 5$.

(b) If $x = \dfrac{2t}{1 + t^2}$ and $y = \dfrac{1 - t^2}{1 + t^2}$, find the value of $x^2 + y^2$ in its simplest form.

ULCI

40 If $\log_x M = p$ and $\log_y M = p$, prove that

$$\frac{p + q}{p - q} = \frac{\log_z y + \log_z x}{\log_z y - \log_z x}$$

Verify this result when $x = 9$, $y = 27$, $z = 3$, and $M = 81$, without using tables.

MID-CHESHIRE

Chapter Four
Logarithms to Other Bases than 10

Relation between the logarithms of a number to different bases
Hitherto in using logarithms the student has used only those which are calculated to base 10, although it should be noted that the laws which govern the use of logarithms have been shown to be true whatever the base.

But the student will find that the logarithms which he has to use, especially in engineering, are not always referred to 10 as a base. Consequently it is necessary to consider in these cases what relation such logarithms bear to those calculated to base 10, which he finds in his tables.

In simple cases, the logarithms to other bases can easily be determined. For example, since $5^2 = 25$, the logarithm of 25 to base 5 is 2, or, with the usual notation,

$$\log_5 25 = 2$$
Similarly
$$\log_4 64 = 3$$

For actual computations the logarithm tables are calculated to base 10, and in working it is not usual to specify this base, as it is taken for granted that logarithms to base 10 are being used. If, however, any other base is employed, it is most important that it should be specified.

To obtain logarithms to bases other than 10
In cases such as we have considered above, where the index or logarithm is a positive integer, we can often obtain this by inspection.

For example, if we require $\log_6 216$, by definition, if x is the required logarithm, then we must have $6^x = 216$.

Hence we can see that $x = 3$, since $6^3 = 216$, and

$$\log_6 216 = 3$$

Now consider a more difficult case: to find $\log_3 10$.
If x be the required logarithm, then, by definition, since $x = \log_3 10$

$$3^x = 10$$

Taking logs to base 10,

$$x \log_{10} 3 = \log_{10} 10$$
$$\therefore \quad x \log_{10} 3 = 1$$

$$\therefore \quad x = \frac{1}{\log_{10} 3} = \frac{1}{0 \cdot 4771}$$

$$\therefore \quad \log_3 10 = 2 \cdot 096$$

To evaluate $\log_3 100$ we could proceed as above, or, employing the third law of indices, we could use the following method.

Since $100 = 10^2$, then $2 = \log_{10} 100$

But, as found above, $10 = 3^{2 \cdot 096}$.

$$\therefore \quad 100 = 10^2 = (3^{2 \cdot 096})^2$$
$$\therefore \quad \log_3 100 = \log_3 (3^{2 \cdot 096})^2$$
$$= 2 \times 2 \cdot 096$$
$$= \log_{10} 100 \times \log_3 10$$

We now proceed to give a general treatment of this.

To find the relation between the logarithms of a number to two different bases

Let N be any number and a and b two bases.

Let
$$N = b^y$$
$$\therefore \quad y = \log_b N$$
$$\therefore \quad \log_a N = \log_a (b^y)$$
$$= y \log_a b$$
$$= \log_b N \times \log_a b$$

This important result should be carefully memorised:

$$\log_a N = \log_b N \times \log_a b$$

It will be noticed that the factor $\log_a b$ is always the same, whatever the value of N; therefore, if we know the log of N to a base b, and we require the log to base a, we multiply the known log of N to the base b, by the log of the base b to the new base a. This constant multiplier is termed a *modulus*.

Since $\log_a N = \log_b N \times \log_a b$ for all values of N, let $N = a$.

Then
$$\log_a a = \log_b a \times \log_a b$$
But
$$\log_a a = 1$$
$$\therefore \quad \log_b a \times \log_a b = 1$$

and
$$log_a b = \frac{1}{\log_b a}$$

which is a useful result.

The student might be helped in remembering the formula

$$\log_a N = \log_b N \times \log_a b \quad \cdot \quad \cdot \quad \cdot \quad \cdot \quad \cdot \quad \text{(i)}$$

by noticing the sequence of the letters in

$$\frac{N}{a} = \frac{N}{b} \times \frac{b}{a}$$

where the numerators represent the appropriate numbers in (i), and the denominators represent the corresponding bases.

Naperian or natural logarithms

Logarithms were first discovered by Lord Napier in the sixteenth century. As he used them, they were not calculated to base 10, but to a base denoted by the letter 'e', where

$$e = 1 + \frac{1}{1} + \frac{1}{1 \cdot 2} + \frac{1}{1 \cdot 2 \cdot 3} + \frac{1}{1 \cdot 2 \cdot 3 \cdot 4} + \ldots \text{ ad infinitum.}$$

In this series, the terms are continually diminishing and ultimately become indefinitely small. To calculate the value of e to any required degree of accuracy, we take as many terms as may be required for the purpose. Its value correct to five places of decimals is 2·718 28.

To the student who has not progressed very far in mathematics, the choice of such a base is difficult to understand, but, when his studies are sufficiently advanced, he will discover that logarithms calculated to this base enter naturally into the higher branches of mathematics, and indeed are used almost exclusively except when computations are necessary. Hence the term 'natural logarithms'. They are also termed 'hyperbolic logarithms'. The student will also find that these logarithms enter considerably into his work in engineering and physics, and consequently a knowledge of their properties is essential.

For numerical computations, Naperian logarithms are inconvenient, and it is best to use the common logarithms, since the base 10 to which they are calculated is also the base of the common scale of notation. One advantage is that numbers with the same significant figures have the same mantissa or decimal part, the characteristic or integral part varying with the position of the decimal point in the number. With this the student is already acquainted, and can work out the proof for himself.

It was Henry Briggs, a Professor of Mathematics at Oxford, who saw the great advantage of common logarithms in calculations, and it was he who first compiled a table of logarithms calculated to base 10.

To convert Naperian logarithms to common logarithms

The student will frequently need, in the course of his work, to find the numerical value of the Naperian logs of numbers, since tables of these are not usually accessible. In order to do this, he must know the relation between these and the common logarithms of his tables. The conversion is effected by using the formula derived above:

$$\log_a N = \log_b N \times \log_a b$$

Adapting this, we have, replacing a by e and b by 10

$$\log_e N = \log_{10} N \times \log_e 10$$

Also, using the result

$$\log_a b = \frac{1}{\log_b a}$$

we have $\log_e 10 = \dfrac{1}{\log_{10} e}$

Consequently, to make the change, we use either,

$$\log_e N = \log_{10} N \times \log_e 10$$

or
$$\log_e N = \log_{10} N \times \frac{1}{\log_{10} e}$$

Now, from the tables,

$$\log_{10} e = 0.4343$$

$$\therefore \quad \log_e 10 = \frac{1}{0.4343} = 2.3026$$

Hence

$$\log_e N = \log_{10} N \times 2.3026$$
or
$$\log_e N = \log_{10} N \div 0.4343$$

Examples
4.1 Evaluate $\log_e 50$.

From tables	$\log_{10} 50 = 1.6990$
$\therefore$	$\log_e 50 = 1.6990 \times 2.3026$
Let	$x = 1.6990 \times 2.3026$
then	$\log x = \log 1.6990 + \log 2.3026$
	$= 0.2303 + 0.3622$
	$= 0.5925$
	$= \log 3.913$
	$x = 3.913$
or	$\log_e 50 = 3.913$

As $\log_{10} 2.3026$ will constantly be used in such calculations, the student should note that $\log_{10} 2.3026 = 0.3622$

4.2 Find the value of Q from the formula $\quad Q = \log_e \dfrac{T}{461} + \dfrac{1420}{T} - 0.65$

when $\qquad\qquad T = 600$

$$\log_e \tfrac{600}{461} = \log_{10} (\tfrac{600}{461}) \times 2.3026$$
$$= (\log_{10} 600 - \log_{10} 461) \times 2.3026$$
$$= (2.7782 - 2.6637) \times 2.3026$$
$$= 0.1145 \times 2.3026$$
$$= 0.2637$$

$$\frac{1420}{T} = \frac{1420}{600} = 2\cdot3667$$

$$\therefore \quad Q = 0\cdot2637 + 2\cdot3367 - 0\cdot65$$
$$= 2\cdot6304 - 0\cdot65$$
$$= 1\cdot9804$$

4.3 Express as simply as possible (a) $b^{\log_b 3}$, (b) $b^{2\,\log_b 3}$.   NCTEC

(a) Let $x = b^{\log_b 3}$
Taking logs to base b,

$$\log_b x = \log_b 3$$
$$\therefore \quad x = 3$$

(b) Let $x = b^{2\,\log_b 3}$
Taking logs to base b,

$$\log_b x = 2\log_b 3$$
$$\therefore \quad \log x = \log (3)^2$$
$$\therefore \quad x = 9$$

Use of natural, Naperian or hyperbolic logarithm tables
Often the Naperian logarithms of numbers from $1\cdot0$ to $10\cdot0$ are available in mathematical tables.

Thus $\log_e 5\cdot501 = 1\cdot7049$ immediately

To find $\log_e 55\,010$

$$\log_e 55\,010 = \log_e (5\cdot501 \times 10\,000) = \log_e (5\cdot501 \times 10^4)$$
$$= \log_e 5\cdot501 + \log_e 10^4$$
$$= \log_e 5\cdot501 + 4\log_e 10$$

At this stage the tables may be used.

Thus, $\log_e 55\,010 = 1\cdot7049 + 4 \times 2\cdot3026$
$$= 1\cdot7049 + 9\cdot2104$$
$$= 10\cdot9153$$

To find $\log_e 0\cdot000\,055\,01$

$$\log_e 0\cdot000\,055\,01 = \log_e (5\cdot501 \times 0\cdot000\,01) = \log_e (5\cdot501 \times 10^{-5})$$
$$= \log_e 5\cdot501 - 5\log_e 10 = 1\cdot7049 - 5 \times 2\cdot3026$$
$$= 1\cdot7049 - 11\cdot5130$$
$$= -9\cdot8081$$

The student is advised to always work in this way, using the laws of logarithms. He will thus formulate for himself the following rule:

Naperian logarithm tables may be used in the same manner as *common logarithm tables*, provided that a factor of $2\cdot3026$ is introduced with the characteristic.

Example
4.4

$$\log_e 955{\cdot}2 = 2(2{\cdot}3026) + 2{\cdot}2567$$
$$= 4{\cdot}6052 + 2{\cdot}2567 = 6{\cdot}8619$$
$$\log_e 0{\cdot}046\ 25 = \bar{2}(2{\cdot}3026) + 1{\cdot}5315$$
$$= -4{\cdot}6052 + 1{\cdot}5315 = -3{\cdot}0737$$
$$\operatorname{antilog}_e 8{\cdot}2031 = \operatorname{antilog}_e (3 \times 2{\cdot}3026 + 8{\cdot}2031 - 3 \times 2{\cdot}3026)$$
$$= \operatorname{antilog}_e (3 \times 2{\cdot}3026 + 8{\cdot}2031 - 6{\cdot}9078)$$
$$= \operatorname{antilog}_e (3 \times 2{\cdot}3026 + 1{\cdot}2953)$$
$$= 257{\cdot}0$$
$$\operatorname{antilog}_e - 0{\cdot}0108 = \operatorname{antilog}_e (-1 \times 2{\cdot}3026 + 1 \times 2{\cdot}3026 - 0{\cdot}0108)$$
$$= \operatorname{antilog}_e (-1 \times 2{\cdot}3026 + 2{\cdot}2918) = 0{\cdot}9893$$

Thus, to find $\operatorname{antilog}_e (N)$, multiples of $2{\cdot}3026$ are added or subtracted from N until a number results which is between 0 and $2{\cdot}3026$ and can thus be located in the tables. The characteristic is used as usual to adjust the decimal point.

Transposition of logarithmic formulae

Examples
4.5 (*a*) Make W the subject of the formula $\log_e \left(\dfrac{T_1 + W}{T_2} \right) = \mu\theta$

(*b*) Make x the subject of the formula $y = \dfrac{1 - e^{-2x}}{1 + e^{-2x}}$ using the substitution $z = e^{-2x}$, or otherwise.

(*a*)
$$\log_e \frac{T_1 + W}{T_2} = \mu\theta \qquad \therefore \quad e^{\theta\mu} = \frac{T_1 + W}{T_2}$$

$$\therefore \quad T_2 e^{\theta\mu} = T_1 + W \qquad \therefore \quad W = T_2 e^{\theta\mu} - T_1$$

(*b*) If
$$y = \frac{1 - e^{-2x}}{1 + e^{-2x}}, \text{ let } z = e^{-2x}$$

then
$$y = \frac{1 - z}{1 + z}$$

$$\therefore \quad y + yz = 1 - z \qquad \therefore \quad yz + z = 1 - y$$

$$\therefore \quad z(1 + y) = 1 - y \qquad \therefore \qquad z = \frac{1 - y}{1 + y}$$

i.e.
$$e^{-2x} = \frac{1 - y}{1 + y} \qquad \therefore \qquad e^{2x} = \frac{1 + y}{1 - y}$$

$$\therefore \qquad 2x = \log_e \frac{1 + y}{1 - y} \qquad \therefore \qquad x = \tfrac{1}{2} \log_e \frac{1 + y}{1 - y}$$

4.6 (*a*) Make n the subject of the formula $Q = \lambda\sqrt{\dfrac{2gh}{n^2 - 1}}$

(*b*) Make v the subject of the formula

$$\frac{1}{2\lambda}\left(\log_e \frac{\lambda + v}{\lambda - v} - \log_e \frac{\lambda + V_1}{\lambda - V_1}\right) = \lambda t$$

(*c*) Determine t when $v = \dfrac{\lambda}{2}$ and $V_1 = \dfrac{\lambda}{4}$ in (*b*)

(*a*) $$\frac{Q}{\lambda} = \sqrt{\left(\frac{2gh}{n^2 - 1}\right)} \qquad \therefore \quad \frac{Q^2}{\lambda^2} = \frac{2gh}{n^2 - 1}$$

$$\therefore \quad n^2 - 1 = \frac{2gh\lambda^2}{Q^2} \qquad \therefore \quad n^2 = \frac{2gh\lambda^2}{Q^2} + 1$$

$$\therefore \quad n^2 = \frac{2gh\lambda^2 + Q^2}{Q^2} \qquad \therefore \quad n = \frac{\sqrt{(2gh\lambda^2 + Q^2)}}{Q}$$

(*b*) $\dfrac{1}{2\lambda}\left(\log_e \dfrac{\lambda + v}{\lambda - v} - \log_e \dfrac{\lambda + V_1}{\lambda - V_1}\right) = \lambda t$

$$\therefore \quad \log_e \frac{\lambda + v}{\lambda - v} - \log_e \frac{\lambda + V_1}{\lambda - V_1} = 2\lambda^2 t$$

$$\therefore \quad \log_e \frac{\lambda + v}{\lambda - v} \cdot \frac{\lambda - V_1}{\lambda + V_1} = 2\lambda^2 t$$

$$\therefore \quad e^{2\lambda^2 t} = \frac{\lambda + v}{\lambda - v} \cdot \frac{\lambda - V_1}{\lambda + V_1}$$

$$\therefore \quad \frac{\lambda + V_1}{\lambda - V_1} e^{2\lambda^2 t} = \frac{\lambda + v}{\lambda - v}$$

Write z for $\dfrac{\lambda + V_1}{\lambda - V_1}e^{2\lambda^2 t}$, then $z = \dfrac{\lambda + v}{\lambda - v}$ and $z\lambda - zv = \lambda + v$

$$\therefore \quad z\lambda - \lambda = v + zv \qquad \therefore \quad \lambda(z - 1) = v(z + 1)$$

$$\therefore \quad v = \lambda\left(\frac{z - 1}{z + 1}\right)$$

Substituting for z, $v = \lambda \dfrac{\left(\dfrac{\lambda + V_1}{\lambda - V_1}e^{2\lambda^2 t} - 1\right)}{\dfrac{\lambda + V_1}{\lambda - V_1}e^{2\lambda^2 t} + 1}$ . . . . (i)

(c) When $\frac{\lambda}{2}$ and $V_1 = \frac{\lambda}{4}$ then

$$\frac{1}{2\lambda}\left(\log_e \frac{\lambda + \lambda/2}{\lambda - \lambda/2} - \log_e \frac{\lambda + \lambda/4}{\lambda - \lambda/4}\right) = \lambda t$$

$$\therefore \ \frac{1}{2\lambda}\left(\log_e 3 - \log_e \frac{5}{3}\right) = \lambda t$$

$$\therefore \ \frac{1}{2\lambda^2}\log_e \frac{3 \times 3}{5} = t$$

i.e. $$t = \frac{1}{2\lambda^2}\log_e \frac{9}{5} \quad . \quad . \quad . \quad . \quad . \ \text{(ii)}$$

The student should note that we can test the consistency of result (i) in

(b) since, when $v = \frac{\lambda}{2}$ and $V_1 = \frac{\lambda}{4}$, from (ii) $2\lambda^2 t = \log_e \frac{9}{5}$ and $e^{2\lambda^2 t} = \frac{9}{5}$.

In (i) L.H.S. $= v = \frac{\lambda}{2}$

$$\text{R.H.S.} = \lambda\frac{\left(\frac{\lambda + \lambda/4}{\lambda - \lambda/4} \times \frac{9}{5} - 1\right)}{\left(\frac{\lambda + \lambda/4}{\lambda - \lambda/4} \times \frac{9}{5} + 1\right)} = \lambda\frac{\left(\frac{5}{3} \times \frac{9}{5} - 1\right)}{\left(\frac{5}{3} \times \frac{9}{5} + 1\right)}$$

$$= \lambda\frac{(3 - 1)}{(3 + 1)} = \frac{\lambda}{2}$$

which is consistent for the particular values considered.

EXERCISE 4.1

1 Transpose the formula $y = \lambda(1 - e^{-At})$ and prove that $t = \frac{1}{A}\log_e\frac{\lambda}{\lambda - y}$.

2 Given that $t = A - \frac{L}{R}\log_e (E - Ri)$ and that $i = 0$ when $t = 0$, find A and prove that

(a) $t = \frac{L}{R}\log_e \frac{E}{E - Ri}$

(b) $i = \frac{E}{R}\left(1 - e^{-Rt/L}\right)$

3 Given that $(\phi_2 - \phi_1) = \frac{R}{J}\log_e\frac{V_2}{V_1}$, prove that

$$v_1 = v_2\, e^{\,J(\phi_1 - \phi_2)/R}$$

4 Given that $\phi_2 - \phi_1 = C_v \log_e \dfrac{T_2}{T_1} + \dfrac{R}{J} \log_e \dfrac{V_2}{V_1}$, prove that

(a) $e^{\phi_2 - \phi_1} = \left(\dfrac{T_2}{T_1}\right)^{C_v} \left(\dfrac{V_2}{V_1}\right)^{R/J}$

(b) $\dfrac{V_2}{V_1} = \left[e^{\phi_2 - \phi_1}\left(\dfrac{T_1}{T_2}\right)^{C_v}\right]^{R/J}$

(c) $V_2 = V_1 e^{(\phi_2 - \phi_1)J/R}\left(\dfrac{T_1}{T_2}\right)^{C_v J/R}$

5 (a) Find the values of x which satisfy the equation $\log \frac{1}{2}x = 2\log(2x - 5)$.
  (b) Express the law $i = ae^{Kt}$ in its 'straight line form'.
  (c) It is found that t and D are connected by the equation $\log t = b + a \log D$. Express t as a function of D free from all logarithms. (Assume that the logarithms are to the base 10.)   NCTEC

6 (a) If $P = P_0 e^{rt/100}$, calculate the value of t when $r = 3$ and $P = 2P_0$.
  (b) If the value found for t in (a) above is t_1, what is the ratio P/P_0 when $t = nt_1$, where n is any positive number and $r = 3$, as before?
  (c) If $e^{x^2} = 5^{3(2x-2)}$, what is the value of x?   NCTEC

EXERCISE 4.2 (miscellaneous)

1 Write down, by inspection, the value of
  (a) $\log_4 64$ (c) $\log_2 64$ (e) $\log_{16} 1024$
  (b) $\log_5 125$ (d) $\log_3 27$ (f) $\log_{81} 9$
2 (a) What power of 5 is 50? Find the value of $\log_5 50$.
  (b) Find $\log_{6.5} 100$.
3 Calculate the value of
  (a) $\log_e 4.6$ (c) $\log_e 9.6$ (e) $\log_e 56$
  (b) $\log_e 7.5$ (d) $\log_e 40$ (f) $\log_e 0.062$
4 Show that
  (a) $\log_e 160 = \log_e 8 + \log_e 5 + 2\log_e 2$
  (b) $\log_e 96 = \log_e 2.4 + \log_e 4 + \log_e 100 - \log_e 10$
5 Find the numbers whose logs to base e are
  (a) 1.39 (b) 1.86 (c) 2.205 (d) $\bar{1}.1560$
6 Given a table of ordinary logarithms, show how you would find the logarithm of any number N to any base b. Hence calculate $\log_e 2.5$, where $e = 2.718$.   UEI
7 (a) Find the value of T_1 when $T_2 = 30$, $\mu = 0.5$, $\theta = 1.2$ radians and $\log_e \dfrac{T_1}{T_2} = \mu\theta$.

  (b) Simplify (i) $e^{\log_e x}\,x$, (ii) $\dfrac{1}{e^{-2\log_e x}}$, (iii) $\dfrac{1}{\log_e a}\,e^{\log_e a}$

  (c) Evaluate (i) $\sqrt{(0.1624)}$, (ii) $0.1624^{-1.5}$, (iii) $\dfrac{1}{0.1624^3}$

          COVENTRY

8 Evaluate (a) $\log_3 81$,   (b) $\log_5 20$,   (c) $\log_e 63$.   RUGBY

9 (a) Evaluate, showing full logarithm details, $(0.569)^{0.73}$

(b) It is known that $\phi = \log_e \dfrac{t}{273}$. If $t = 273 + \theta$, find ϕ when $\theta = 150$.

DUDLEY

10 If P_m is the mean pressure of steam in a cylinder of an engine, P the pressure of steam on admission to the cylinder, and r the ratio of expansion, then (assuming hyperbolic expansion and exhaust at atmospheric pressure)

$$P_m = P\,\frac{1 + \log_e r}{r}$$

Calculate P_m when $P = 8.0$ bar and $r = 2$.   WORCESTER

11 (a) Find the values of (i) $\log_7 87$ and (ii) $\log_3 0.4178$.

(b) Use logarithms to evaluate $(0.2137)^{0.7863}$.

(c) If $A = 5.27 - 1.3 \log_e \dfrac{(x + 5)}{(x - 2)}$, find the value of A when $x = 4.6$.

SW ESSEX

12 If $2 \log_e x = 5.4$, find x. Solve the equation $2^x = 100$.   COVENTRY

13 When a metal cools its temperature (T) at a time (t) is given by the formula $T = R + (T_0 - R)e^{-kt}$, where T_0 is the initial temperature, and R is the room temperature.

If, when the room temperature is $10°C$ and the initial temperature is $510°C$, the metal cools to $100°C$ in 1 h, find the constant k and the additional time taken for the metal to cool to $50°C$. ($e = 2.718$)

COVENTRY

14 The insulation resistance R megohms of length l m of a certain wire is given by

$$R = \frac{0.42\rho}{l} \times \log_e \frac{d_2}{d_1}.$$

where d_1 and d_2 are the inside and outside diameters of the insulating material and ρ is the resistivity. Find l to the nearest metre when $\rho = 1000$ megohm metres, $R = 0.44$ megohm, $d_2 = 3.0$ mm and $d_1 = 1.6$ mm. Check your result by reverse working to find R.

15 Show that   (a) $\log_b a \cdot \log_c b \cdot \log_a c = 1$,   (b) $\log_a A \cdot \log_b B = \log_b A \cdot \log_a B$.

16 In a calculation on the dryness of steam, the following formula was used.

$$\frac{qL}{T} = \frac{q_1 L_1}{T_1} + \log_e \frac{T_1}{T}$$

Use it to find q when $L_1 = 850$, $L = 1000$, $T_1 = 780$, $T = 650$ and $q_1 = 1$.

17 If $m = \dfrac{P(1 + \log_e r)}{r} - B$, find m when $P = 120$, $r = 2 \cdot 5$ and $B = 16$.

18 (a) If $x \log_{10} e = \log_{10} 1 - \log_{10} 148 \cdot 4$, what is the value of x?

    (b) Calculate the value of $x^{\sqrt{x}}$ when $x = 0 \cdot 834$.   NCTEC

19 (a) If $T = 15 \cdot 25$ and $e = 2 \cdot 718$, find $\log_e T$, correct to three significant figures.

    (b) If $pv^K = R^{0 \cdot 352}$, by taking logs find the value of K when $p = 0 \cdot 675$, $v = 3 \cdot 992$, $R = 0 \cdot 875$, as accurately as tables will allow.   NCTEC

20 (a) Evaluate $(0 \cdot 0645)^{0 \cdot 475}$.

    (b) If $PA^n = \dfrac{x(A^n - 1)}{A - 1}$, calculate x to three significant figures, when $P = 5000$, $A = 1 \cdot 05$ and $n = 25$.

    (c) Evaluate $\dfrac{1 - e^{-2x}}{1 + e^{-2x}}$ when $x = 0 \cdot 75$ and $e = 2 \cdot 7183$.

                                        WEST RIDING

21 (a) The formula $\theta = \theta_1{}^{-kt}$ is the equation for Newton's law of cooling, where θ is the difference between the temperature of the body and that of its surroundings after t seconds and θ_1 is the initial temperature difference. If when $t = 0$, $\theta = 84°C$, and when $t = 10$, $\theta = 30°C$, determine the value of k. ($e = 2 \cdot 718$.)

    (b) If $2^{x^2} = 4^{(4-x)}$, find the two values of x.   WEST RIDING

22 If $\dfrac{qL}{T} = \dfrac{q_1 L_1}{T_1} + \log_e \dfrac{T_1}{T}$ and $\dfrac{q_1 L_1}{T_1} = \dfrac{qL}{3T}$, prove that $T_1 = Te^{2qL/3T}$.

Chapter Five
The Slide Rule

Consider the C and D scales, which are used for multiplication and division and are calibrated for numbers between 1 and 10 (Fig. 5.1).

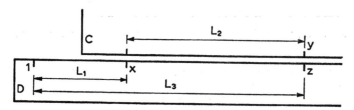

FIG. 5.1   Diagram for the multiplication of x and y to give the product z

Let $L_1 = K \log_{10} x$, $L_2 = K \log_{10} y$ and $L_3 = K \log_{10} z$, where K is a scale constant which we shall take to be 100 mm.

From the diagram,

$$L_1 + L_2 = L_3$$
$$\therefore \quad K \log_{10} x + K \log_{10} y = K \log_{10} z$$

i.e.
$$\log_{10} x + \log_{10} y = \log_{10} z$$

But, by the first rule of logarithms,

$$\log_{10} x + \log_{10} y = \log_{10} xy$$

hence
$$\log_{10} z = \log_{10} xy \quad \text{or} \quad z = xy$$

i.e.
z represents the product of x and y

Since the C and D scales can only be used directly for numbers between 1 and 10, if the product of 0·037 and 13 500 is required we in effect convert the numbers to standard form, as $3·7 \cdot 10^{-2}$ and $1·35 \cdot 10^4$, so that $(3·7 \cdot 1·35) \cdot 10^2$ is required. The slide rule gives $3·7 \cdot 1·35 = 5·00$, so the answer is $(5·00) \cdot 10^2$, which is written down as 500 (Fig. 5.2). In practice only the significant figures of x, y and z are considered, and the decimal point is inserted afterwards.

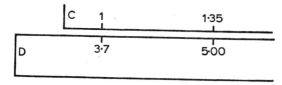

FIG. 5.2   Diagram for the multiplication of 0·037 and 13 500 to give the product 500

Procedure for multiplication of x and y

Align 1 on C with the number x on D (Fig. 5.3). Move the cursor to the number y on C, and read off the product number z on D. Adjust the decimal point as required.

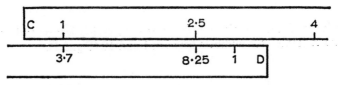

FIG. 5.3 Diagram for the multiplication of 3·7 and 2·5 to give the product 8·25

It may be noted that, if the product 3·7 . 4·0 were required, with the slide in the position of the last example, 4·0 on the C scale is off the D scale. In this case the slide is moved to the left, as illustrated in Fig. 5.4, giving 3·7 . 4 = 14·8.

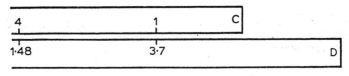

FIG. 5.4

EXERCISE 5.1

Work through the flow chart in Fig. 5.5 for various values of x and y, checking the product by another method of calculation.

Division

Division being the inverse process to multiplication, instead of proceeding to z via x times y, we proceed to x via z divided by y (Fig. 5.6). y on scale C is aligned with z on scale D, and x on scale D is read off in alignment with 1 on scale C.

It will be noted that for multiplication it is immaterial which factor of the product is x and which is y. In division, however, y must be the divisor, and x is then the quotient. x will always be accessible at either the left- or the right-hand 1 on scale C.

Construction of a slide rule with scales C and D

It is instructive to construct a slide rule with graph paper, using a scale factor of 100 mm. The following table is first compiled for the integers from 1 to 10, to show the procedure. Intermediate values could be included for greater detail.

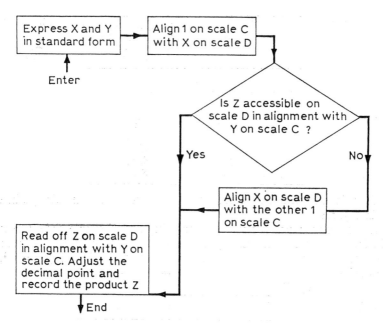

FIG. 5.5   Flow diagram for multiplication procedure $xy = z$ by slide rule

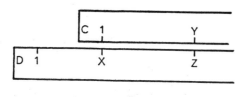

FIG. 5.6 $x = \dfrac{z}{y}$ since $xy = z$

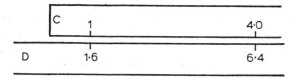

FIG. 5.7   Diagram to illustrate the division of 64 by 4000 to obtain the quotient 0·016

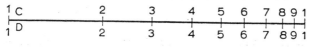

FIG. 5.8

Number	Logarithm	Length in mm measured from 1 mark with 100 mm scale factor
1	0·0000	0
2	0·3010	30·1
3	0·4771	47·7
4	0·6021	60·2
5	0·6990	69·9
6	0·7782	77·8
7	0·8451	84·5
8	0·9031	90·3
9	0·9542	95·4
10	1·0000	100·0

These lengths with the corresponding numbers, are marked along a line as in Fig. 5.8.

If the graph paper is now cut along this line, a simple slide rule is obtained, with which the multiplication and division of integers can be illustrated. Further points on the scales can be located by using $\frac{9}{5} = 1\cdot8$, $\frac{8}{5} = 1\cdot6$, $\frac{7}{5} = 1\cdot4$, etc., and then $1\cdot4 \times 2 = 2\cdot8$, $1\cdot4 \times 3 = 4\cdot2$, etc.

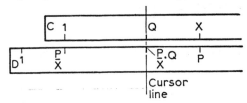

FIG. 5.9   Diagram to show $\frac{PQ}{X}$ obtained on scale D as $\frac{P}{X}$ multiplied by Q

Extended multiplication and division

An expression of the type $\dfrac{P \cdot Q \cdot R \cdot \text{etc.}}{X \cdot Y \cdot Z \cdot \text{etc.}}$ is evaluated as

$$\left[\left(\frac{P}{X} \cdot Q/Y\right)R\right]/Z$$

in which division of P by X is followed by multiplication by Q and division by Y, and so on; the reason being that multiplication by Q can follow division by X without movement of the C scale slide (see Figs 5.9 and 5.10).

FIG. 5.10   Diagram to show $\frac{PQ}{X}$ divided by Y and multiplied by R giving $\frac{P}{X} \cdot \frac{Q}{Y} \cdot R$ on scale D

Examples

5.1 Evaluate $\dfrac{2 \cdot 18 \cdot 3 \cdot 14 \cdot 0 \cdot 07}{21 \cdot 4 \cdot 72 \cdot 1 \cdot 0 \cdot 003}$.

By the above procedure the significant figures 1036 are obtained.

The expression is approximately $\dfrac{2 \cdot 3 \cdot 70}{20 \cdot 60 \cdot 3} \simeq 0 \cdot 1$.

Hence $0 \cdot 1036$ is the required result: say $0 \cdot 104$ correct to three significant figures.

5.2 Evaluate $\dfrac{113 \cdot 0 \cdot 0 \cdot 0712 \cdot 14\,600}{0 \cdot 000\,812 \cdot 0 \cdot 0123}$.

In this example a slight variation is to reduce the numbers to standard form, thus

$$\frac{(1 \cdot 13 \cdot 10^2) \cdot (7 \cdot 12 \cdot 10^{-2}) \cdot (1 \cdot 46 \cdot 10^4)}{(8 \cdot 12 \cdot 10^{-4}) \cdot (1 \cdot 23 \cdot 10^{-2})}$$

$$= \frac{1 \cdot 13 \cdot 7 \cdot 12 \cdot 1 \cdot 46}{8 \cdot 12 \cdot 1 \cdot 23} \cdot \frac{10^{2-2+4}}{10^{-4-2}} = 1 \cdot 176 \cdot 10^{10}$$

EXERCISE 5.2

1 Verify the multiplication tables from the 'twice times' to the 'twelve times', using the slide rule.

2 Divide each of the integers from 1 to 9 by 1, 2, . . ., 9 successively, and check the results mentally.

3 (*a*) Evaluate $\dfrac{24 \cdot 18 \cdot 25}{6 \cdot 30 \cdot 15}$ by slide rule, and check your result by cancelling the expression.

   (*b*) Evaluate $\dfrac{24 \cdot 1 \cdot 18 \cdot 2 \cdot 24 \cdot 9}{6 \cdot 2 \cdot 30 \cdot 1 \cdot 14 \cdot 8}$ by slide rule.

4 (*a*) Evaluate $\dfrac{28 \cdot 13 \cdot 75}{7 \cdot 39 \cdot 5}$ by slide rule, and check your result by cancelling the expression.

   (*b*) Evaluate $\dfrac{27 \cdot 9 \cdot 13 \cdot 1 \cdot 74 \cdot 8}{6 \cdot 9 \cdot 38 \cdot 7 \cdot 4 \cdot 9}$ by slide rule, making use of 4 (*a*).

5 Verify the following, using repeated multiplication by slide rule.
  (*a*) $2^3 = 8,\ 2^4 = 16,\ 2^5 = 32$
  (*b*) $3^2 = 9,\ 3^3 = 27,\ 3^4 = 81,\ 3^5 = 243$
  (*c*) $7^2 = 49,\ 7^3 = 343$

6 Evaluate 4^3 by slide rule as   (*a*) $4 \cdot 4 \cdot 4$,   (*b*) $\dfrac{12 \cdot 12 \cdot 12}{3 \cdot 3 \cdot 3}$,

  (*c*) $\dfrac{20 \cdot 20 \cdot 20}{5 \cdot 5 \cdot 5}$,   (*d*) $\dfrac{28 \cdot 28 \cdot 28}{7 \cdot 7 \cdot 7}$. Check by long multiplication.

Log-log scales

Consider scale C (Fig. 5.11) in relation to the log-log scale LL below scale D. K is the scale factor for the LL scale.

Let Q be the number on the LL scale aligned with 1 on scale C. Let n be the number on scale C aligned with P on the LL scale. Let L_1, L_2, L_3 be the lengths corresponding to Q, n and P on the appropriate scales.

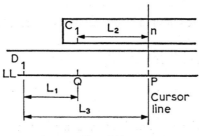

FIG. 5.11

Then
$$L_1 + L_2 = L_3$$
i.e.
$$K \log (\log Q) + K \log n = K \log (\log P)$$
$$\therefore \quad \log (\log Q) + \log n = \log (\log P)$$
$$\therefore \quad \log (n \log Q) = \log (\log P)$$
$$\therefore \quad \log (\log Q^n) = \log (\log P)$$
$$\therefore \quad \log Q^n = \log P$$
$$\therefore \quad Q^n = P$$

Thus, if Q on the LL scale is aligned with 1 on the C scale, then Q^n is on the LL scale, aligned with n on scale C.

Note that the following examples refer to a slide rule with three log-log scales scaled from 1·01 to 22 000.

Example
5.3 Find $5^{2·98}$.

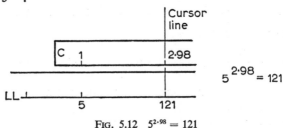

FIG. 5.12 $5^{2·98} = 121$

Determination of the nth root of P

Since $Q^n = P$, it follows that $Q = P^{1/n}$: thus, if n on the C scale is aligned with P on the LL scale, then $P^{1/n}$ on the LL scale is aligned with 1 on the C scale.

Example

5.4 Find $63^{1/5\cdot9}$

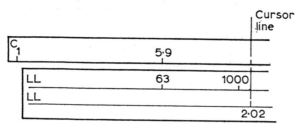

FIG. 5.13

In this case (Fig. 5.13) the left-hand 1 on scale C is off the LL scale. The right-hand 1 on scale C is aligned with a number in excess of 1000 on the LL scale. The result, 2·02, is found on an adjacent LL scale. *Unlike on the C and D scales, numbers such as* 1·4, 14, 140, 1400, 14 000 *are located in different positions on the LL scales.* Thus a slide rule may have about five parallel LL scales with a length of approximately 500 mm in total. It is useful to estimate the approximate value of the required result, or to make a check using logarithms. In this example $63^{1/5\cdot9} \simeq 64^{1/6} = 8^{1/3} \simeq 2$, which directs the attention to the appropriate LL scale and the result 2·02. Using four-figure logarithms, 2·018 is obtained.

No.	Log	Log-log
63	1·7993	0·2551
	5·9	0·7709
2·018	0·3049	$\bar{1}$·4842

Examples

5.5 It is required to evaluate $4\cdot28^{7\cdot68}$.

(*a*) Make a rough approximation to the required value, (*b*) evaluate by slide rule, (*c*) check by logarithms.

(*a*) Say $4\cdot28^{7\cdot68} \simeq 4^8 = 16^4 = 256^2 \simeq 200 \cdot 300$ or 60 000

(*b*) On setting 4·28 on the appropriate LL scale, 7·68 on scale C is off the LL scale, but

$$4\cdot28^{7\cdot68} = 4\cdot28^3 \cdot 4\cdot28^{4\cdot68} = 78\cdot5 \cdot 900 \text{ (using the LL scale)}$$
$$= 70\ 650$$

(*c*)

No.	Log	Log-log
4·28	0·6314	$\bar{1}$·8003
	7·68	0·8854
70 790	4·850	0·6857

Using four-figure logarithms, 70 790 is obtained. It must be expected that the number of correct significant figures in such evaluations will fall when the more crowded right-hand end of the log-log scale is approached.

5.6 It is required to evaluate $1.03^{18.2}$.

(a) Make a rough approximation to the required value, (b) evaluate by slide rule, (c) evaluate by logarithms.

(a) $1.03^{18.2} \simeq 1.03^{20} \simeq 1 + 20 \times 0.03$, since $(1 + x)^n \simeq 1 + nx$, if x is small enough $= 1.6$

(b) On setting 1.03 on the appropriate LL scale, the result is 1.714 on the appropriate LL scale, under 1.82 on scale C.

(c)

No.	Log	Log-log
1·03	0·0128	$\bar{2}$·1072
	18·2	1·2601
1·710	0·2330	$\bar{1}$·3673

Using four-figure logarithms, 1.710 is obtained.

5.7 It is required to evaluate $328^{1/5.8}$.

(a) Make a rough approximation to the required value, (b) evaluate by slide rule, (c) evaluate by logarithms.

(a) $328^{1/5.8} \simeq 300^{1/6} \simeq 18^{1/3} \simeq 2.5$, since $2^3 = 8$ and $3^3 = 27$.

(b) On setting 328 on the LL scale and aligning 5·8 on scale C, the result is 2.715 on the appropriate LL scale.

(c)

No.	Log	Log-log
328	2·5159	0·4007
	5·8	0·7634
2·715	0·4338	$\bar{1}$·6373

Using four-figure logarithms, 2.715 is obtained.

5.8 Evaluate $\left(\dfrac{V_2}{V_1}\right)^{n-1/1}$, where $V_2 = 150$, $V_1 = 9$ and $n = 1.35$.

(a) Make a rough approximation to the required value, (b) evaluate by slide rule, (c) evaluate by logarithms.

(a) The expression is $(\frac{150}{9})^{0.35/1.35} \simeq (16)^{3.5/13.5} \simeq 16^{1/4} = 2$.

(b) Align 9 on scale C with 15 on scale D, giving 16·65. Set 16·65 on LL, align 1·35 on scale C, move the cursor to 3·5 on scale C: the result is 2·07 on the appropriate LL scale.

(c)

No.	Log	Log-log
150	2·1761	
	0·9542	
	1·2219	0·0871
	0·35	$\bar{1}$·5441
		$\bar{1}$·6312
	1·35	0·1303
2·074	0·3169	$\bar{1}$·5009

Using four-figure logarithms, 2·074 is obtained.

The necessity of a rough calculation must be emphasised. Also it must be remembered that, although 0·14 and 14·0 may be represented by 1·4 on scales C and D, this is not so on the log-log scales. 1·4 and 14·0 are in different positions on the log-log scales, whilst only the *reciprocal* of 0·14 can be located on the log-log scales *unless special scales of the LL_0 type are available.*

EXERCISE 5.3

Evaluate the following by slide rule and by logarithms, after making a rough approximation to the result.

1 $4·9^{3·12}$ **2** $71^{1/6·3}$ **3** $5·12^{7·42}$ **4** $1·02^{21·1}$

5 $296^{1/5·2}$ **6** $\left(\dfrac{V_2}{V_1}\right)^{(n-1)/n}$, where $V_2 = 9$, $V_1 = 140$ and $n = 1·4$

EXERCISE 5.4

Evaluate the following by slide rule after making a rough approximation to the result.

1 $\left(\dfrac{6·2}{3·0}\right)^{0·286}$ **2** $\left(\dfrac{5·88}{1·47}\right)^{3/13}$ **3** $60 \cdot \left(\dfrac{50}{13}\right)^{1·3}$

4 $560 \cdot \left(\dfrac{6·50}{0·147}\right)^{0·395/1·395}$ **5** $15 \cdot \left(\dfrac{173}{10·8}\right)^{1·4}$

6 $450 \cdot \left(\dfrac{1·84}{17·3}\right)^{1·4}$ **7** $\left[1 - \left(\dfrac{35}{7}\right)^{\frac{1}{5}}\right]^4$ **8** $\left[1 + \dfrac{0·87.34·8}{590}\right]^{3·5}$

9 $\left[\dfrac{2}{\gamma + 1}\right]^{\gamma/(\gamma-1)}$ where $\gamma = 1·25$.

10 $\sqrt{\left(64·4 \cdot \dfrac{1·135}{0·135} \cdot 144 \cdot 120 \cdot 3·73 \left[1 - \left(\dfrac{3·5}{6}\right)\left(\dfrac{0·135}{1·135}\right)\right]\right)}$

11 $\dfrac{2944 \cdot (\frac{7}{12})^{1/1·135}}{144 \cdot 7·46}$ **12** $\dfrac{600 \cdot 144 \cdot 2}{0·8}\left[1 - \left(\dfrac{4}{20}\right)^{0·2/1·2}\right]$

Chapter Six
Quadratic Equations

Simultaneous quadratic equations
Simultaneous quadratic equations are simultaneous equations which contain the unknown quantities in the second, but no higher, degree. Only a few special cases can be dealt with here.

When one equation is linear
Using the linear equation, we can express one unknown quantity in terms of the other and then substitute that value in the other (quadratic) equation. This will give a quadratic with one variable only, which can be solved in the usual way. The other variable can then be found by substitution, as in the case of simple simultaneous equations.

Examples
6.1 Solve
$$x + y = 6 \qquad \text{. . . . . (i)}$$
$$x^2 + 3xy - y^2 = 36 \qquad \text{. . . . . (ii)}$$

From (i), $x = 6 - y$

Substituting this value of x in (ii), we get
$$(6 - y)^2 + 3(6 - y)y - y^2 = 36$$
$$36 - 12y + y^2 + 18y - 3y^2 - y^2 = 36$$
$$-3y^2 + 6y = 0$$
$$y^2 - 2y = 0$$
$$\therefore \quad (y)(y - 2) = 0 \qquad\qquad \therefore \quad y = 0 \text{ or } 2$$

Substituting $y = 0$ in (i), $x = 6$
Substituting $y = 2$ in (i), $x = 4$
$\therefore$ The required solutions are

$$x = 6 \qquad y = 0$$
and $$\qquad x = 4 \qquad y = 2$$

6.2 Solve
$$x^2 + 2y^2 = 17 \qquad \text{. . . . . (i)}$$
$$x + 3y = 9 \qquad \text{. . . . . (ii)}$$

From (ii)
$$x = 9 - 3y$$

Substituting this value for x in (i), we get

$$(9 - 3y)^2 + 2y^2 = 17$$
$$81 - 54y + 9y^2 + 2y^2 = 17$$
$$11v^2 - 54v + 64 =$$

$$y = \frac{54 \pm \sqrt{\{54^2 - (4 \times 11 \times 64)\}}}{22}$$

$$y = \frac{54 \pm 10}{22} = \tfrac{64}{22} \text{ or } \tfrac{44}{22}$$

$$\therefore \quad y = \tfrac{32}{11} \text{ or } 2$$

Substituting $y = \tfrac{32}{11}$ in (ii)

$$x + \tfrac{96}{11} = 9$$
$$\therefore \quad x = \tfrac{3}{11}$$

Substituting $\qquad\qquad y = 2$ in (ii)

$$x + 6 = 9$$
$$\therefore \quad x = 3$$

∴ the solutions are

$$x = \tfrac{3}{11}, \qquad y = \tfrac{32}{11}$$

and $\qquad\qquad x = 3, \qquad y = 2$

When both equations are in the second degree
Solve $\qquad\qquad\qquad\qquad\qquad xy = 18$
$$x^2 - xy + y^2 = 67$$

In this case, substitution as before would give an equation involving the fourth power of the variable, which would render the calculation more difficult and lengthy and usually not capable of solution. An easier method is as follows.

Since $\qquad\qquad x^2 - 2xy + y^2 = (x - y)^2$
and $\qquad\qquad x^2 + 2xy + y^2 = (x + y)^2$

it is easier to obtain the values of these expressions and so find the values of $x - y$ and $x + y$.

Thus, $\qquad\qquad \left. \begin{aligned} x^2 - xy + y^2 &= 67 \\ xy &= 18 \end{aligned} \right\}$ subtract

$$\overline{x^2 - 2xy + y^2 = 49}$$

$$\therefore \quad (x - y)^2 = 49$$

$$\therefore \quad x - y = \pm 7 \quad \cdots \cdots \cdots \text{ (i)}$$

Again, $\qquad\qquad \left. \begin{aligned} x^2 - xy + y^2 &= 67 \\ 3xy &= 54 \end{aligned} \right\}$ add

$$\overline{x^2 + 2xy + y^2 = 121}$$

$$\therefore \quad (x + y)^2 = 121$$

$$\therefore \quad x + y = \pm 11 \quad \cdots \cdots \cdots \text{ (ii)}$$

Using (i) and (ii) in all possible ways

(a) $x - y = + 7$
       $x + y = +11$
∴ $2x\qquad = 18$

∴ $x = 9$ and $y = \dfrac{18}{x} = 2$

(b) $x - y = -7$
       $x + y = +11$
∴ $2x\qquad = 4$

∴ $x = 2$ and $y = \dfrac{18}{x} = 9$

(c) $x - y = + 7$
       $x + y = -11$
∴ $2x\qquad = -4$

∴ $x = -2$ and $y = \dfrac{18}{x} = -9$

(d) $x - y = -7$
       $x + y = -11$
∴ $2x\qquad = -18$

∴ $x = -9$ and $y = \dfrac{18}{x} = -2$

Thus the solutions are

$$x = 9,\qquad y = 2$$
$$x = 2,\qquad y = 9$$
$$x = -2,\qquad y = -9$$
and $$x = -9,\qquad y = -2$$

As will be seen, the device used is suggested by the particular form of the expressions occurring. Many difficulties attending straightforward substitutions may similarly be evaded by an alert worker.

Solve $3x^2 - 2xy + 4y^2 = 31$ . . . . . . (i)
         $x^2 + xy + y^2 = 19$ . . . . . . (ii)

Here by division and cross-multiplication a new equation can be obtained without the introduction of a higher power of the variable

$$\frac{3x^2 - 2xy + 4y^2}{x^2 + xy + y^2} = \frac{31}{19}$$

$$57x^2 - 38xy + 76y^2 = 31x^2 + 31xy + 31y^2$$
$$26x^2 - 69xy + 45y^2 = 0$$
$$(13x - 15y)(2x - 3y) = 0$$

$$x = \frac{15y}{13} \quad \text{or} \quad \frac{3y}{2}$$

Substituting these values for x in equation (ii) and solving for y gives

$$y = \pm\frac{13}{\sqrt{31}} \quad \text{or} \quad y = \pm 2$$

Substituting these numerical values for y, in turn gives

$$x = \pm\frac{15}{\sqrt{31}} \quad \text{or} \quad x = \pm 3$$

EXERCISE 6.1

Solve

1 $x - y = 1$, $x^2 + y^2 = 61$
2 $3x - 2y = 7$, $x^2 - 3xy + y^2 = -19$
3 $4x + 5y = 0$, $2x^2 + xy - y^2 = 14$
4 $x + y + 1 = 0$, $3x^2 - 5y^2 - 7 = 0$
5 $x + y = 1$, $3x^2 - xy + y^2 = 37$
6 $2x + 3y = 14$, $4x^2 + 2xy + 3y^2 = 60$
7 $x - 2y = 2$, $xy = 12$
8 $xy = 4$, $x^2 + y^2 = 17$
9 $2xy = 80$, $x^2 + y^2 = 89$
10 $xy + 3x = 15$, $2y + 3xy = 22$
11 $xy - 6y = 1$, $2xy + 4x = 10$
12 $x^2 + 2y = 1\frac{3}{4}$, $x^2 - x + y = \frac{1}{2}$
13 Solve the following equations.
   (a) $6x^2 + 8x - 9 = 0$

   (b) $\dfrac{2}{x-1} + \dfrac{1}{x-3} = \dfrac{2}{x}$

   (c) $x^2 - 9y^2 = 24$
   $\ x - 3y = 8$

ULCI

14 Solve the equations
   (a) $5x^2 + 7x + 1 = 0$

   (b) $\dfrac{x+4}{2x+4} - \dfrac{x+1}{x+3} = 1$

CHELTENHAM

15 (a) Simplify $\dfrac{(xy - 2y^2)^2}{x^2 - xy - 2y^2} \times \dfrac{x+y}{x^2 - 4y^2}$.

   (b) Simplify $\dfrac{x^{1\cdot7} \cdot y^{-0\cdot6}}{x^{-0\cdot3} \cdot y^{0\cdot4}}$.

   (c) Solve the simultaneous equations $x^2 - xy + y^2 = 13$, $x + y + 5 = 0$.

16 (a) Solve the equation $\sqrt{(3x + 3)} - \sqrt{(x - 3)} = 3$.

(b) If $x = 2$ is a root of the equation

$$6x^3 - px^2 - 14x + 24 = 0,$$

find p and the other roots.   HANDSWORTH

17 (a) Solve the equation $\dfrac{1}{x - 1} = \dfrac{3}{x + 2} - \dfrac{2}{x - 3}$.

(b) Make A the subject of the formula

$$E = 100\left(\frac{c\sqrt{A}}{b} + d\right)$$

(c) Given $\dfrac{3}{x} - \dfrac{2}{y} = 17$, and $\dfrac{4}{x} + \dfrac{1}{y} = 8$, find x and y.   RUGBY

18 Find the roots of the equation $x^2 + 2x - 6 = 0$ graphically, and check by calculation. Take values of x between -4 and $+2$.   RUGBY

19 (a) Solve $3x - \dfrac{4}{x} = 11$.

(b) If $w = \dfrac{AB}{r}\sqrt{\left(\dfrac{C^2 + d^2}{2d^2}\right)}$, make d the subject of the formula.

(c) Solve the equations $3x + 2y = 1$, $x = \dfrac{y}{2} + 5$.   RUGBY

Graphical solution of simultaneous quadratic equations
In considering this method of solving simultaneous quadratics, it must be remembered that the coordinates of every point on a graph must satisfy the equation of that graph. Therefore, if two graphs intersect, the co-ordinates of the point (or points) of intersection, since they lie on both graphs, must satisfy both equations simultaneously. If, therefore, we are seeking those x and y values which satisfy two equations at the same time, that is, we are seeking the roots of a pair of simultaneous equations, we must find the coordinates of the points of intersection of their graphs.

Consider a simple example.

Solve graphically $\qquad 4x - y + 15 = 0$ . . . . (i)

$\qquad\qquad\qquad\qquad\quad x^2 - y = 0$ . . . . (ii)

Equation (i) can be written $y = 4x + 15$: its graph is a straight line. Equation (ii) can be written $y = x^2$: its graph is a parabola. The roots of the simultaneous equation given above will be the coordinates of the points where the straight line intersects the parabola.

Draw the graph of $y = x^2$ when x varies from -8 to $+8$.

Tabulating the corresponding values of x and y,

x	-8	-6	-4	-2	0	2	4	6	8
$y = x^2$	64	36	16	4	0	4	16	36	64

The graph is shown in Fig. 6.1.

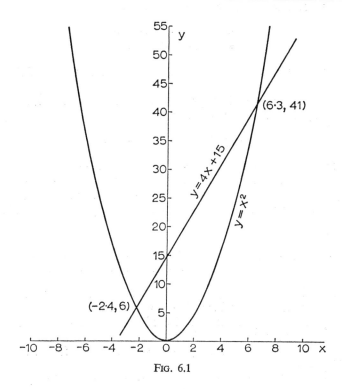

FIG. 6.1

The line $y = 4x + 15$ can be drawn by taking two suitable points, such as:

x	-2	$+2$
$y = 4x + 15$	$+7$	$+23$

As seen in the diagram, these graphs intersect at (approx.) $x = -2.4$, $y = 6$ and $x = 6.3$, $y = 41$.

∴   The roots of the simultaneous equations $4x - y + 15 = 0$, $x^2 - y = 0$ are $(-2.4, 6)$ and $(6.3, 41)$.

Checking algebraically, we get

$$y = x^2 = 4x + 15$$

For the x roots of the equation,

$$x^2 - 4x - 15 = 0$$

$$\therefore x = \frac{4 \pm \sqrt{(16 + 60)}}{2} = 6\cdot36 \quad \text{or} \quad -2\cdot36$$

and by substitution of these values in $y = 4x + 15$

$$y = 4(6\cdot36) + 15 = 40\cdot44$$
or
$$y = 4(-2\cdot36) + 15 = 5\cdot56$$

These are values which agree (approximately) with those read off from the graph (Fig. 6.1).

It will also be noticed in the algebraic solution used as a check above that solving the simultaneous quadratic

$$4x - y + 15 = 0, \quad x^2 - y = 0,$$

is the same problem as solving $x^2 = 4x + 15$ or $x^2 - 4x - 15 = 0$. We could have obtained the roots by drawing the graph of $x^2 - 4x - 15$ and noting the x values of the points where $x^2 - 4x - 15$ cuts the x-axis; that is, the values of x when $x^2 - 4x - 15 = 0$. This graph is shown in Fig. 6.2.

Examples

6.3 Solve graphically

$$xy = 20 \quad \ldots \ldots \ldots \text{(i)}$$
$$x^2 + y^2 = 64 \quad \ldots \ldots \ldots \text{(ii)}$$

Before tabulating x and y values and drawing the graphs from them, consider equation (ii), in which $x^2 + y^2 =$ a constant.

If AB (Fig. 6.3) represents a portion of a curve, and P_1 is any point on the curve, then the coordinates of P_1 are x_1 and y_1 (say), where $x_1^2 + y_1^2 =$ a constant (in this case 64). But $x_1^2 + y_1^2 = (OP_1)^2$, therefore $(OP_1)^2 = 64$ and $OP_1 = \pm 8$.

Similarly, if P_2 is another point on the same curve, its coordinates (x_2, y_2) must also satisfy the equation $x_2^2 + y_2^2 = 64$.

But $\quad x_2^2 + y_2^2 = (OP_2)^2 \quad \therefore \quad (OP_2)^2 = 64 \quad \therefore \quad OP_2 = \pm 8.$

Similarly for all other points on the curve: they will all lie on a curve whose distance from O is ± 8. Therefore the graph of $x^2 + y^2 = 64$ is a circle, whose centre is at O, and whose radius is 8 units.

To draw the graph of $x^2 + y^2 = 64$, we draw this circle of radius 8 units with centre at the origin, which is much easier and quicker than tabulating and plotting corresponding values of x and y.

To draw the graph of $xy = 20$,

write as
$$y = \frac{20}{x}$$

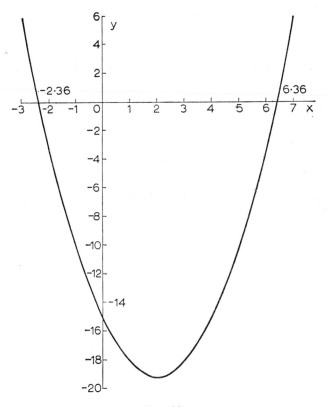

FIG. 6.2

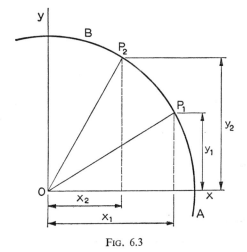

FIG. 6.3

Tabulating values from this, we get

x	-8	-6	-4	-2	0	2	4	6	8
$y = \dfrac{20}{x}$	$-2\frac{1}{2}$	$-3\frac{1}{3}$	-5	-10	$-\infty$	10	5	$3\frac{1}{3}$	$2\frac{1}{2}$

On drawing this graph, which is called a hyperbola, we find it cuts the circle at four points A, B, C, D (see Fig. 6.4).

At point A, $x = 2 \cdot 8$ $y =$ $7 \cdot 4$
　　　,,   B, $x = 7 \cdot 4,$ $y =$ $2 \cdot 8$
　　　,,   C, $x = -7 \cdot 4,$ $y = -2 \cdot 8$
　　　,,   D, $x = -2 \cdot 8,$ $y = -7 \cdot 4$

The roots found in the usual algebraic way agree very nearly with these values.

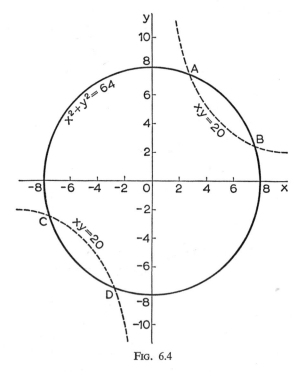

FIG. 6.4

6.4 Solve graphically $xy = 10$ . . . . (i)
　　　　　　　　　　　$y = x^2 + 6$ . . . . (ii)

Plotting the graphs in the usual way, we find that (i) is a curve of the same type as (i) in the last example (a curve known as a hyperbola), whilst (ii) is a parabola. In this case there is only one point of intersection: its coordinates are $x = 1·35$, $y = 7·75$. Thus the roots of the simultaneous quadratics have these values, as nearly as can be read directly from the graph (see Fig. 6.5).

(If more accurate values were required, it would be necessary to plot the quadrant containing the point of intersection on a much larger scale, showing the portions of the curves lying in that quadrant.)

It will also be noticed that we have solved graphically an equation which would have proved difficult to solve algebraically. This is often the case: sometimes the graphical method is the *only* method of obtaining a solution.

The student should now solve, by the graphical method, a few of the equations given in Exercise 6.1, as well as those of Exercise 6.2 which follows.

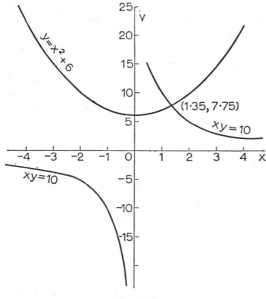

FIG. 6.5

EXERCISE 6.2

1 Find the three values of x which satisfy the equation

$$3x^3 + 7x^2 - 4 = 0.$$ COVENTRY

2 On the same reference axes, for the range between 0° and 120°, plot graphs of the functions (a) $\sin 2x + 3$,   (b) $\tan x - 2$.
Hence solve the equation $\sin 2x - \tan x + 5 = 0$.   WORCESTER

3 Solve the equations

(a) $\sqrt{(y + 3)} + \sqrt{y} = \dfrac{6}{\sqrt{(y + 3)}}$

(b) $\dfrac{5}{x} - \dfrac{3}{y} = 2$, $xy + 15 = 0$ BURTON UPON TRENT

4 Draw a graph of $y = 6x - 4 - x^3$, and use it to solve the equations

(a) $6x - 4 - x^3 = 0$,   (b) $6x - 4 - x^3 = e^{-x}$.

Take values of x between -3 and $+3$.   COVENTRY

5 Draw the graph of $y = 3x - \dfrac{x^3}{2}$ from $x = -2$ to $x = 2$. Use the graph
to find a value of x satisfying $x^3 - 6x + 2 = 0$.
   By drawing a straight line on the graph, solve the equation
$x^3 - 8x + 4 = 0$.   HANDSWORTH

6 Find, to three significant figures, the values of x between $+3$ and -3
which satisfy the equation $x^3 - 6x + 3 = 0$.   RUGBY

7 Draw the graph $y = x^3 - 4x + 1$ for values of x from -3 to $+3$, and
use the graph to solve $x^3 - 4x = 2$.
   By drawing a line on the graph, solve the equation $x^3 - 6x = 3$.
                                   HANDSWORTH

8 Using the same scales and axes, draw the curve whose equation is
$x^2 + y^2 = 4$, and the line whose equation is $y = 2x + 1\cdot5$. State the
values of x and y for the points where the curve and line intersect.

9 Solve graphically the equation $x^3 = 3x^2 - 2$, and check your solution
algebraically.   COVENTRY

10 (a) Solve the equation $4x^2 - 5x - 21 = 0$ by the two-graph method,
   and verify your result by calculation.
   (b) An open cistern has a square base, and the area of the sides and
   base is 10 m². If x m is a side of the base, and y m is the height,
   show that $x^2 + 4xy = 10$, and that the volume $V = x^2y$
   $= \dfrac{x}{4}(10 - x^2)$ m³.   HANDSWORTH

11 Plot the above expression for V between $x = 0$ and $x = 3$, and hence
find x when the volume is a maximum.   HANDSWORTH

Polynomial functions

An expression of the type $Ax^n + Bx^{n-1} + Cx^{n-2} + \ldots + Jx + K$,
where A is not zero, is called a polynomial function of degree n in x, where n
is a positive integer (whole number).

The remainder theorem

If $f(x)$ is a polynomial function, then $f(a)$ is the remainder when $f(x)$ is
divided by $x - a$.

Since $\dfrac{f(x)}{x - a} = \phi(x) + \dfrac{R}{(x - a)}$ . . . . .   (i)

where if $f(x)$ is of degree n, $\phi(x)$ is of degree $n - 1$ and R is a constant remainder term,

then $\qquad\qquad f(x) = \phi(x)(x - a) + R \quad \cdot \ \cdot \ \cdot \ \cdot \ \cdot \ \cdot$ (ii)

Put $x = a$, then $\qquad f(a) = \phi(a)(a - a) + R$

i.e. $\qquad\qquad\qquad f(a) = R$

$\phi(x)$ and R could be obtained by long division.

The factor theorem

If, in the above, $f(a) = 0$, then $R = 0$; i.e. from (ii), $f(x) = \phi(x)(x - a)$, and $(x - a)$ is a factor of $f(x)$. If a is substituted for x in a polynomial function $f(x)$, and $f(a) = 0$, then $x - a$ is a factor of $f(x)$.

Examples
6.5 Find the remainder when $x^4 - x^3 - 3x^2 + 3x + 4$ is divided by $x - 2$.

If $\qquad\qquad f(x) = x^4 - x^3 - 3x^2 + \ ^\cdot x + 4$

then $\qquad\qquad f(2) = 2^4 - 2^3 - 3 \cdot 2^2 + 3 \cdot 2 + 4$

$\qquad\qquad\qquad = 16 - 8 - 12 + 6 + 4 = 6$

i.e. 6 is the remainder.

We have previously obtained the identity

$$\frac{x^4 - x^3 - 3x^2 + 3x + 4}{x - 2} \equiv x^3 + x^2 - x + 1 + \frac{6}{(x - 2)}$$

(see question 1, Exercise 1.3)

or $\quad (x^4 - x^3 - 3x^2 + 3x + 4) \equiv (x^3 + x^2 - x + 1)(x - 2) + 6$

showing that 6 is the remainder.

6.6 Find the remainder when $x^{200} + 3x^{105} + 5$ is divided by $x + 1$.

$$f(x) = x^{200} + 3x^{105} + 5$$
$$f(-1) = (-1)^{200} + 3(-1)^{105} + 5$$
$$= +1 - 3 + 5 = 3$$

i.e. 3 is the remainder.

Examples
6.7 (*a*) Show that $x - y$ is a factor of $x^3 - y^3$, and complete the factorisation.
(*b*) Show that $x - y$ is a factor of $x^5 - y^5$, and complete the factorisation.
(*c*) Show that $x + y$ is a factor of $x^3 + y^3$, and complete the factorisation.

(*a*) In $x^3 - y^3$ put $x = y$, then $f(x, y) = y^3 - y^3 = 0$. $\quad \therefore \quad x - y$ is a factor. Consider $(x - y) \times$ (a second-degree expression in x and y). We see immediately that the first term is x^2 and the last term is $+y^2$, i.e. consider $(x - y)(x^2 \qquad + y^2)$. But we now have the terms x^3, $- yx^2$, xy^2 and $-y^3$: we must insert a term $+xy$ which will give rise to terms x^2y and $-xy^2$, and so cancel the terms $-yx^2$ and xy^2.

Thus $(x^3 - y^3) \equiv (x - y)(x^2 + xy + y^2)$
which can be verified in the usual ways.

(b) In $x^5 - y^5$ put $x = y$, then $f(x, y) = y^5 - y^5 = 0$. $\therefore$ $x - y$ is a factor.

Consider $(x - y)(x^4 \qquad + y^4)$: to cancel $-x^4y$ introduce $+x^3y$.

Consider $(x - y)(x^4 + x^3y \qquad + y^4)$: to cancel $-x^3y^2$ introduce $+x^2y^2$.

Consider $(x - y)(x^4 + x^3y + x^2y^2 \qquad + y^4)$: to cancel $-x^2y^3$ introduce $+ xy^3$.

Consider $(x - y)(x^4 + x^3y + x^2y^2 + xy^3 + y^4)$: to cancel $-xy^4$ there is a term $+xy^4$, and the factorisation is complete.

i.e. $x^5 - y^5 \equiv (x - y)(x^4 + x^3y + x^2y^2 + xy^3 + y^4)$

(c) In $x^3 + y^3$ put $x = -y$, then $f(x, y) = (-y)^3 + y^3 = 0$. $\therefore x + y$ is a factor.

Consider $(x + y)(x^2 \qquad + y^2)$: to cancel x^2y introduce a term $-xy$

$$(x + y)(x^2 - xy + y^2).$$

It can be seen that all terms except x^3 and y^3 cancel, and the factorisation is complete.

i.e. $x^3 + y^3 \equiv (x + y)(x^2 - xy + y^2)$

6.8 (a) Find the value of a if $(x + 2)$ is a factor of $x^3 - ax^2 + 7x + 10$.

(b) Factorise completely (i) $x^3 - 2x^2 - 5x + 6$, (ii) $\pi d^3 - \dfrac{\pi}{27}$.

NCTEC

(a) If $f(x) = x^3 - ax^2 + 7x + 10$, put $x = -2$.

$$f(-2) = (-2)^3 - a(-2)^2 + 7(-2) + 10$$
$$= -8 - 4a - 14 + 10 = -12 - 4a$$

If $f(-2) = 0$,   then $-12 - 4a = 0$

$$\therefore\quad 4a = -12$$

$$\therefore\quad a = -3$$

We can check this result thus: if $a = -3$,

$$f(x) = x^3 + 3x^2 + 7x + 10 \equiv (x + 2)(x^2 + x + 5)$$

(b) (i) Consider $x^3 - 2x^2 - 5x + 6$
We can try the integral factors of 6, i.e. 1, 6, 2, 3, with plus or minus signs.

If $f(x) = x^3 - 2x^2 - 5x + 6$,
        $f(1) = 1 - 2 - 5 + 6 = 0$ $\therefore$ $x - 1$ is a factor

$$x^3 - 2x^2 - 5x + 6 \equiv (x - 1)(x^2 - x - 6)$$
$$\equiv (x - 1)(x - 3)(x + 2)$$

Check

When $x = 2$, L.H.S. $= 2^3 - 2 \times 2^2 - 5 \times 2 + 6 = 8 - 8 - 10 + 6 = -4$

R.H.S. $= (2 - 1)(2 - 3)(2 + 2) = 1 \times (-1) \times 4 = -4$

i.e. L.H.S. $= -4 = $ R.H.S.

(ii) $\pi d^3 - \dfrac{\pi}{27} = \pi \left(d^3 - \dfrac{1}{27} \right) = \pi \left[d^3 - \left(\dfrac{1}{3} \right)^3 \right]$

Compare with $x^3 - y^3 = (x - y)(x^2 + xy + y^2)$

Replacing x by d and y by $\dfrac{1}{3}$,

$$d^3 - \left(\dfrac{1}{3} \right)^3 = \left(d - \dfrac{1}{3} \right) \left(d^2 + d \left(\dfrac{1}{3} \right) + \left(\dfrac{1}{3} \right)^2 \right) = \left(d - \dfrac{1}{3} \right) \left(d^2 + \dfrac{d}{3} + \dfrac{1}{9} \right)$$

$$\therefore \quad \pi d^3 - \dfrac{\pi}{27} \equiv \pi \left(d - \dfrac{1}{3} \right) \left(d^2 + \dfrac{d}{3} + \dfrac{1}{9} \right)$$

Check

When $d = 1$, L.H.S. $= \pi \times 1^3 - \dfrac{\pi}{27} = \dfrac{26\pi}{27}$

R.H.S. $= \pi \left(1 - \dfrac{1}{3} \right) \left(1 + \dfrac{1}{3} + \dfrac{1}{9} \right)$

$$= \pi \left(\dfrac{2}{3} \right) \left(\dfrac{9 + 3 + 1}{9} \right) = \dfrac{26\pi}{27}$$

i.e., when $d = 1$, L.H.S. $= \dfrac{\pi \times 26}{27} = $ R.H.S.

6.9 (*a*) Factorise the polynomial function $x^3 - 3x^2 - 4x + 12$.
(*b*) Show that, if a is a very large positive number, when $x = a$ the function takes the value a^3 approximately, and when $x = -a$ the function takes the value $-a^3$ approximately.
(*c*) Make a rough sketch of the function.

(*a*) We first try the integral factors of 12, i.e. 1, 2, 3, 4, 6, 12, with plus or minus signs.

$$f(1) = 1 - 3 - 4 + 12 = 6$$
$$f(2) = 2^3 - 3 \times 2^2 - 4 \times 2 + 12$$
$$= 8 - 12 - 8 + 12 = 0$$

$\therefore \quad x - 2$ is a factor

$\therefore \quad x^3 - 3x^2 - 4x + 12 \equiv (x - 2)(x^2 - x - 6)$
$$\equiv (x - 2)(x + 2)(x - 3)$$

(b)
$$f(a) = a^3 - 3a^2 - 4a + 12$$

$$= a^3 \left[1 - \frac{3}{a} - \frac{4}{a^2} + \frac{12}{a^3} \right] \simeq a^3$$

since the last three terms in the bracket are negligible in comparison with 1 if a is very large.

$$f(-a) = -a^3 - 3a^2 + 4a + 12$$

$$= -a^3 \left[1 + \frac{3}{a} - \frac{4}{a^2} - \frac{12}{a^3} \right] \simeq -a^3$$

(c) Since
$$f(x) = (x - 2)(x + 2)(x - 3)$$
$$f(x) = 0, \quad \text{when} \quad x = 2, \ x = -2 \quad \text{or} \quad x = 3$$

also
$$f(0) = 0 - 0 - 0 + 12 = +12$$

We plot $f(0) = +12, f(1) = 6, f(2) = 0, f(-2) = 0, f(3) = 0, f(-1) = 12$

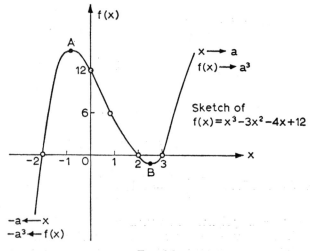

Fig. 6.6

The points A and B (Fig. 6.6) (known as turning points) on the graph of the function can be obtained by the calculus: see later work on maxima and minima.

EXERCISE 6.3

1 (a) Calculate the remainder when $x^3 - 2x^2 - x + 2$ is divided by
  (i) $x + 5$, (ii) $x - 1$.
  (b) Factorise $x^3 - 2x^2 - x + 2$, and check.

2 Show that $(x - a)$ is a factor of $x^3 - (a + b + c)x^2 + (ab + bc + ca) x - abc$. Complete the factorisation, and check.

3 Show that $(t + p)$ is a factor of $t^3 + (p - q + r)t^2 + (pr - rq - qp)$ $t - pqr$. Complete the factorisation, and check.

4 Show that $(y + 1)$ and $(y - 1)$ are factors of $y^4 + y^3 - 7y^2 - y + 6$. Complete the factorisation, and check.

5 With reference to the function $-x^3 - 3x^2 + x + 3$, show that (a) $(x + 3)$ is a factor; (b) if a is a very large positive number, when $x = a$ the function takes the value $-a^3$ approximately, and when $x = -a$ the function takes the value $+a^3$ approximately; (c) make a rough sketch of the function.

6 Show that $(x + y - 2)$ is a factor of $x^3 + y^3 + 6xy - 8$. Complete the factorisation, and check.

7 Show that $(a - b)$ and $(b - c)$ are factors of $ab^2 + bc^2 + ca^2 - a^2b$ $- b^2c - c^2a$, and complete the factorisation.

8 If $j^2 = -1$ show that (a) $(x + j)$ is a factor of $x^2 + 1$, and complete the factorisation, (b) $\left(x + \frac{1}{2} + \frac{\sqrt{3}j}{2}\right)$ is a factor of $x^2 + x + 1$, and complete the factorisation.

9 If $(x + 5)$ is a factor of $x^3 + 4x^2 + Px - 10$, determine the value of P, and complete the factorisation.

10 If $(x + 3)$ and $(x - 2)$ are factors of $x^4 + x^3 + Ax^2 - x + B$, show that $9A + B = -57$ and $4A + B = -22$. Solve for A and B, and complete the factorisation.

11 If the remainder when $x^3 + 6x^2 + Px + 6$ is divided by $x + 4$ is -6, determine the value of P, and factorise the cubic polynomial function.

Chapter Seven
The Plotting of More Difficult Graphs

Recapitulation
The graphs of $y = ax + b$, $y = ax^2 + bx + c$ and $y = a/x$ were dealt with in previous work.

It will be remembered that $y = ax + b$ represents straight lines which differ in slope and position according to the values borne by a and b. The value of a denotes the gradient of the line, and the value of b the intercept on the y-axis.

The graph of $y = ax^2 + bx + c$ was found to be a curve with one turning-point (a parabola): a reference to Chapter Six of this volume will serve as a useful reminder.

The graph of $y = a/x$ was found to be a curve of the hyperbola type.

In this chapter it is proposed to deal with a few graphs of harder types.

Graphs of cubic expressions
A cubic expression is one in which the highest power of the variable is the third: x^3, $3x^3 + 7$, $5x^3 - 2x + 3$, and $7 + 2x - 5x^2 - 4x^3$ are cubic expressions, or cubic functions of x.

The graph of $y = x^3$
Connected values of x and y are tabulated below.

x	-4	-3	-2	-1	0	1	2	3	4
$y = x^3$	-64	-27	-8	-1	0	1	8	27	64

An examination of these values reveals that

(a) when x is negative, y is negative,

(b) when x is zero, y is zero,

(c) when x is positive, y is positive,

(d) as x increases, y increases,

(e) when x is infinitely large (positive or negative), y is infinitely large (positive or negative respectively).

(f) for each numerically equal pair of positive and negative values of x, there are numerically equal pairs of positive and negative values of y: there are two symmetrical (but relatively reversed) portions of the curve.

On plotting the curve (Fig. 7.1), the above conclusions are verified. We also note that the origin, O, is the *centre of symmetry*: here the curve changes from concave downwards to concave upwards, the x-axis being tangential to both portions. For this reason, O is called a *point of inflection*.

The graph of $y = -x^3$
Tabulating and plotting as before, we obtain

x		-4	-3	-2	-1	0	1	2	3	4
$y = -x^3$		$+64$	$+27$	$+8$	$+1$	0	-1	-8	-27	-64

The graphs of $y = x^3$ (Fig. 7.1) and $y = -x^3$ (Fig. 7.2) may be looked upon as 'reflections' of each other.

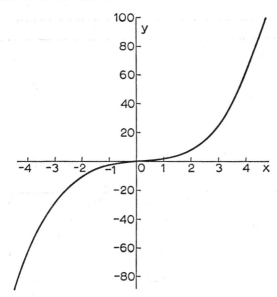

FIG. 7.1 Graph of $y = x^3$

The graph of $y = ax^3$ (where a is constant)
The student will easily see that
   (a) the introduction of the constant a simply multiplies the y value by a, for each value of x,
   (b) when $x = 0$, $y = 0$,
   (c) when $x \longrightarrow \infty$, $y \longrightarrow \infty$,
   (d) in general form, the graph is similar to that of $y = x^3$,
   (e) when a is negative, the graph will be similar to that of $y = -x^3$.

The graph of $y = ax^3 + b$
The effect of adding the constant b (which may be positive or negative) is to increase or decrease the value of y by the value of b. The graph will be the same shape as the graph of $y = ax^3$, but its position will be different, since it no longer passes through the origin, as when $x = 0$, $y = b$.
   The student will now be able to visualise graphs such as $10 + x^3$, $7 - x^3$, $2x^3 - 8$, etc.

The graph of $y = ax^3 + bx + c$

Here we find the variable in the first as well as the third degree.

Let $a = 3$, $b = 5$ and $c = 10$. The graph of $y = 3x^3 + 5x + 10$ can be drawn by tabulating and plotting in the usual way.

x		-3	-2	-1	0	1	2	3
$3x^3$		-81	-24	-3	0	3	24	81
$5x$		-15	-10	-5	0	5	10	15
10		10	10	10	10	10	10	10
$y = 3x^3 + 5x + 10$		-86	-24	2	10	18	44	106

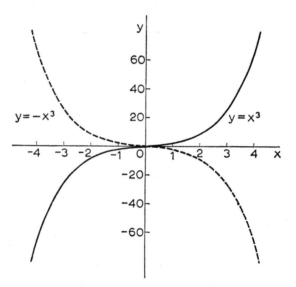

FIG. 7.2 Graphs of $y = x^3$ and $y = -x^3$

Here we notice that, as x increases from -3 through 0 to $+3$, y increases from -86 to $+106$, passing through the value 0 when x is between -2 and -1 (probably nearer to -1 than -2).

From the graph (Fig. 7.3) it is seen that $y = 0$ only when $x = -1\cdot1$ (approx). The graph is a cubic curve similar to $y = x^3$, having a point of inflection but no turning-points.

We also note that the equation $3x^3 + 5x + 10 = 0$ has one real root which is $x = -1\cdot1$ (approx). If a more accurate value of this root were required, it would be necessary to plot the graph from $x = -1\cdot5$ to $x = -0\cdot5$ on a large scale.

We could also use this graph to solve the equation

$$3x^3 + 5x + 10 = 50$$

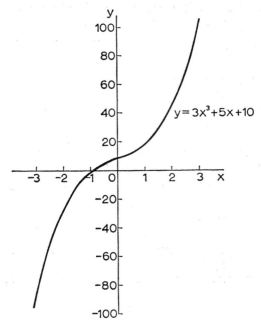

Fig. 7.3 Graph of $y = 3x^3 + 5x + 10$

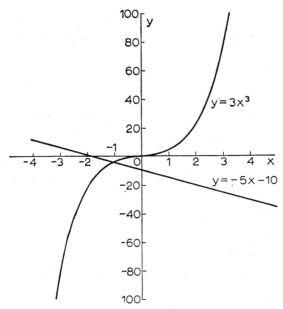

Fig. 7.4 Intersection of $y = 3x^3$ and $y = -5x - 10$

by drawing a horizontal line through $y = 50$ and reading off the x value of the point of intersection. It appears to be 2·15 approximately.

Since $3x^3 + 5x + 10 = 50$ can be written as $3x^3 + 5x - 40 = 0$, the real root of this equation is also 2·15.

Alternative method of solving $3x^3 + 5x + 10 = 0$

Since $3x^3 + 5x + 10 = 0$ may be written as $3x^3 = -5x - 10$, we may proceed to draw the graphs of $y = 3x^3$ and $y = -5x - 10$, and then find their point of intersection.

$y = 3x^3$ may be plotted by using the values of $3x^3$ shown in the table used for the last graph, and, since $y = -5x - 10$ is a straight line, we need only two pairs of corresponding values of x and y, such as

x	0	4
$y = -5x - 10$	-10	-30

When both graphs are plotted on the same axes and using the same scales, we obtain Fig. 7.4.

The graphs intersect at a point whose x value is $-1·1$ (approx). At this point

$$3x^3 = -5x - 10$$
or
$$3x^3 + 5x + 10 = 0$$

∴   the root of $3x^3 + 5x + 10 = 0$ is $x = -1·1$ (approx).

These methods are both of use in a general way, and by their application (whichever may seem preferable) it is often possible to solve an equation which might be much more difficult to solve algebraically.

Graph of $y = ax^3 + bx + c$, where b is negative

Suppose we require the graph of $y = 2x^3 - 7x - 3$.

Tabulating values as before, we obtain

x	-2	$-1·5$	-1	$-0·5$	0	0·5	1	1·5	2	2·5
$2x^3$	-16	$-6·75$	-2	$-0·25$	0	0·25	2	6·75	16	31·25
$-7x$	$+14$	$+10·5$	$+7$	$+3·5$	0	$-3·5$	-7	$-10·5$	-14	$-17·5$
-3	-3	-3	-3	-3	-3	-3	-3	-3	-3	-3
y	-5	0·75	2	0·25	-3	$-6·25$	-8	$-6·75$	-1	10·75

Here we note that
(a) y changes sign in three places:

    (i) between $x = -2$ and $x = -1·5$, the change in y being from negative to positive,

    (ii) between $x = -1$ and $x = 0$, y changing from positive to negative,

    (iii) between $x = 2$ and $x = 2·5$, y changing from negative to positive,

(b) These results suggest two turning-points on the curve:
  (i) between $x = -1.5$ and $x = -0.5$, which will be of the type known as a *maximum*,
  (ii) between $x = 0.5$ and $x = 1.5$, a point of the type known as a *minimum*.

Other points might be plotted to confirm this and obtain the turning-points more exactly.

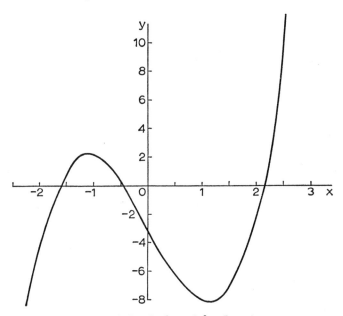

Fig. 7.5 Graph of $y = 2x^3 - 7x - 3$

On drawing the curve (Fig. 7.5), we find it crosses the x-axis at the points where $x = -1.6$, -0.45 and 2.15, reaching a maximum value of 2.1 when $x = -1.1$, and a minimum value of -8.1 when $x = 1.1$. We also note that $x = -1.6$, -0.45 and 2.15 are the roots of the equation $2x^3 - 7x - 3 = 0$.

Alternative Method of solving $2x^3 - 7x - 3 = 0$

Writing $2x^3 = 7x + 3$, we may use the method of the previous example: by drawing the graphs of $y = 2x^3$ and $y = 7x + 3$, we can find their points of intersection, whose x values will be the above roots (Fig. 7.6).

We find that the straight line cuts the cubic curve in three points, whose x values are -1.6, -0.45 and 2.15. These agree with the values previously found for the roots of the equation.

In this particular case we find three real roots of the equation, but in other positions of the straight line, i.e. for other values of b and c, it might cut the cubic curve in only *one* point. There would then be only *one real* root of the equation: the two others would be imaginary.

Thus a cubic equation of the type $ax^3 + bx + c = 0$ may, in general, have three real roots or one real and two imaginary roots.

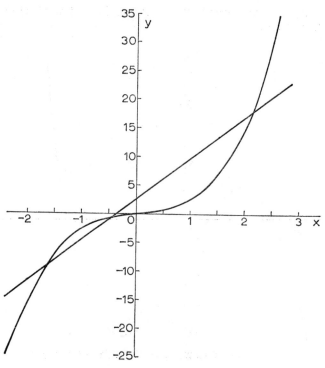

Fig. 7.6 Intersection of $y = 2x^3$ and $y = 7x + 3$

The graph of $y = x(x - a)(x - b)$

When the function is given in this form, we can get a clear idea of the shape of the curve by inspection.

Suppose we require the graph of $y = 4x(x - 1)(x - 2)$. By inspection we note that

(a) when $x = 0, 1$ or 2, $y = 0$, $\therefore$ the curve cuts the x-axis at these points,

(b) when x is negative, both $(x - 1)$ and $(x - 2)$ are negative, $\therefore y$ is *negative* for all negative values of x,

(c) when x lies between 0 and 1, x is positive, while $(x - 1)$ and $(x - 2)$ are both negative, $\therefore y$ is *positive*,

(d) when x lies between 1 and 2, x is positive, $(x - 1)$ is positive, but $(x - 2)$ is negative; $\therefore y$ is *negative*,

(e) when x is greater than 2, x, $x - 1$ and $x - 2$ are all positive, ∴ y is *positive*.

(f) Since y is positive for values of x between 0 and 1, and zero at these points, the curve (if continuous) must pass through a turning-point which is clearly a *maximum point* between these two points.

(g) Similarly, the graph will have a turning-point between $x = 1$ and $x = 2$, and, since y is negative between these points, the turning-point will be a *minimum point*.

(h) As x approaches infinity (either positive or negative) y approaches infinity (positive or negative, respectively).

It should be noted that the coefficient 4 does not affect the above arguments; it merely multiplies the y values.

We will now draw the graph of $y = 4x(x - 1)(x - 2)$ in the usual way.

x	−0·25	0	0·25	0·5	0·75	1	1·25	1·5	1·75	2	2·25
$y = 4x(x - 1)(x - 2)$	−2·81	0	1·31	1·5	0·94	0	−0·94	−1·5	−1·31	0	2·81

The graph obtained (Fig. 7.7) verifies the above general observations. The equation $4x(x - 1)(x - 2) = 0$ is shown to have three real roots—namely, $x = 0$, 1 and 2.

The maximum and minimum values of the expression $4x(x - 1)(x - 2)$ are given by the values of y at the turning-points: they are

(a) maximum value, 1·55, when $x = 0.45$,
(b) minimum value, −1·55, when $x = 1.55$.

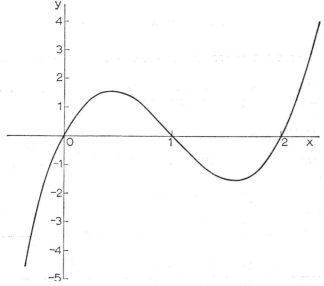

FIG. 7.7 Graph of $y = 4x (x - 1) (x - 2)$

The graph of $y = ax^3 + bx^2 + cx + d$
Lastly we will consider the graph of an expression of the type in which the variable x occurs in the first, second and third degrees.

Let $a = 2$, $b = -9$, $c = 3$ and $d = 14$.

The graph of $y = 2x^3 - 9x^2 + 3x + 14$ is shown in Fig. 7.8. It is a curve with two turning-points, and cuts the x-axis at $x = -1$, 2 and 3·5.

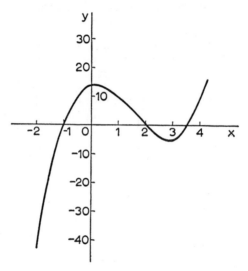

FIG. 7.8 Graph of $y = 2x^3 - 9x^2 + 3x + 14$

Hence the roots of the equation $2x^3 - 9x^2 + 3x + 14 = 0$ are $x = -1$, 2 or 3·5.
Thus we see how a cubic equation may be solved or a cubic expression factorised generally, by means of its graph.

Alternative method of solving $2x^3 - 9x^2 + 3x + 14 = 0$
Writing $2x^3 = 9x^2 - 3x - 14$, we can draw the graphs of (a) $y = 2x^3$, (b) $y = 9x^2 - 3x - 14$, and find their points of intersection as in previous examples.

x	-2	-1	0	1	2	3	4
$2x^3$	-16	-2	0	2	16	54	128
$9x^2$	36	9	0	9	36	81	144
$-3x$	6	3	0	-3	-6	-9	-12
-14	-14	-14	-14	-14	-14	-14	-14
$9x^2 - 3x - 14$	28	-2	-14	-8	16	58	188

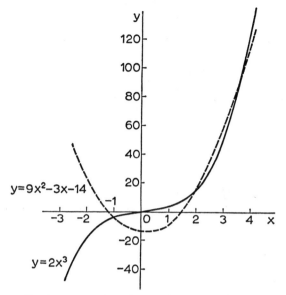

FIG. 7.9 Intersection of $y = 2x^3$ and $y = 9x^2 - 3x - 14$

Figure 7.9 shows the points of intersection, which are three in number, when $x = -1$, 2 and 3·5. These are therefore the roots of the equation $2x^3 - 9x^2 + 3x + 14 = 0$ (as found in the first method). Values obtained by this alternative method of solution should be checked for numerical accuracy. The points of intersection of the two curves are not well defined.

In this case we note that the graphs are the cubic curve (as before) and a parabola. If the parabola had its vertex (turning-point) *above* 0 on the y-axis, instead of *below*, the parabola would cut the cubic curve only *once*: thus we should find *one real* root instead of three as above (two roots would be imaginary).

This conclusion could also be verified by reference to Fig. 7.10, for if the graph A with a different value of d were transferred vertically upwards to position C, the minimum point would lie above the x-axis, which would then be cut by the curve in only *one* point, instead of the three previously obtained. (See Fig. 7.8.)

As a particular case, we might imagine the minimum point lying *on* the x-axis, in which case the two points of intersection of the curve and the x-axis would coincide (B). There would thus be three roots, two being coincident in value.

Other types of graphs commonly used
Two other types of graphs which the student will often meet are

$$(a)\ y = ax^n, \qquad (b)\ y = ae^{bx}.$$

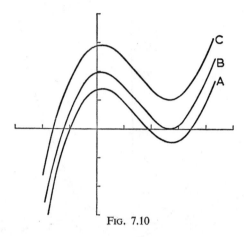

Fig. 7.10

Examples
7.1 Draw the graph of $y = 3 \cdot 5x^{2 \cdot 8}$.

A graph of this type needs a rather more difficult calculation than the cubic types, as the values of $x^{2 \cdot 8}$ can be plotted only by using logarithms.

Taking logs of both sides,

$$\log y = \log 3 \cdot 5 + 2 \cdot 8 \log x$$

Our table of values could be set out as follows.

x	0	0·5	1	1·5	2	2·5	3
$\log x$	$-\infty$	$\overline{1}\cdot6990$ $= -0\cdot3010$	0	0·1761	0·3010	0·3979	0·4771
$2\cdot8 \log x$	$-\infty$	$-0\cdot8428$	0	0·4931	0·8428	1·1141	1·3359
$\log 3\cdot5$	0·5441	0·5441	0·5441	0·5441	0·5441	0·5441	0·5441
$\log y$	$-\infty$	$\overline{1}\cdot7013$	0·5441	1·0372	1·3869	1·6582	1·800
y	0	0·5027	3·500	10·89	24·37	45·52	75·86

The graph is shown in Fig. 7.11.

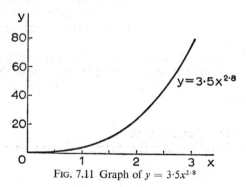

$y = 3 \cdot 5x^{2 \cdot 8}$

Fig. 7.11 Graph of $y = 3 \cdot 5x^{2 \cdot 8}$

7.2 Draw the graph of $y = 3e^{2x}$.

The same method of calculation is used in this case. Taking logs of both sides,

$$\log y = \log 3 + 2x \log e$$

Tabulating as before, and noting that $\log_{10} e = 0{\cdot}4343$

x	0	0·25	0·5	0·75	1	1·25	1·5	1·75	2
$2x \log e$	0	0·2172	0·4343	0·6513	0·8686	1·0858	1·3029	1·5201	1·7372
$\log 3$	0·4771	0·4771	0·4771	0·4771	0·4771	0·4771	0·4771	0·4771	0·4771
$\log y$	0·4771	0·6943	0·9114	1·1286	1·3457	1·5629	1·7800	1·9972	2·2143
y	3	4·946	8·155	13·45	22·17	36·56	60·26	99·36	163·8

The graph is shown in Fig. 7.12.

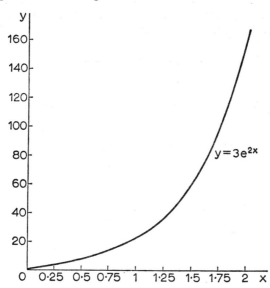

FIG. 7.12 Graph of $y = 3e^{2x}$

EXERCISE 7.1

1 Draw the graph of $y = 6x^2 + 17x - 45$ from $x = -5$ to $x = 4$. Use it to solve the equations (a) $6x^2 + 17x - 45 = 0$, (b) $6x^2 + 17x = 25$.

What is the minimum value of $6x^2 + 17x - 45$?

2 Draw the graphs of $6x^2$ and $(45 - 17x)$, using the same scales and axes. What are the values of x where the graphs intersect? How do these values compare with the roots of $6x^2 + 17x - 45 = 0$ in exercise 1 above?

3 Draw the graph of $y = 7{\cdot}2x - 1{\cdot}5x^2$ from $x = 0$ to $x = 4$. Find the maximum value of $7{\cdot}2x - 1{\cdot}5x^2$, and give the corresponding value of x.

4 Given that $Y = 3M + 75/M$, find a value of M which makes Y a minimum. What is the minimum value of Y? (Draw the graph from $M = 0$ to $M = 5$.)

5 Given that $P = \dfrac{36R}{(R + 0.32)^2}$ plot the graph from $R = 0.1$ to $R = 0.6$, and find what value of R gives a maximum value for P.

6 Solve graphically the equation $x(x - 3)(2x + 8) = 0$, and find the maximum and minimum values of the function $x(x - 3)(2x + 8)$.

7 Draw the graph of $y = 1.7x^{1.6}$, and hence solve the equation $1.7x^{1.6} = 8.8$. (Draw the graph from $x = 0$ to $x = 4$.)

8 Draw the graph of $2.2x^{1.2}$ from $x = 0$ to $x = 5$. Use it to solve $2.2x^{1.2} - 10 = 0$.

9 Graph the functions $1/x$ and $[0.4(x + 1)(3 - x)]$ for values of x from -2 to $+4$, using the same scales and reference axes for both. By means of the graphs, estimate to within ± 0.05 the roots of the equation $2x(x + 1)(3 - x) = 5$.   NCTEC

10 The volume in litres of a gas storage cylinder is given by
$$V = \frac{11x}{75}\left(21 - \frac{x^2}{900}\right)$$
where x is in millimetres. Taking values of x from 25 to 100, calculate V and tabulate. Show on a diagram how V varies as x varies from 25 to 100. From your diagram read off (a) the value of x that gives a maximum volume, (b) the maximum volume.   UEI

11 The cost per hour of running a ship at a speed of c knots is £$\left(5 + \dfrac{c^3}{1350}\right)$. For a voyage of 2700 nautical miles, express the total cost over a range of values of c. Tabulate your results and plot a graph of total cost against speed from $c = 5$ to $c = 25$ knots. Hence find the value of c which makes the total cost as small as possible. (1 knot is a speed of 1 international nautical mile per hour.)   EMEU

12 Plot the curve of the equation $y = 2x^3 - 5x^2 + 2$ between the values $x = -2$ and $x = 3$, and from your curve find the roots of the equation $2x^3 - 5x^2 + 2 = 0$ which lie within the range given.   ULCI

13 Find, to two significant figures, the values of x between $+3$ and -3 which satisfy the equation $x^3 - 6x + 3 = 0$.   ULCI

14 Graph the function $0.1(x - 1)(2x + 3)(2x - 7)$ for all values of x from -2 to $+4$, using 20 mm as the unit along each axis. By means of the graph, estimate to within ± 0.05 the roots of the equation $(x - 1)(2x + 3)(2x - 7) = 5$.   NCTEC

15 Graph the function $\frac{1}{5}(x - 4)^2(x + 2)$ for values of x from -2 to 6. By means of the graph, solve each of the equations (a) $(x - 4)^2(x + 2) = 10$, (b) $(x - 4)^2(x + 2) = 17$.   NCTEC

16 On the same axes, and to the same scales, plot graphs of the functions $x^3 + 5x^2 + 3$ and $17 - x$ for values of x between -5 and $+2$. Hence solve the equation $x^3 + 5x^2 + x - 14 = 0$.   UEI

Solution of equations involving exponential, logarithmic and circular functions

A solution of the equation $F(x) = f(x)$, where $F(x)$ and $f(x)$ may be exponential, logarithmic, or circular (trigonometrical) functions, may be obtained by finding the value of x corresponding to the intersection of the graphs of $F(x)$ and $f(x)$. Figure 7.13 shows the graphs of $F(x)$ and $f(x)$ plotted over a certain range of values of x which includes the point where $x = b$, a point of intersection B of the graphs of $F(x)$ and $f(x)$.

In the diagram M is the point $[a, F(a)]$, N is the point $[a, f(a)]$ and NM represents the difference between the two functions, i.e. $F(a) - f(a)$. At the point of intersection B it can be seen that this difference has become zero: i.e. $F(b) - f(b) = 0$, or $F(b) = f(b)$. Considered as a point on $f(x)$, B is the point $[b, f(b)]$, but, considered as a point on $F(x)$, B is the point $[b, F(b)]$: i.e. CB in the diagram represents both $f(b)$ and $F(b)$, and again $F(b) = f(b)$. We can therefore say that $x = b$ is *the solution of the equation $F(x) = f(x)$, since $F(b) = f(b)$.*

Example

7.3 On the same axes plot the curves $y = \log_e x^2$ and $y = 3e^{-x/2}$ for x between 1 and 5. Hence determine an approximate solution of $\log_e x^2 = 3e^{-x/2}$. Explain briefly how an approximate solution of the equation $\log_e x^2 - 3e^{-x/2} = 1$ can be found.   ULCI specimen

Using the appropriate mathematical tables and the fact that $\log_e x^2 = 2 \cdot \log_e x$,

x	1	2	3	4	5
$\log_e x$	0	0·6931	1·0986	1·3863	1·6094
$2 \cdot \log_e x$, i.e., $\log_e x^2$	0	1·39	2·20	2·77	3·22
$\dfrac{-x}{2}$	−0·5	−1·0	−1·5	−2·0	−2·5
$e^{-x/2}$	0·6065	0·3679	0·2231	0·1353	0·0821
$3e^{-x/2}$	1·82	1·10	0·67	0·41	0·25

On plotting the graphs of $3e^{-x/2}$ and of $\log_e x^2$ (see Fig. 7.14), an approximate solution of $\log_e x^2 = 3e^{-x/2}$ is seen to be $x = 1\cdot8$. The solution can be refined by replotting in the neighbourhood of $x = 1\cdot8$, if more extensive tables are available, giving $x = 1\cdot83$, and so on.

An approximate solution of $\log_e x^2 - 3e^{-x/2} = 1$, i.e. $\log_e x^2 = 3e^{-x/2} + 1$, may be found by plotting the function $3e^{-x/2}$ with each ordinate increased by 1 unit. The required solution will be given by the intersection of the graph of the function $3e^{-x/2} + 1$ with the graph of the function $\log_e x^2$.

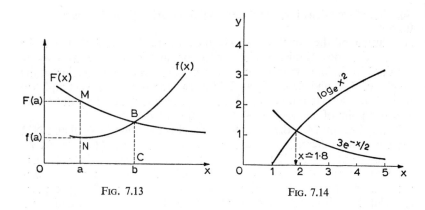

FIG. 7.13   FIG. 7.14

EXERCISE 7.2

1 Using the same axes plot the functions $\sin 2x$ and $x^2 + 0.25$ for values of x from 0 to 1·4 radians. Hence determine the approximate solutions of $x^2 + 0.25 = \sin 2x$ within this range.

2 Plot the graph of $y = e^{-x} . \sin x$ for values of x from $x = 0$ to $x = 1.2$ at intervals of 0·2 units. Hence solve the equation $e^{-x} . \sin x = 0.1 + x/4$.

3 Plot the graphs of $y = e^{-x} . \cos x$ and $y = \sin x$, using the same axes, for values of x from $x = 0$ to $x = 0.8$ at intervals of 0·1 units. Hence determine an approximate solution of   (a) $e^{-x} . \cos x = \sin x$, (b) $e^{-x} . \cos x = 0.4$, (c) $\sin x = 0.4$.

4 Plot the graph of $y = \log_e (1 + \sin x)$ from $x = 0$ to $x = 1$ at intervals of 0·1 units. Use your graph to solve approximately (a) $e^{7x/10} = 1 + \sin x$, (b) $\log_e (1 + \sin x) = 0.4 - x$.

5 Plot the graph of $y = \log_e (2 - \cos x)$ for values of x from $x = 0$ to $x = 1$ at intervals of 0·1 units. Hence obtain approximate solutions of the equation $\frac{18}{7} \log_e (2 - \cos x) = x - 0.1$.

6 Plot the graph of $y = \log_e (1 + 2x)$ and of $y = 1.08x + 0.1$ for values of x from $x = 0$ to $x = 0.9$. Hence obtain approximate solutions of the equation $\log_e (1 + 2x) = 1.08x + 0.1$.

7 Plot the graph of $y = e^{-x}$ and $y = \sin x/2$ for values of x from $x = 0$ to $x = 1$. Hence obtain an approximate solution of the equation $e^{-x} = \sin x/2$.

Chapter Eight
Determination of Laws

Having considered the methods of drawing various graphs from given equations (or given functions of x), we will now consider the converse type of problem.

In this case we are given the graph, or more usually sets of corresponding values of the two variables, and we are required to find the relation, or equation, connecting them. Usually the sets of corresponding values are the result of experiment, and it may be very useful to find a law connecting them. For example, the student may have measured the different values of an electric current I in a circuit containing a source of constant voltage when the resistance R in the circuit was repeatedly altered. By tabulating the corresponding values of current I and resistance R, he is able to find how they are related to each other or, as the problem is usually stated, 'to determine the law' connecting I and R. Once the law has been determined, it serves as a useful basis for further calculations.

The linear law, $y = ax + b$
This law has already been dealt with in previous work, where it was shown that all straight lines may be represented generally by the equation $y = ax + b$, the gradient of the line and its position with respect to the axes depending on the values of a and b respectively. Thus $y = 3x + 2$, $y = 5 - 4x$, and $y = x/3 - 6$ are straight lines in which the gradients are 3, -4 and $\frac{1}{3}$, and the intercepts on the y-axis are 2, 5 and -6 respectively. If we plot a given set of corresponding values of two variables (say x and y) and we obtain a straight line, then we know that x and y (or other variables taking their places) are connected by a law of the above type. To determine the law is merely to find the particular values of a and b in the equation $y = ax + b$.

Example
8.1 Given the following values of x and y, find the law connecting them.

x	-3	-2	-1	0	2	3	4
y	-17	-13	-9	-5	3	7	11

On plotting these values in the usual way, a straight line is obtained (Fig. 8.1). Therefore the equation of the line (or the law connecting x and y) is of the form $y = ax + b$.
The values of a and b can now be found in one of two ways.

(a) By drawing two lines, such as **AB** and **BC**, we find that the line has risen 24 units (as measured on the y-scale) in a horizontal distance of 6 units (measured on the x-scale)

∴   the gradient of the line $= a = \frac{24}{6} = 4.$

Since the line crosses the y-axis at -5,
$$b = -5$$

∴   the equation of the line is
$$y = 4x - 5$$

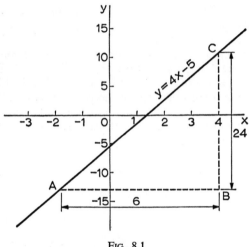

FIG. 8.1

(b) Alternatively, by choosing values from the given table, such as $x = -2$ when $y = -13$, and $x = 3$ when $y = 7$, we can substitute in the equation $y = ax + b$ and so obtain the simultaneous equations

$$7 = 3a + b. \quad . \quad . \quad . \quad . \quad . \quad \text{(i)}$$
$$-13 = -2a + b \quad . \quad . \quad . \quad . \quad . \quad \text{(ii)}$$

whence
$$20 = 5a$$
$$\therefore \quad a = 4$$

Substituting $a = 4$ in the equation (i),

$$7 = 12 + b$$
$$\therefore \quad b = -5$$

∴   the equation of the line (or the law connecting x and y) is, as previously found,
$$y = 4x - 5$$

In dealing with sets of related quantities found by actual experiment, the same method is adopted. The values are plotted and, if the graph is a straight line (allowance being made for slight divergences owing to experimental error), the values plotted vertically (in the place of y) are connected with those plotted horizontally (in the place of x) by the linear law $y = ax + b$, adapted by putting the variables used in the place of y and x respectively.

If we have plotted a series of values of M vertically and a series of values of N horizontally and the result is a straight line, we know that the law connecting M and N is of the form $M = aN + b$.

Laws other than linear

Only occasionally do the values give a straight line: generally the graph is curved. A reference to the graphs shown in Figs 7.9, 7.11 and 7.12 in the previous chapter will convince the student that there is sufficient similarity between the graphs of $y = 2x^3$, $y = 9x^2 - 3x - 14$, $y = 3{\cdot}5x^{2{\cdot}8}$ and $y = 3e^{2x}$ (or for such portions as can be drawn with the limits of the paper) to make it impossible by mere inspection to state definitely the type of law to which a curve, as seen in the first quadrant, belongs, for it must be remembered that usually only that part of the curve which lies in the first quadrant will be obtained by plotting experimental data.

When the graph is a straight line, the form of the law can be definitely stated, but a curved graph might represent any one of a large number of laws, such as $y = ax^2 + b$, $y = ax^3 + bx^2 + cx + d$, $y = ax^n$, $y = ae^{bx}$, etc. To decide the type of law on which the curved graph is based, we must adopt some method of 'reducing the curve to a straight line'. We will illustrate possible methods by examples.

Consider the graph of $y = 3x^2 - 10$. We already know that this is a parabola, but between $x = 0$ and $x = 3$ we have a curve (Fig. 8.2) which

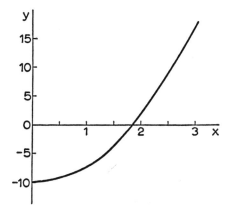

Fig. 8.2 Graph of $y = 3x^2 - 10$

we could not identify merely by inspection *if we did not previously know the type of equation.* It might satisfy reasonably well one of several other types of laws such as have been dealt with in previous chapters.

But we already know in this particular case that y and x^2 are the forms in which the variables occur in the equation, so let us try the effect of plotting the y values against those of x^2 (instead of x).

Tabulating these values, we get

x	0	$\frac{1}{2}$	1	$1\frac{1}{2}$	2	$2\frac{1}{2}$	3
x^2	0	$\frac{1}{4}$	1	$2\frac{1}{4}$	4	$6\frac{1}{4}$	9
$y = 3x^2 - 10$	-10	$-9\frac{1}{4}$	-7	$-3\frac{1}{4}$	2	$8\frac{3}{4}$	17

When these are plotted, we obtain Fig. 8.3.

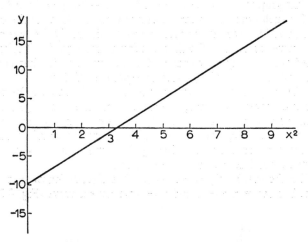

FIG. 8.3 Result of plotting y against x^2

The graph is a straight line

$\therefore$ y is connected with x^2 by the linear form of law

$$\therefore\quad y = ax^2 + b$$

(a result which we already knew, since $a = 3$ and $b = -10$).

Thus, in future, if we can obtain a straight line by plotting some *functions* of y and x instead of the actual y and x values given, we shall be able to deduce the law.

For example, if we obtain a straight line by plotting M^3 (say) vertically against N^2 horizontally, we infer that

$$M^3 = aN^2 + b$$

In general, if we plot any particular function of a variable θ vertically against a function of another variable ϕ horizontally and so obtain a straight line, we know that the two functions are connected by the linear law. Thus

$$\text{the function of } \theta = a \times (\text{the function of } \phi) + b.$$

The problem now resolves itself into finding what functions of the variables must be plotted so as to give the straight line.

Let us consider the following typical examples.

Examples
8.2 Type of law $y = ax^2 + b$
The following values of R and V are possibly connected by a law of the type $R = aV^2 + b$. Test if this is so, and find the law (i.e. find a and b).

V	12	16	20	22	24	26	30
R	6·44	7·56	9	9·84	10·76	11·76	14

If these values are plotted as given, a curve will result, which tells us nothing except that they are *not* connected by the linear law $y = ax + b$ or $R = aV + b$. This is a valueless result in view of the suggestion offered in the problem.

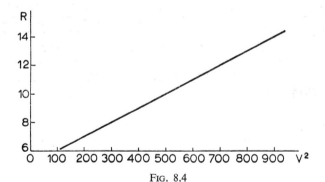

Fig. 8.4

But if R is plotted against V^2, a straight line will result if the supposition made is correct. The values of R and V^2 are therefore tabulated and plotted (Fig. 8.4).

V^2	144	256	400	484	576	676	900
R	6·44	7·56	9	9·84	10·76	11·76	14

The straight line obtained shows that $R = aV^2 + b$.

We can find the values of a and b (and so determine the actual law) by either of the two methods given previously.

Using the simultaneous equations method, we choose two suitable points not too close together, such as (256, 7·56) and (676, 11·76).

Thus $11·76 = a \times 676 + b$ . . . . . (i)
                        $7·56 = a \times 256 + b$ . . . . . (ii)

or $11·76 = 676a + b$
                        $7·56 = 256a + b$

By subtraction,

$$4·20 = 420a$$

$$\therefore \quad a = \frac{4·2}{420} = 0·01$$

By substitution of $a = 0·01$ in equation (ii),

$$7·56 = (0·01 \times 256) + b$$
$$\therefore \quad 7·56 = 2·56 + b$$
$$\therefore \quad b = 5$$

i.e. The law is $R = 0·01V^2 + 5$.

The other method of finding the gradient and the intercept on the y-axis could be used as a check on this result.

In practice it often happens that the points do not give a perfectly straight line, owing to the fact that experimental readings involve slight errors. In such cases, a line should be drawn intermediate between the points and the simultaneous equations be derived from values chosen from the line, rather than from the slightly erroneous values given. Of course, the student must first satisfy himself that the points *do* represent a straight line with slight divergencies.

8.3 Type of law, $y = a/x + b$

In measuring the resistance, R ohms, of a carbon-filament lamp at various voltages, V, the following results were obtained.

V (volts)	60	70	80	90	100	120
R (ohms)	70	67·2	65	69·3	62	60

Show that the law connecting R and V is of the form $R = a/V + b$ and then find it.

By comparing $y = ax + b$ with $R = a/V + b$ (which, as we have seen earlier in the course, is a hyperbola) we find that R occupies the place of y and $1/V$ occupies the place of x in the equation.

We therefore plot R values vertically and values of $1/V$ horizontally.

Tabulating and plotting, we obtain

V	60	70	80	90	100	120
$1/V$	0·0167	0·0143	0·0125	0·0111	0·01	0·0083
R	70	67·2	65	63·3	62	60

A straight line is obtained (Fig. 8.5)

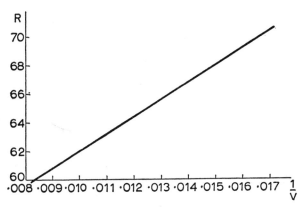

FIG. 8.5 Graph of $R = a/V + b$

$$\therefore \quad R = a \times \frac{1}{V} + b$$

or $\qquad\qquad R = \frac{a}{V} + b$ is the form of the law

The values of a and b may be found by selecting $1/V = 0·0143$ when $R = 67·2$, and $1/V = 0·0083$ when $R = 60$.

Substituting in $R = \frac{a}{V} + b$, we obtain

$$67·2 = a \times 0·0143 + b$$
$$60 \ \ = a \times 0·0083 + b$$

or
$$67·2 = 0·0143a + b \ \ . \quad . \quad . \quad . \quad . \quad . \quad \text{(i)}$$
$$60 \ \ = 0·0083a + b \ \ . \quad . \quad . \quad . \quad . \quad . \quad \text{(ii)}$$

whence
$$7·2 = 0·0060a$$

$$\therefore \quad a = \frac{7·2}{0·006} = 1200$$

Substituting $a = 1200$ in equation (ii)

$$60 = (0·0083 \times 1200) + b$$
$$\therefore \quad 60 = 9·96 + b$$
$$\therefore \quad b = 60 - 9·96 = 50·04 = 50 \text{ (approx)}$$

$\therefore$ The law is $\qquad R = \dfrac{1200}{V} + 50$

8.4 Type of law, $y = ax^n$
Find the law connecting H and v (it is probably of the form $H = av^n$).
Note that a and n are constants.

v	15	18	20	22	24	25	27
H	354	623	863·4	1160	1519	1724	2190

If the law connecting H and v is of the type suggested, we know that, by plotting the given values, we should get a curve (of the type shown in Fig. 7.9), but, as before, we cannot distinguish such a curve from curves depending on other types of laws. Thus the curve obtained does not indicate the type of law applicable.

To obtain a straight line in the given type of equation $H = av^n$, take logs of both sides,

$$\log H = \log a + n \log v$$
or
$$\log H = n \log v + \log a$$

By comparison with the linear law $y = ax + b$, we find that the two variables $\log H$ and $\log v$ are connected in the same way as y and x.

$\log H = \log v \times (\text{a constant}, n) + (\text{a constant}, \log a)$ corresponding to

$$y = x \times (\text{a constant}, a) + (\text{a constant}, b)$$

$\therefore$ if $\log H$ is plotted against $\log v$, a straight line should be obtained if the suggested law holds good.

We therefore tabulate $\log H$ and $\log v$ and plot in the usual way.

$\log v$	1·1761	1·2553	1·3010	1·3424	1·3802	1·3979	1·4314
$\log H$	2·5490	2·7945	2·9362	3·0645	3·1816	3·2365	3·3404

The graph is a straight line (Fig. 8.6).

$$\therefore \quad \log H = n \log v + \log a$$

where n and a are constants.

$$\therefore \quad H = av^n \text{ is the type of law}$$

The constants a and n are found in the usual way. Choosing $\log H =$

3·2365 when log $v = 1·3979$, and log $H = 2·7945$ when log $v = 1·2553$, and substituting in log $H = n$ log $v + $ log a, we obtain

$$3·2365 = n \times 1·3979 + \log a \quad . \quad . \quad . \quad . \quad . \quad \text{(i)}$$
$$2·7945 = n \times 1·2553 + \log a \quad . \quad . \quad . \quad . \quad . \quad \text{(ii)}$$

whence $\quad\quad 0·4420 = 0·1426n$

$$\therefore \quad n = \frac{0·4420}{0·1426} = 3·1$$

Substituting this value of n in equation (ii),

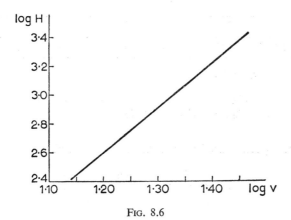

FIG. 8.6

$$2·7945 = (3·1 \times 1·2553) + \log a$$
$$\therefore \quad 2·7945 = 3·8914 + \log a$$
$$\therefore \quad \log a = 2·7945 - 3·8914$$
$$\therefore \quad \log a = \bar{2}·9031$$

whence $\quad\quad\quad a = 0·08$

$\therefore$ The law connecting H and v is

$$H = 0·08v^{3·1}$$

8.5 Type of law, $y = ae^{bx}$
Show that the following values of x and y are connected by a law of the form $y = ae^{bx}$, and find the constants a and b.

x	2	2·5	3	3·5	4	4·5	5
y	30·26	47·44	74·47	116·7	182·8	287·1	449·8

If $\quad\quad\quad y = ae^{bx}$

then $\quad\quad\quad \log y = \log a + bx \log e$

Let $\qquad b \log e = B$, since both b and e are independent of x,

$$\therefore \quad \log y = \log a + Bx$$

Comparing this with $y = a + bx$, we find that

$$\log y = \log a + Bx$$

is the linear law, where $\log y$ and x are the variables, $\log a$ and B being constants.

$\therefore$ If we plot $\log y$ against x (not $\log x$), we shall get a straight line if the suggested law applies.

Tabulating and plotting as below,

x	2	2·5	3	3·5	4	4·5	5
$\log y$	1·4809	1·6762	1·872	2·067	2·262	2·458	2·653

The graph is a straight line (Fig. 8.7),

$$\therefore \quad \log y = \log a + Bx \text{ (where } B = b \log e)$$
$$\therefore \quad \log y = \log a + bx \log e$$
$$\therefore \quad \quad y = ae^{bx}$$

Substituting selected values of x and $\log y$ in the equation,

$$\log y = \log a + Bx$$
$$2 \cdot 4580 = \log a + B \times 4 \cdot 5 \quad . \quad . \quad . \quad . \quad . \quad \text{(i)}$$
$$1 \cdot 6762 = \log a + B \times 2 \cdot 5 \quad . \quad . \quad . \quad . \quad . \quad \text{(ii)}$$

whence $\qquad 0 \cdot 7818 = 2B$

$$\therefore \quad B = \frac{0 \cdot 7818}{2} = b \log e$$

$$\therefore \quad b = \frac{0 \cdot 7818}{2 \log e} = 0 \cdot 9001 = 0 \cdot 9 \text{ (approx)}$$

To find a, substitute $B = \dfrac{0 \cdot 7818}{2}$ in equation (ii)

$$\therefore \quad 1 \cdot 6762 = \log a + \frac{0 \cdot 7818}{2} \times 2 \cdot 5$$

$$\therefore \quad 1 \cdot 6762 = \log a + 0 \cdot 9772$$
$$\therefore \quad \log a = 1 \cdot 6762 - 0 \cdot 9772$$
$$\therefore \quad \log a = 0 \cdot 6990$$
$$\therefore \quad \quad a = 5 \cdot 00$$

$\therefore$ The required constants are

$$a = 5, \quad b = 0 \cdot 9$$

or the law is

$$y = 5e^{0 \cdot 9x}$$

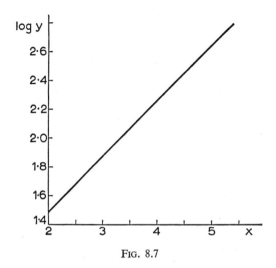

Fig. 8.7

Summary

We have examined five types of laws and the methods of determining the actual laws by finding the constants.

(a) $y = ax + b$,
(b) $y = ax^2 + b$ (of which $R = aV^2 + b$ is a special case),
(c) $y = a/x + b$ (of which $R = a/V + b$ is a special case),
(d) $y = ax^n$ (of which $H = av^n$ is a special case),
(e) $y = ae^{bx}$.

If a suggested type of law be compared with $y = ax + b$, the functions which occupy the places of y and x, respectively, when plotted as x and y coordinates, will give a straight line if the law is as supposed. Conversely, if by trial we obtain a straight line by plotting a function of y against a function of x, then these functions are connected by the linear law.

In every case except the first the curve was reduced to a straight line, each type requiring its own particular treatment, but the methods used will suggest ways of dealing with other forms as they arise.

EXERCISE 8.1

1 Prove that the following values of x and y are connected by the law $y = ax + b$. Find the constants a and b.

x	-4	-2	0	2	4	6	8
y	$-19\cdot2$	$-14\cdot6$	-10	$-5\cdot4$	$-0\cdot8$	$3\cdot8$	$8\cdot4$

2 The following are values of L, the latent heat of steam, at corresponding temperatures $\theta\,°C$. Find the law connecting L and θ in the form $L = a + b\theta$, and find the latent heat of steam at $100\,°C$

$\theta\,°C$	70	80	90	110	130	140
L kJ	2340	2300	2280	2220	2160	2130

3 Complete the following table of values, given that y and x are connected by a law of the form $y = ax + b$, where a and b are constants.

x	-1	0		1·4	3	5	7
y			0		12		32

NCTEC

4 The following table gives the result of an experiment. Plot the load-effort diagram and determine the equation (of the form $E = mW + c$) which most nearly accords with the results.

W (load N)	7	14	21	28	35	42	49	56
E (effort N)	3	6·5	9·5	12·5	16	18·75	22	25

ULCI

5 By experiment, the following relation is found between two quantities R and D.

R	10·5	15·8	21·3	26·6	31·6	36·4	42
D	0	1	2	3	4	5	6

It is believed that this relation can be expressed by an equation of the form $R = mD + c$. Plot the results and from your graph determine the values of m and c.

ULCI

6 Complete the following table, given that y and x are connected by a law of the form $\dfrac{y}{a} = \dfrac{1}{x + b}$.

x	-1			2	3	6
y		4	2·5	2		1

NCTEC

7 The following table gives related values of x and y. Determine whether these values are connected by an equation of the form $y = ax^2 + b$, where a and b are constants, and, if so, find the values of a and b.

x	4	5	6	7	8	9
y	14·3	18	22·5	28	34·5	41·5

ULCI

8 The following values of R and V satisfy a law of the form $R = a + bV^2$, where a and b are constants.

V	20	25	30	35	40	45
R	52	58	67	76	88	100

Plot suitable quantities to obtain a straight-line graph, and obtain the values of a and b.   ULCI

9 In an experiment, the following values of N and D were obtained. There is reason to suspect that they are connected by a formula of the type $N = a + b/D$. Test if this is so, and find a and b.

N	0·322	0·317	0·313	0·311	0·309	0·308
D	0·45	0·6	0·75	0·9	1·05	1·2

10 A certain type of vessel needs power P kW to drive it at a speed v knots. Find the relation between P and v using the following values.

v	10	12	14	15	17
P	1500	2300	3400	4000	5700

(The relation is believed to be of the form $P = mv^3 + a$.)

11 The law connecting the coefficient of friction μ between a belt and a pulley, with the velocity v of the pulley in m/s, is believed to be of the form

$$\mu = a\sqrt{v} + b$$

Test if the following values of μ and v agree with this law, and find a and b.

μ	0·210	0·223	0·235	0·245	0·262	0·277
v	2	3	4	5	7	9

12 Air is compressed adiabatically, and the pressure and temperature are measured. The law connecting them is believed to be $T = ap^n$, where T is the absolute temperature and p is the pressure in bars. From the following numbers, obtain the best values of a and n.

p bars	1	1·7	3·1	4·8	6·2
T K	500	580	685	775	837

13 The following values of H and Q are connected by a law of the type $Q = aH^n$. Find a and n.

H	1·2	1·6	2·0	2·2	2·5	3·0
Q	6·087	6·751	7·316	7·571	7·927	8·467

14 Given that $T = ae^{\mu\theta}$, find a and μ.

θ	0·35	0·698	1·047	1·396	1·745
T	17·77	19·74	21·91	24·32	27

Additional questions may be found in Exercise 8.2, which follows.

EXERCISE 8.2. (miscellaneous)

In many of these miscellaneous examples, and in others which have gone before, the student is called upon to determine constants such as a, p, m, n, etc. The information given and the methods available usually leave the student no alternative but to determine these constants, and to state them, as *numbers*. In fact, however, as examination of the equations in which they occur will show, these constants have physical dimensions: they are lengths, areas, times, velocities, etc., or powers or products or ratios of these. Thus the constants will differ according to the units of measurement employed in stating the laws, etc. It is important therefore that no such constant should be used outside the immediate context of the question asked.

1 Draw a graph of $y = 4 - x^2 + \dfrac{8}{x + 5}$ from $x = -4$ to $x = +4$. Use the graph to solve the equation $4 - x^2 + \dfrac{8}{x + 5} = 0$.

What is the maximum value of y?   WEST RIDING

2 On the same axes plot the graphs of $y = \cos\theta$ and $y = \sin 2\theta$ for values of θ from $0°$ to $360°$. Hence solve the equation $\sin 2\theta - \cos\theta = 0$, giving roots between $0°$ and $360°$.   SURREY

3 The tabulated values of x and y are thought to be related by a law of the type $y = ax + bx^2$. Verify graphically that this is so, and determine the values of a and b.

x	1	2	3	4	5
y	3	9·2	18·6	31·2	47

NUNEATON

4 The following values of H and Q are connected by a law of the type $Q = aH^n$, where a and n are constants.

H	1·2	1·6	2·0	2·2	2·5	3·0
Q	6·087	6·751	7·316	7·571	7·927	8·467

Plot suitable quantities to obtain a straight-line graph and from it determine the values of a and n.   RUGBY

5 The following table gives corresponding values of x and y which are connected by a relationship of the type $y = a + bx^2$, a and b being constants.

x	0·5	1·5	2·5	3·0	3·5	4·0	5·0
y	6·3	8·4	12·0	14·6	17·5	21·0	29·2

By plotting a suitable graph, or by any other method, find the values of a and b, and hence state the relationship between x and y.
                                                                  HANDSWORTH

6 Find the maximum value of y, if $y = 2x - 3x^2 + 2$. Also sketch the graph of y for values of x from -2 to $+3$.   RUGBY

7 It is believed that the following values of e and l obey a law of the form $e = a/l + b$. Verify, by means of a suitable graph, that this is so, and determine from your graph the best values for a and b.

l	2	4	6	8	10	12	14
e	64·2	41·6	34·3	30·4	28·2	26·8	25·6

CANNOCK

8 It is believed that the following values of w and I obey the law $w = b + a/I$.

I	36·8	31·5	26·3	21·0	15·8	12·6	8·4
w	12·5	12·9	13·1	13·3	14·1	14·5	16·3

Verify the law by means of a suitable graph, and determine the best values for a and b.   CANNOCK

9 Draw on the same axes, the graphs of $y = \sin x$ and $y = e^{-x}$ for values of x between 0 and π.

From your graph, find the solutions of the equation $\sin x - e^{-x} = 0$.   COVENTRY

10 The tension T in a rope wrapped around a rough curved block is found at points in the rope defined by an angle θ. The corresponding values obtained are

θ (radians)	0·2	0·4	0·7	0·9	1	1·3
T (newtons)	10·29	10·58	11·05	11·34	11·50	12

Test these values for a law of the type $T = Ae^{\mu\theta}$, and determine the most suitable values for the constants A and μ.   SUNDERLAND

11 The table gives values of u and p. Verify graphically that u and p are connected by a law of the form $u = ap^2 + bp$, and find approximate values of the constants a and b.

p	1	2	3	4	5	6	8
u	0·11	0·37	0·795	1·36	2·1	3·0	4·0

NUNEATON

12 η is the coefficient of viscosity of glycerine at various temperatures $\theta\,°C$. Find a formula of the type $\eta = Ce^{-k\theta}$ by drawing a suitable straight-line graph.

θ	0	10	20	30
η	46	21	8·5	3·5

COVENTRY

13 If the cost of a voyage, £C per hour, is given by

$$C = 16 + \frac{v^3}{1000}$$

where v is the speed of the ship in knots, find the total cost of a voyage L international nautical miles long. Use a graphical method to find the most economical speed for this journey.

(Consider speeds between 10 and 30 knots at intervals of 2 knots.)
                                                                    SUNDERLAND

14 Plot the graph of $y = \frac{1}{2}x^2 - 3x$ for values of x between -1 and 7. Using your graph, state the values for which y is 'zero'. Read also the minimum value of y, and the value of x for which it occurs.
                                                                    COVENTRY

15 In the following table x and y are approximately connected by a law of the form $y = a + b/x$. Find a and b from a graph.

x	10	12	15	18	20	22	25
y	3·120	2·985	2·850	2·755	2·710	2·675	2·628

NUNEATON

16 On the same axes, draw the graphs $y = x^2 - 4x$ and $y = 3\log_{10} x$, between $x = 0$ and $x = 5$. Hence find as accurately as possible the two roots of the equation $x^2 - 4x = 3\log_{10} x$.   WEST RIDING

17 Prove that the area of a segment subtending an angle θ radians at the centre of a circle of unit radius is $\frac{1}{2}(\theta - \sin\theta)$.

If this area is equal to that of a quadrant of the same circle, verify that θ satisfies the equation $\theta - \sin\theta = 1.5708$, and solve this equation by a suitable graphical method.

Find the perimeter of the segment for this value of θ. (Take $\pi = 3.1416$.) $\hfill$ SUNDERLAND

18 One side of a rectangle with an area of 10 m² is x m. If the total distance round the rectangle is y m, show that

$$y = 2\left(x + \frac{10}{x}\right)$$

Plot a graph of y against x from $x = 1$ to $x = 7$. From the graph read off the least value of y for a rectangle of area 10 m², and give the length of its sides. $\hfill$ HANDSWORTH

19 (a) A pinion is rotating at 8 revolutions a minute. Find (i) the speed in radians per second, (ii) the time taken to turn through an angle of 1 radian.

(b) Calculate the value of (i) $\sin(100\pi t - 0.08)$ when $t = 0.025$, the angle being in radians, (ii) t when $\sin(100\pi t - 0.08)$ has its maximum value. $\hfill$ NCTEC

20 A gas-engine test gave the following relationship between I (indicated power) and S (shaft power).

S kW	18	39	57	100	141
I kW	45	67	90	135	180

By plotting the values, find a law of the form $S = mI + b$ connecting them. $\hfill$ COVENTRY

21 What physical meanings do you attach to the constants m and b determined in answering the last question?

22 Taking $e = 2.718$, draw the graph of $y = \log_e(x + 3)$ between $x = -2.5$ and $x = 3$. With the same axes and scales, draw the graph of $5y = x^3$, and use these graphs to solve approximately the equation

$$(x + 3)^5 = e^{x^3} \hfill \text{HANDSWORTH}$$

23 The values of T and θ in the table below are thought to be connected by a law of the form $T = ae^{b\theta}$, where a and b are constants and $e = 2.7183$. Show, by drawing a graph, that T and θ are connected by a law of this form, and use your graph to find the values of a and b.

θ	1	2	3	4	5
T	165	272	448	739	1220

$\hfill$ WEST RIDING

24 H and V are connected by law of the form $H = aV^n$. Find this law if corresponding values of H and V are as follows.

V	8·04	11·67	14·43	17·41	19·9
H	3·03	6·11	9·07	12·21	15·62

SHREWSBURY

25 The following values were obtained in an experiment on bending moments.

x	8	10	12	14	16	18
w	48	38·2	30·8	26·1	22	19·4

Show, by means of a suitable graph, that w and x are related by an equation of the form

$$w = \frac{a}{x} + b$$

and find the values of the constants a and b. COVENTRY

26 In an experiment to find the luminosity I of a lamp for varying voltage V, the following measurements were made.

V(V)	60	80	100	120	140	160
I(lx)	10	31·6	88	184	322	580

Show that the law is of the type $I = aV^n$, where a and n are constants, and find the values of a and n. EMEU

27 A body is raised to a temperature of $20\,°C$ and then allowed to cool: $100\,s$ afterwards its temperature is found to be $11°C$. If θ is the temperature at the time t and it is known that θ and t are connected by an equation of the form ae^{bt}, where a and b are constants, find the values of a and b. WORCESTER

Chapter Nine
Progressions

Meaning of a series
A sequence of quantities connected by a definite law is called a 'series'. Each of the quantities is called a term of the series, and the law connecting the terms is called the 'law of formation'.

Simple examples of numbers 'in series' are

(a) 1, 2, 3, 4, 5, . . .
(b) 1, 3, 5, 7, 9, . . .
(c) 2, 4, 6, 8, 10, . . .
(d) 1, 4, 9, 16, 25, . . . (squares of the natural numbers)
(e) 1, 8, 27, 64, 125, . . . (cubes of the natural numbers)
(f) 1, 5, 9, 13, 17, . . . (add 4 to each in succession)
(g) 5, 10, 20, 40, 80, . . . (multiply by 2)
(h) 4, 5·3, 6·6, 7·9, . . . (add 1·3)
(i) 1, −1, 1, −1, . . . (multiply by −1)
(j) 100, −50, 25, −12·5, . . . (multiply by −0·5).

Arithmetic progressions
An *arithmetic progression* is a series in which each term is formed from the preceding one by the addition (or subtraction) of the same quantity. Thus the terms increase (or decrease) by a fixed amount, called the *common difference*.

Examples of arithmetic series
(a) 1, 5, 9, 13, 17, . . . (common difference is +4)
(b) 16, 10, 4, −2, −8, . . . (common difference is −6)
(c) $x, x + y, x + 2y, x + 3y, . . .$ (common difference is +y)
(d) yearly salaries rising by fixed annual increments
(e) successive yearly values of a car if it depreciates by a constant amount annually
(f) the heights of the supports of a sloping hand-rail when equally spaced out and resting on a level base

General expression for the terms in an A.P.
Let the first term be a, and the common difference d, then the series can be expressed by

$$a, a + d, a + 2d, a + 3d, a + 4d, . . .$$

By noting that

> the first term is a
> the second term is $a + (2 - 1)d$
> the third term is $a + (3 - 1)d$
> the fourth term is $a + (4 - 1)d$

we see that

the twelfth term would be $a + (12 - 1)d = a + 11d$
the twentieth term would be $a + (20 - 1)d = a + 19d$
and thus any term, the nth, could be written

$$n\text{th term} = a + (n - 1)d$$

Examples

9.1 Write down the tenth and the twentieth terms of 1, 6, 11 . . .

The nth term $= a + (n - 1)d$
where $a = 1, \; d = 5$
$\therefore$ tenth term $= a + 9d$
$= 1 + (9 \times 5)$
$= 46$
twentieth term $= a + 19d$
$= 1 + (19 \times 5)$
$= 96$

9.2 If the first term of an A.P. is 5 and the seventh is 29, find the eighteenth term.

First term $= a = 5$
seventh term $= a + 6d = 29$
$\therefore$ By subtraction, $6d = 24$
$\therefore \;\; d = 4$
But   eighteenth term $= a + 17d$
$\therefore$ eighteenth term $= 5 + (17 \times 4)$
$= 73$

9.3 Which term of the series 2·3, 4·2, 6·1 . . . is 36·5?

Let the nth term $= 36\cdot5$

$\therefore \;\; a + (n - 1)d = 36\cdot5$
But $a = 2\cdot3$ and $d = 1\cdot9$
$\therefore \;\; 2\cdot3 + (n - 1)1\cdot9 = 36\cdot5$
$\therefore \;\; 2\cdot3 + 1\cdot9n - 1\cdot9 = 36\cdot5$
$\therefore \;\; 1\cdot9n = 36\cdot1$
$\therefore \;\; n = 19$
$\therefore \;\; 36\cdot5$ is the nineteenth term of the series.

9.4 Insert six arithmetic means between 10 and 25·4.

This means that we must put six numbers between 10 and 25·4, such that 10, ?, ?, ?, ?, ?, ?, 25·4 form an arithmetic progression.

∴   25·4 must be the eighth term and 10, the first.

$$\text{eighth term} = a + 7d = 25 \cdot 4$$

and   first term $= a = 10$

∴   By subtraction, $7d = 15 \cdot 4$

$$\therefore \quad d = 2 \cdot 2$$

∴   The series will be

$$10, \ 12 \cdot 2, \ 14 \cdot 4, \ 16 \cdot 6, \ 18 \cdot 8, \ 21, \ 23 \cdot 2, \ 25 \cdot 4$$

and the required arithmetic means will be

$$12 \cdot 2, \ 14 \cdot 4, \ 16 \cdot 6, \ 18 \cdot 8, \ 21 \ \text{and} \ 23 \cdot 2$$

9.5 A machine is bought for £500. If its value depreciates at the rate of £30 per year, what is its value at the end of 9 years?

At the end of: 1 yr, 2 yrs, 3 yrs, ... 9 yrs.
Value: 470   440   410 ... ?

It is plain that we require the ninth term of the series 470, 440, 410 . . .

where $a = 470$ and $d = -30$
$$\text{ninth term} = a + 8d$$
$$= 470 - (8 \times 30)$$
$$= 470 - 240$$
$$= £230$$

Questions suitable for practice may be picked out from Exercise 9.1.

Geometric series
A series of terms, each of which is formed by multiplying the term which precedes it by a constant factor, is called a *geometric series* or *geometric progression*. The constant factor is called the *common ratio* of the series.

Examples
(a) 4, 8, 16, 32, . . . are in G.P. (common ratio is 2)
(b) 1, $\frac{1}{4}$, $\frac{1}{16}$, . . . are in G.P. (common ratio is $\frac{1}{4}$)
(c) 9, −27, +81, . . . are in G.P. (common ratio is −3)
(d) x, rx, r^2x . . . are in G.P. (common ratio is r)
(e) If a ball bounces up to a height equal to $\frac{5}{6}$ of the height from which it fell, the successive heights to which it rises form a G.P. If h is the initial height from which it drops, the series will be

$$h, \ (5/6)h, \ (5/6)^2h, \ (5/6)^3h, \ \text{etc.}$$

General expression for a series in G.P.

Let the first term be denoted by a, and the common ratio by r.
   Then the series can be expressed by

$$a, ar, ar^2, ar^3, \ldots$$

   The relation between the index of r and the number of the corresponding term should be carefully noted:

$$\text{first term} = a$$
$$\text{second term} = ar \text{ (index of } r = 1)$$
$$\text{third term} = ar^2 \text{ (index of } r = 2)$$
$$\text{seventh term} = ar^6 \text{ (index of } r = 6)$$
$$\text{tenth term} = ar^9 \text{ (index of } r = 9)$$
$$n\text{th term} = ar^{n-1} \text{ (index of } r = n - 1)$$

   It will be seen that the index of r is always *one less* than the number of the term in the series. We can now write down any required term in a given series, viz.

$$n\textbf{th term} = ar^{n-1}$$

Examples
9.6 Write down the sixth term of 6, 18, 54, . . .

Here
$$\text{first term } a = 6 \text{ and } r = 3$$
$$\therefore \quad \text{sixth term} = ar^5$$
$$= 6 \times 3^5$$
$$= 6 \times 243$$
$$= 1458$$

9.7 Write down the seventh term of 6, -4, $+2\frac{2}{3}$, . . .

In this case
$$a = 6$$
and
$$r = \frac{-4}{6} = -\frac{2}{3}$$
$$\therefore \quad \text{seventh term} = ar^6$$
$$= 6 \times (-\tfrac{2}{3})^6$$
$$= 6 \times \tfrac{64}{729}$$
$$= \frac{128}{243}$$

9.8 Find the first five terms of a series in G.P. whose second term $= 24$ and whose sixth term $= 121\frac{1}{2}$.

$$\text{Second term} = ar = 24$$
$$\text{sixth term} = ar^5 = 121\frac{1}{2}$$

Dividing the sixth term by the second,

$$\frac{ar^5}{ar} = \frac{121\frac{1}{2}}{24}$$

$$\therefore \quad r^4 = \frac{243}{48} = \frac{81}{16}$$

$$\therefore \quad r = \sqrt[4]{\frac{81}{16}}$$

$$\therefore \quad r = \frac{3}{2}$$

$$\therefore \quad r = 1\cdot5$$

Now, substituting $1\cdot5$ for r in $ar = 24$,

$$\therefore \quad a \times 1\cdot5 = 24$$

$$\therefore \quad a = \frac{24}{1\cdot5} = 16 = \text{first term}$$

$\therefore$ The first five terms of the series are

$$16, \ 24, \ 36, \ 54, \ 81$$

9.9 Insert four geometric means between 100 and 10.

This is equivalent to putting four terms between 100 and 10 such that all the six numbers form a G.P.

$\therefore$ 100 ? ? ? ? 10 form a G.P.

$$\therefore \quad \text{sixth term} = ar^5 = 10$$
$$\text{first term} = a = 100$$

$$\therefore \quad \frac{ar^5}{a} = \frac{10}{100}$$

$$\therefore \quad r^5 = 0\cdot1$$

$$\therefore \quad r = \sqrt[5]{0\cdot1}$$

$$\therefore \quad r = 0\cdot6310$$

$$\therefore \quad \text{first term} = 100$$
$$\text{second term} = 100 \times 0\cdot6310 = 63\cdot10$$
$$\text{third term} = 63\cdot10 \times 0\cdot6310 = 39\cdot81$$
$$\text{fourth term} = 39\cdot81 \times 0\cdot6310 = 25\cdot12$$
$$\text{fifth term} = 25\cdot12 \times 0\cdot6310 = 15\cdot85$$
$$\text{sixth term} = 10$$

$\therefore$ The required geometric means are $63\cdot10, 39\cdot81, 25\cdot12, 15\cdot85$.

9.10 Which term of the series $1\cdot01, 0\cdot909, 0\cdot8181 \ldots$ is $0\cdot4827$?

Let the required term be the nth term.

Then $a = 1\cdot01$ and $r = \dfrac{0\cdot909}{1\cdot01} = 0\cdot9$

and the nth term, $ar^{n-1} = 1\cdot01 \times (0\cdot9)^{n-1} = 0\cdot4827$

$$\therefore \quad (0\cdot9)^{n-1} = \frac{0\cdot4827}{1\cdot01}$$

Taking logs of both sides,

$$(n-1) \log 0{\cdot}9 = 0{\cdot}4827 - \log 1{\cdot}01$$
$$\therefore \quad (n-1) \times \bar{1}{\cdot}9542 = \bar{1}{\cdot}6836 - 0{\cdot}0043$$

$$\therefore \quad n-1 = \frac{\bar{1}{\cdot}6836 - 0{\cdot}0043}{\bar{1}{\cdot}9542}$$

$$= \frac{\bar{1}{\cdot}6793}{\bar{1}{\cdot}9542}$$

$$= \frac{-0{\cdot}3207}{-0{\cdot}0458}$$

$$= \frac{0{\cdot}3207}{0{\cdot}0458}$$

$$= 7$$
$$\therefore \quad n = 8$$
$$\therefore \quad 0{\cdot}4827 \text{ is the eighth term.}$$

Questions suitable for practice may be picked out from Exercise 9.1

Progressions compared
Series of *numbers* are of great interest and importance to the mathematician. It is, in fact, only through the study of infinite series that the columns in the logarithm tables and the tables of trigonometrical ratios have been computed. Progress in higher algebra demands a ready appreciation of the properties of series and some skill in their manipulation.

The *quantities* whose magnitudes are expressed by some of the number series are of no less importance in practical matters. Arithmetical progressions do not often occur in the studies of physics and engineering, or even in commerce—one of the few examples is that the velocity of a falling body has the same increment second after second. But quantities in geometric progression are of great practical importance—in fact geometric progression seems to be an expression of natural growth.

Once the law of a series has been laid down, we may wish to make calculations which concern large numbers of terms—the sum of n terms, the magnitude of the nth term, and so on. Often, however, our interest is limited to a few terms, and concentrates on the relation between one term and another.

Ordinary weights and measures, for example a box of gramme 'weights' (10–5–2–2–1), are designed to enable a series of quantities with a *common difference* of 1 unit to be weighed simply and quickly.

Engineers use drills ranging in diameter from less than a millimetre to 50 or more millimetres. The different sizes of small drills have to be quite close together, but it would be needless and, incidentally, impossibly costly to stock large drills differing by amounts measured in tenths of a millimetre. Here again common sense leads to something like a geometrical progression of sizes—the progression cannot be exact because the terms must be simply expressed sizes in general use, and not recurring decimals.

Now no one ever decided that he would, as a policy, arrange them to form a geometrical progression. As we have said, common sense made the decision. Later, looking at the result, we say that this is a geometrical progression. Can we now seek to understand how common sense has worked?

Suppose that we have an article, and have to decide upon a larger size. There are many factors to be taken into account. There may be a high cost of tooling for each size of an engineering product. There is always the matter of capital locked up as stock. There is consumers' preference, and so on. But one factor in deciding the increase—the magnitude of the step to be taken—*will always be the size of the article we have.* Other factors may decide *r*, the common ratio, but a new term in the series is obtained by multiplying the *term we have* by *r*. The rate of growth depends upon the value already attained: the actual *increase* in magnitude, instead of being a constant 'difference', *d*, is $(r - 1)$ times the last term.

This is a natural law of growth. It is the *compound interest* law, which equally concerns students of commerce and science.

Example from engineering

As an example of figures from practice which often approximate closely to geometrical progressions, consider the revolution speeds of a lathe. The machine is required to turn work whose diameter ranges from 10 to 150 mm. The surface speed of the work is to be as nearly as possible 35 m/min for all diameters. Eight speeds of revolution are to be made available: to what values should they approximate?

It is required that for all diameters of work the maximum divergence between the attainable cutting speed and the intended cutting speed of 35 m/min, expressed as a ratio, should be the same. Now, for this to be the case, the speeds must be in geometrical progression.

The lowest rev/min must give 35 m/min at the circumference of a 150 mm circle. That is

$$\text{lowest speed} = \frac{35 \times 1000}{\pi \times 150} \text{ rev/min}$$

$$= 74 \text{ rev/min}$$

The ratio of the highest speed to the lowest will be $\frac{150}{10} = 15$. The logarithm of 15 is 1·1761. There are seven steps to go from the lower to the higher. Each will be taken by multiplying the rev/min by the common ratio *r* to bring the speed up to the next value. It is clear that

$$7 \log r = \log 15 = 1 \cdot 1761$$

and
$$\log r = \frac{1 \cdot 1761}{7} = 0 \cdot 1680$$

Logarithm	Rev/min	Speed no.
1·8692 add 0·1680	74	1
2·0372 add 0·1680	108·9 from antilog tables	2
2·2052 add 0·1680	160·4 „ „ „	3
2·3732 add 0·1680	236·1 „ „ „	4
2·5412 add 0·1680	347·7 „ „ „	5
2·7092 add 0·1680	511·9 „ „ „	6
2·8772 add 0·1680	753·7 „ „ „	7
3·0452	1110 „ „ „	8

Checking by multiplication, speed no. 8 is 15 times speed no. 1 = 15 ×
74 = 1110.

The accuracy is better than could be guaranteed with the use of four-
figure logarithms.

The eight speeds provided will be as near to the geometrical progression
74, 108·9, 160·4, etc. as can be arranged without unreasonable complication
in the gearbox.

For practice questions see Exercise 9.2 (miscellaneous).

To find the sum of n terms of an A.P.

Let S_n denote the sum of n terms of a series in A.P. whose first term $= a$,
and whose common difference $= d$.

Then

$$S_n = a + (a + d) + (a + 2d) + \ldots + \{a + (n - 1)d\}$$

Now let l denote the nth or last term, $a + (n - 1)d$

Then

$$S_n = a + (a + d) + (a + 2d) + \ldots + (l - 2d) + (l - d) + l$$

By writing the series in reverse order,

$$S_n = l + (l - d) + (l - 2d) + \ldots + (a + 2d) + (a + d) + a$$

By adding these two series together,

$$\therefore \quad 2S_n = (a + l) + (a + l) + (a + l) + \ldots$$
$$+ (a + l) + (a + l) + (a + l)$$
$$= (a + l) \ldots \text{to } n \text{ terms}$$
$$= n(a + l)$$
$$\therefore \quad S_n = \frac{n}{2}(a + l)$$

But $l = a + (n - 1)d$

$$\therefore \quad S_n = \frac{n}{2}\{a + a + (n - 1)d\}$$

$$\therefore \quad S_n = \frac{n}{2}\{2a + (n - 1)d\}$$

Examples
9.11 Find the sum of $3 \cdot 2 + 4 \cdot 1 + 5 \cdot 0 \ldots$ to ten terms

$$a = 3 \cdot 2, \ d = 0 \cdot 9 \text{ and } n = 10$$
$$\therefore \quad S_{10} = \frac{10}{2}\{6 \cdot 4 + (9 \times 0 \cdot 9)\}$$
$$= 5\{6 \cdot 4 + 8 \cdot 1\}$$
$$= 5 \times 14 \cdot 5$$
$$\therefore \quad S_{10} = 72 \cdot 5$$

9.12 How many terms of the series $2\frac{1}{2}$, 5, $7\frac{1}{2}$, 10 $\ldots$ will give a sum of $137\frac{1}{2}$?

Here $a = 2\frac{1}{2}$, $d = 2\frac{1}{2}$ and $S_n = 137\frac{1}{2}$. It is required to find n.

$$\therefore \quad 137\frac{1}{2} = \frac{n}{2}\{5 + (n - 1)2\frac{1}{2}\}$$

$$= \frac{n}{2}\{2\frac{1}{2} + 2\frac{1}{2}n\}$$
$$\therefore \quad 275 = n\{2\frac{1}{2} + 2\frac{1}{2}n\}$$
$$= 2\frac{1}{2}n + 2\frac{1}{2}n^2$$

Dividing both sides by $2\frac{1}{2}$,

$$110 = n + n^2$$

$\therefore$ Solving for n,

$$n^2 + n - 110 = 0$$
$$(n + 11)(n - 10) = 0$$
$$\therefore \quad n = -11 \text{ or } 10$$

As $n = -11$ is evidently meaningless in this case, $n = 10$.
$\therefore$ The number of terms to give a sum of $137\frac{1}{2}$ is 10.

9.13 Find the number of terms in an A.P., whose third term is 2, whose sixth term $= -13$ and whose sum $= -105$.

$$S_n = -105$$
$$\text{sixth term} = a + 5d = -13$$
$$\text{third term} = a + 2d = 2$$

Subtracting $\therefore \quad 3d = -15$
                              $\therefore \quad d = -5$

Substituting $d = -5$ in $a + 2d = 2$
                    $a - 10 = 2$
                    $\therefore \quad a = 12$

Now substituting $a = 12$, $d = -5$, $S_n = -105$, in the formula

$$S_n = \frac{n}{2}\{2a + (n-1)d\}$$

we obtain

$$-105 = \frac{n}{2}\{24 + (n-1)(-5)\}$$
$$= \frac{n}{2}\{24 - 5n + 5\}$$
$$\therefore \quad -210 = 29n - 5n^2$$
$$\therefore \quad 5n^2 - 29n - 210 = 0$$
$$\therefore \quad (5n + 21)(n - 10) = 0$$
$$\therefore \quad n = -\frac{21}{5} \quad \text{or} \quad 10$$

Rejecting the negative value as meaningless in this case, we have the number of terms required is 10.

9.14 From a piece of wire 2 m long, a short piece is cut, then a piece 5 mm longer than the first, then another 5 mm longer than the second, and so on. If 30 pieces so cut exactly use up the wire, find the length of the first piece.

Let the length of the first piece $= a$ mm.
Then the lengths of the other pieces are

$$a + 5 \text{ mm}, \ a + 10 \text{ mm} \ldots$$
$$\therefore \quad \text{If the total length of 30 pieces} = S_n$$
$$S_n = \frac{30}{2}\{2a + (29 \times 5)\}$$
$$\therefore \quad 3 \text{ m} = 3000 \text{ mm} = 15\{2a + (29 \times 5)\}$$
$$3000 = 30a + 2175$$
$$30a = 825$$
$$a = \frac{825}{30} = 27\cdot5 \text{ mm}$$

i.e.   Length of first piece $= 27\cdot5$ mm.

To find the sum of n terms of a G.P.

Let S_n denote the sum of n terms of a series in G.P. whose first term is a, and whose common ratio is r.

Then

(i) $S_n = a + ar + ar^2 + \ldots + ar^{n-2} + ar^{n-1}$.

Multiplying both sides by r,

(ii) $\therefore \quad rS_n = ar + ar^2 + \ldots \qquad\qquad + ar^{n-1} + ar^n$.

Subtracting the second series from the first,

$$S_n - rS_n = a - ar^n$$
$$\therefore \quad S_n(1 - r) = a - ar^n$$
$$\therefore \quad S_n = \frac{a - ar^n}{1 - r}$$
$$\therefore \quad S_n = \frac{a(1 - r^n)}{1 - r} \quad \cdots \cdots \cdots \text{(i)}$$

or by subtracting the first series from the second,

$$S_n = \frac{a(r^n - 1)}{r - 1} \quad \cdots \cdots \cdots \text{(ii)}$$

It is better to use (i) when r is a proper fraction or negative.

Infinite geometric series

The approach to infinity

When the common ratio of a geometric series is numerically greater than unity, as in

$$1, 2, 4, 8, \ldots$$
$$2 \cdot 5, 7 \cdot 5, 22 \cdot 5, \ldots$$

the terms increase in magnitude. The sum of n terms will, therefore, also increase as n increases.

If the number of terms increases without limit, i.e. n is greater than any number we may select, however great, then the sum of these terms will also increase without limit.

This we may express by saying that as n, the number of terms, approaches infinity, S_n, the sum of these terms, also approaches infinity. This may also be expressed in the following notation.

If $\qquad\qquad n \longrightarrow \infty$, then $S_n \longrightarrow \infty$.

A decreasing series

If, however, the common ratio is numerically less than unity, as in such series as

$$1, \tfrac{1}{2}, \tfrac{1}{4}, \tfrac{1}{8} \ldots$$
$$0 \cdot 3, 0 \cdot 03, 0 \cdot 003, 0 \cdot 0003 \ldots$$

then, as the number of terms increases, the terms themselves decrease. Using the terms employed above, we may say that, as n increases without limit, the terms themselves decrease without limit.

We cannot say, however, that the sum of these terms decreases without limit as n increases without limit. This is a matter for further investigation.

Questions suitable for practice may be picked out from Exercise 9.1.

Recurring decimals

Let us consider the case of what is termed a recurring decimal. We know from arithmetic that $\frac{1}{3} = 0\cdot3333\ldots$ to any number of places.

i.e. $\frac{1}{3} = 0\cdot3 + 0\cdot03 + 0\cdot003 + 0\cdot0003 + \ldots$ to any number of terms.

The series on the right-hand side is seen to be a geometric series, with common ratio $\frac{1}{10}$.

There is no limit to the number of terms of this series: at the same time, it is evident that the sum of this series is $\frac{1}{3}$.

Let us now find the sum of a finite number of terms of the series.

For example,
$$S_2 = \tfrac{33}{100}$$
$$S_3 = \tfrac{333}{1000}$$
$$S_4 = \tfrac{3333}{10000}$$

. . . .

From these and other similar examples, it is clear that the difference between $\frac{1}{3}$ and the various sums $S_2, S_3, S_4, \ldots$ is diminishing as the number of terms is increased. We come to the conclusion that, the greater the number of terms we take, the more nearly does S_n approach to equality with $\frac{1}{3}$, and that it is always less than $\frac{1}{3}$.

Using our previous notation, we can express the result thus:

$$n \longrightarrow \infty, \ S_n \longrightarrow \tfrac{1}{3}.$$

There is thus a *limit* to which S_n approaches and which it cannot reach.

Clearly all series representing recurring decimals will lead to similar results.

A graphical illustration

Another special case which serves to illustrate this approach of the sum of a series to a limit may be seen if we represent graphically the series

$$\tfrac{1}{2} + \tfrac{1}{4} + \tfrac{1}{8} + \tfrac{1}{16} + \cdots$$

Let the rectangle ABCD (Fig. 9.1) represent a unit area.

If EF be drawn as shown, bisecting the rectangle, then AEFD represents $\frac{1}{2}$.

Similarly, bisecting EFCB, we get EFHG representing $\frac{1}{4}$. Continuing this process of bisecting the rectangle left over in each case, we get a series of rectangles which represent the terms of the series above, viz.

$$\tfrac{1}{2} + \tfrac{1}{4} + \tfrac{1}{8} + \tfrac{1}{16} + \cdots$$

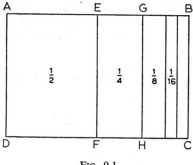

FIG. 9.1

Clearly these rectangles are diminishing as we represent more and more terms of the series in this way.

It is also clear that the sum of all these rectangles, as more and more are taken, is approaching the area of the whole rectangle, i.e. 1, and can never reach this. Consequently, 1 is a limit which the sum of the series approaches as the number of terms is increased without limit, but which it can never reach.

If the series

$$\tfrac{1}{2} + \tfrac{1}{4} + \tfrac{1}{8} + \tfrac{1}{16} + \dots$$

be summed up by using the formula

$$S_n = \frac{a(1 - r^n)}{1 - r}$$

we get

$$S_n = \frac{\tfrac{1}{2}\{1 - (\tfrac{1}{2})^n\}}{1 - \tfrac{1}{2}}$$

$$= \frac{\tfrac{1}{2}\{1 - (\tfrac{1}{2})^n\}}{\tfrac{1}{2}}$$

i.e.

$$S_n = 1 - (\tfrac{1}{2})^n.$$

When $n = 10$, $S_{10} = 1 - (\tfrac{1}{2})^{10} = 1 - \dfrac{1}{2^{10}} = 1 - \dfrac{1}{1025}$

When $n = 20$, $S_{20} = 1 - (\tfrac{1}{2})^{20} = 1 - \dfrac{1}{2^{20}} = 1 - \dfrac{1}{1\,048\,576}$

When $n = 40$, $S_{40} = 1 - (\tfrac{1}{2})^{40} = 1 - \dfrac{1}{2^{40}} = 1 - \dfrac{1}{1\,099\,511\,627\,776}$

Examining these results, we see that the term $(\tfrac{1}{2})^n$ is decreasing as n increases.

By taking a sufficiently large value of n, we can always make $(\tfrac{1}{2})^n$ smaller than any assigned value, no matter how small. Thus, although S_n never reaches the value 1, we can make S_n as close to 1 as we please by taking a sufficiently large value of n. We say that $S_n \longrightarrow 1$, as $n \longrightarrow \infty$.

The sum to infinity

We must now proceed to a general treatment of the problem. Using the formula previously quoted, viz.

$$S_n = \frac{a(1 - r^n)}{1 - r}$$

i.e.

$$S_n = \frac{a - ar^n}{1 - r}$$

we have

$$S_n = \frac{a}{1 - r} - \frac{ar^n}{1 - r}$$

or

$$S_n = \frac{a}{1 - r} - a \cdot \frac{r^n}{1 - r}$$

Consider the second term of the right-hand side. If r be a 'proper' fraction (i.e. it lies between $+1$ and -1), then r^n diminishes as n increases, or, with the previous notation,

as

$$n \longrightarrow \infty, \ r^n \longrightarrow 0$$

Consequently, as

$$n \longrightarrow \infty, \ a \cdot \frac{r^n}{1 - r} \longrightarrow 0.$$

Thus the right-hand side approaches $\frac{a}{1 - r}$ as a limit. This is called the *sum to infinity* of the series.

If it be represented by S_∞,

then

$$S_\infty = \frac{a}{1 - r}$$

It must not be imagined that 'sum' here is used in the sense that we can add term by term and attain this 'sum', but that by adding a sufficient number of terms we can approach this limit as closely as we please, although, no matter how many terms we add, we shall never actually obtain this value. Thus, if n is sufficiently large,

$$S_n \simeq \frac{a}{1 - r}$$

since $\frac{ar^n}{1 - r}$ will be negligible for all practical purposes. The term 'sum to infinity' is an unfortunate one, since it is not possible to add an infinite number of terms: the expression is used conventionally to denote a limit.

Examples

9.15 Sum to infinity $2 + \frac{1}{2} + \frac{1}{8} \ldots$, and find an expression for the amount by which the sum of one hundred terms differs from this value.

Here $a = 2$ and $r = \frac{1}{4}$

$$\therefore \; S_\infty = \frac{2}{1 - \frac{1}{4}}$$

$$= \frac{2}{\frac{3}{4}} = 2\frac{2}{3}$$

$$S_{100} = \frac{2}{1 - \frac{1}{4}} - 2 \cdot \frac{(\frac{1}{4})^{100}}{1 - \frac{1}{4}} = 2\frac{2}{3} - \frac{2}{\frac{3}{4}} \cdot \frac{1}{4^{100}}$$

$$= 2\frac{2}{3} - \frac{2 \times 4}{3 \times 4^{100}} = 2\frac{2}{3} - \frac{2}{3 \times 4^{99}} = 2\frac{2}{3} - \frac{1}{6 \times 4^{98}}$$

The 'sum to infinity' exceeds the sum of one hundred terms by $\dfrac{1}{6 \times 4^{98}}$.

9.16 Find the sum to infinity of the series, $5, -1, \frac{1}{5}, \ldots$

Here
$$a = 5, \; r = -\frac{1}{5}$$

$$\therefore \; S_\infty = \frac{5}{1 - (-\frac{1}{5})}$$

$$= \frac{5}{1\frac{1}{5}} = 4\frac{1}{6}$$

9.17 £500 is invested at 5% per annum, compound interest. What will be the amount of principal and interest together at the end of 10 yrs?

In 1 yr each £1 gains £0·05 interest, i.e. in 1 yr £1 amounts to £1·05.
∴ in 1 yr £500 will amount to £500(1·05).
∴ £500(1·05) is the amount at the end of first year.
During the second year each £1 amounts to £1·05.
∴ during the second year £500(1·05) will amount to £500(1·05)².
Similarly, amount at end of third year is £500(1·05)³, and amount at end of tenth year is £500(1·05)¹⁰,

or
$$A_{10} = 500(1·05)^{10}$$
$$= £814·7$$

9.18 A man pays an insurance premium of £20 at the beginning of each year. At the end of 25 yrs he is to receive all his premiums, together with 3% per annum compound interest. What should he receive, to the nearest £1?

The first £20 earns interest for 25 yrs,

∴ it becomes worth £20$(1 \cdot 03)^{25}$ (see previous example).

The second £20 earns interest for 1 yr,

∴ it amounts to £20$(1 \cdot 03)^{24}$.

Similarly, the third £20 amounts to £20$(1 \cdot 03)^{23}$, etc., etc.

The last £20 paid earns interest for 1 yr,

∴ It amounts to £20$(1 \cdot 03)$.

∴ Total amount = £20$\{1 \cdot 03 \quad 1 \cdot 03^2 + \ldots 1 \cdot 03^{25}\}$

$\qquad\qquad$ = £20 × sum of twenty-five terms of above series

$\qquad\qquad$ = £20 × $\dfrac{1 \cdot 03(1 \cdot 03^{25} - 1)}{1 \cdot 03 - 1}$

$\qquad\qquad$ = £748 *to nearest* £ (within limits of four-figure logs).

9.19 A marble dropped on a stone floor bounces up a distance equal to 0·85 of the height from which it fell. If it was dropped from a height of 3 m, how far would it have travelled altogether in its up-and-down movements, when it reached the floor for the tenth time?

During its first fall it drops 3 m, then it bounces up (0·85 × 3) m and falls down (0·85 × 3) m.

∴ Between its first and second contact with the floor it has travelled 2 × 0·85 × 3 m.

After its second contact with the floor it bounces up 0·85 × 0·85 × 3 m, and then falls down 0·85 × 0·85 × 3 m.

∴ Between its second and third contact with the floor it has travelled 2 × 0·85² × 3 m, and so on.

∴ Between its ninth and tenth contact with the floor it has travelled 2 × 0·85⁹ × 3 m.

∴ Total distance travelled

$\qquad$ = 3 + (2 × 0·85 × 3) + (2 × 0·85² × 3) + … (2 × 0·85⁹ × 3) m

$\qquad$ = 3 + (2 × 3){0·85 + 0·85² + … (0·85)⁹}

$\qquad$ = 3 + 6{4·355}

$\qquad$ = 3 + 26·130

$\qquad$ = 29·13 m

For practice questions see Exercises 9.1 and 9.2.

EXERCISE 9.1

1 Find the ninth term of 17, 13, 9, . . .

2 Find the tenth term of 50, −25, 12½, . . .

3 Find the twenty-fifth term of 0·6, 0·72, 0·84, . . .

4 Find the eighth term of −0·5, +0·15, −0·045, . . .

5 Find the nth term of 2, 7, 12, . . .

6 Find the fifth term of a G.P. whose fourth term is 5, and whose seventh term is 320.

7 The tenth term of a series in A.P. is 18 and the fourteenth is 30. Find the fifth term.

8 Find the seventh term of a geometric series, if the sixth term is 1·09, and the tenth 0·12.

9 Find the tenth term of an A.P. whose third term is 2·9 and whose seventh term is 1·3.

10 Insert three geometric means between 75 and 30·72.

11 Which term of the series $\frac{1}{3}$, $\frac{5}{9}$, $\frac{7}{9}$, . . . is $2\frac{5}{9}$?

12 Find the sum of the first eight terms of 30, -15, $7\frac{1}{2}$, . . .

13 Find the missing terms in ?, ?, ?, 12, ?, ?, 30.

14 Find the sum of the first twelve terms of 4, 5, $6\frac{1}{4}$, . . .

15 Insert three arithmetic means between x and y.

16 Sum $-12 + 9\cdot6 - 7\cdot68$ . . . to six terms.

17 Sum 13, 10, 7 . . . to 10 terms.

18 Sum to infinity $16 - 8 + 4$ . . .

19 Find the sum of twelve terms of -6, -2, $+2$, . . .

20 Evaluate $0\cdot\dot{5}$, $0\cdot3\dot{6}$, and $3\cdot8\dot{3}$.

21 How many terms of the series 11, 15, 19, etc., will give a sum of 341?

22 At the end of a certain year a tree was 10 m high. During the next year it grew 1 m, and in each succeeding year its rate of growth was $\frac{9}{10}$ of what it was in the previous year. Find its greatest height.

23 A parent puts away 25p on his son's first birthday, 50p on his second birthday, 75p on his third, and so on. How much will the boy have when he is 10 yrs old, and how old must he be before the total is £34?

24 A man starts in business and loses £150 in the first year, £120 in the second year and £90 in the third year. If the improvement continues at the same rate, find his total profit or loss at the end of 20 yrs. When would his losses be just balanced by his gains?

25 A contractor agrees to sink a well 80 m deep at a cost of 50p for the first metre, 52p for the second metre and an extra 2p for each additional metre. Find the total cost, and also the cost of the last metre.

26 A clerk in one office commences at £500 a year and receives a yearly rise of £50. Another clerk in another office commences at £500, but receives a rise of £220 after every fourth year. Which receives the greater total salary, and by how much, in (*a*) 32 yrs, (*b*) 36 yrs?

27 A boy builds his bricks into a wall in such a way that each row contains one brick less than the row before it. How many rows can he build with 200 bricks and how many will be left over? (The top row contains only 1 brick.

28 A cyclist's average speed per hour diminishes by $\frac{1}{2}$ km/h during each hour that he rides. If he covers $172\frac{1}{2}$ km in 10 h, find his average speed during the first hour.

29 Find the sum of fourteen terms of the A.P. whose first term is 11 and common difference 9. ULCI

30 Find the thirtieth term and the sum of thirty terms of the series 4, 8, 12, 16. . . . ULCI

31 The yearly output of a silver-mine is found to be decreasing by 25 % of its previous year's output. If in a certain year its output was £25 000, what could be reckoned as its total future output?

32 A heavy ball is suspended at the end of a string 2 m long. It is drawn to one side so that the string makes an angle of 30° with the vertical. When released, it swings to and fro, each swing in one direction being 1 % less than the previous one in the opposite direction. What distance will the ball have travelled before it finally comes to rest?

EXERCISE 9.2 (miscellaneous)

1 (*a*) How many terms must be taken of the series 42, 39, 36 + ... to total 315?

(*b*) Find the sum of seven terms, and the 'sum to infinity' of the series $\frac{1}{2}, \frac{1}{3}, \frac{2}{9} + \ldots$ CANNOCK

2 (*a*) The sum of an arithmetic progression is −105. If the third term is 2 and the sixth term is −13, find the number of terms in the progression.

(*b*) The first term of a geometric progression is 3, and the ratio of the third to the seventh term is 3 : 4. Find the ninth term of the progression, and the sum of 10 terms. CHELTENHAM

3 (*a*) Find the number of terms of the series 27, 25, 23, . . . required to total 192.

(*b*) The pitches of BA threads are in geometric progression. No. 2 is 0·81 mm, No. 3 is 0·729 mm. Find the pitches of No. 1 and No. 4. COVENTRY

4 (*a*) How many terms must be taken of the series 19·25 + 15·75 + 12·25 + ... for the sum to be zero?

(*b*) A small sphere is dropped on to a horizontal table from a height of 1 m and rebounds to a height of 0·75 m. Calculate (i) the total distance the sphere will have travelled just before the fourth impact, (ii) the total distance it will travel from start to finish. CANNOCK

5 (*a*) Find the twelfth term and the sum of the first twenty terms of the series 1, 5, 9, . . .

(*b*) The speeds of a drilling machine are to be five in number, varying from 20 to 200 rev/min. If the speeds are to be in geometrical progression, find the common ratio and the middle term. NUNEATON

6 (*a*) In an A.P. the ratio of the third term to the seventh is 3 : 4. If the common difference is $-\frac{1}{2}$, find the sum of twenty terms.

(*b*) At the beginning of every year a man puts £30 into a society which pays 5 % compound interest per annum. What will be the amount of his savings at the end of 18 yrs? CHELTENHAM

7 (a) Find the number of terms of the series 20, 19, 18, . . . required to make the sum 200, and find the value of the last term.

(b) Find the value of the twentieth term, and the sum to infinity of the series $\frac{2}{3} + \frac{4}{9} + \frac{8}{27} + \ldots$ COVENTRY

8 A firm leases a plot of land at the rate of £1000 for the first year, and for subsequent years at the rate of 90% of the previous year's rent. What is the maximum amount that would be paid in rent, assuming that the land is leased in perpetuity? Find by means of a progression how much rent would be paid in the first 5 years. WORCESTER

9 (a) How many terms of the arithmetical progression 3, 7, 11, . . . must be taken if the sum is 2775?

(b) The ninth term of an A.P. is 13 and the fourteenth term is 3. Find the first term, common difference, and thirtieth term.

NUNEATON

10 (a) How many terms of the series in arithmetic progression 7, $4\frac{1}{2}$, 2 . . . add to $-3\frac{1}{2}$? CHELTENHAM

(b) The third term of a geometric progression is 8, and the tenth term is -1024. Find the sum of the first nine terms.

11 (a) The third term of a geometric series is 5, and the seventh term is 3. Find, as accurately as the log tables allow, the tenth term and the sum of the first ten terms.

(b) Evaluate $\log_e \dfrac{T_1 - t}{T_2 - t}$, given that $T_1 = 450$, $T_2 = 300$ and $t = 273$.

COVENTRY

12 In sinking an oil well it is estimated that the rate of drilling will be 10 m per h for the first hour, but the depth sunk in each successive hour will be 8% less than in the preceding hour. In how many hours will the depth be 58 m? What will be the depth after 5 h?

WORCESTER

13 The second term of a G.P. is 4, and the fifth term is $\frac{32}{27}$. Find the common ratio, first term, and sum to infinity. NUNEATON

14 A firm markets a successful product in two sizes, known respectively as the 100 mm and the 300 mm models. They propose to extend the range by introducing one larger and one intermediate size. They have no direct lead from consumer research as to what these sizes should be. In those circumstances what sizes would you suggest and why?

15 A firm has a machine which cost £1000, and 'writes off' £100 from its book value each year. Another firm has the same machine but prefers to 'write off' 10% of its book value each year. What is the value of the machine at the end of 10 years in the books of the respective firms? Which system of book-keeping would you recommend and why?

Chapter Ten
The Binomial Theorem

The binomial theorem for a positive integral index
By multiplication we can obtain the results

$$(a + x)^2 = a^2 + 2ax + x^2 = a^2 + \frac{2}{1}a^1x + \frac{2 \cdot 1}{1 \cdot 2}x^2$$

$$(a + x)^3 = a^3 + 3a^2x + 3ax^2 + x^3$$

$$= a^3 + \frac{3}{1}a^2x^1 + \frac{3 \cdot 2}{1 \cdot 2}a^1x^2 + \frac{3 \cdot 2 \cdot 1}{1 \cdot 2 \cdot 3}x^3$$

$$(a + x)^4 = a^4 + 4a^3x + 6a^2x^2 + 4ax^3 + x^4$$

$$= a^4 + \frac{4}{1}a^3x + \frac{4 \cdot 3}{1 \cdot 2}a^2x^2 + \frac{4 \cdot 3 \cdot 2}{1 \cdot 2 \cdot 3}ax^3 + \frac{4 \cdot 3 \cdot 2 \cdot 1}{1 \cdot 2 \cdot 3 \cdot 4}x^4$$

From the form of the above results we infer that, when n is a positive integer,

$$(a + x)^n = a^n + \frac{n}{1}a^{n-1}x^1 + \frac{n(n-1)}{1 \cdot 2}a^{n-2}x^2 + \frac{n(n-1)(n-2)}{1 \cdot 2 \cdot 3}a^{n-3}x^3 + \ldots x^n$$

The expansion contains $n + 1$ terms and is true for all values of a and x. It is known as the *binomial theorem* (for a positive integral index), since a term of the form $(a + x)$ is a binomial term. Note that the sum of the indices of a and x in each term is n, and that the factors in the numerator and denominator of the third and subsequent terms decrease and increase by 1 respectively. The proof of the theorem will be given later.

Examples
10.1 Expand $(a + x)^5$ by the binomial theorem.

$$(a + x)^n = a^n + \frac{n}{1}a^{n-1}x^1 + \frac{n(n-1)}{1 \cdot 2}a^{n-1}x^1 + \ldots x^n.$$

Let $n = 5$,

$$(a + x)^5 = a^5 + \frac{5}{1}a^{5-1}x^1 + \frac{5(5-1)}{1 \cdot 2}a^{5-2}x^2$$

$$+ \frac{5(5-1)(5-2)}{1 \cdot 2 \cdot 3}a^{5-3}x^3 + \frac{5(5-1)(5-2)(5-3)}{1 \cdot 2 \cdot 3 \cdot 4}a^{5-4}x^4$$

$$+ \frac{5(5-1)(5-2)(5-3)(5-4)}{1 \cdot 2 \cdot 3 \cdot 4 \cdot 5}a^{5-5}x^5$$

$$= a^5 + 5a^4x + \frac{5 \cdot 4}{1 \cdot 2}a^3x^2 + \frac{5 \cdot 4 \cdot 3}{1 \cdot 2 \cdot 3}a^2x^3$$

$$+ \frac{5.4.3.2}{1.2.3.4}a^1x^4 + \frac{5.4.3.2.1}{1.2.3.4.5}a^0x^5$$

$$= a^5 + 5a^4 + 10a^3x^2 + 10a^2x^3 + 5ax^4 + x^5$$

i.e. $(a + x)^5 \equiv a^5 + 5a^4x + 10a^3x^2 + 10a^2x^3 + 5ax^4 + x^5$

We may test the truth of this identity by inserting convenient values for a and x.

Let $a = 2$ and $x = -1$.
L.H.S.
$= (2 - 1)^5 = 1^5 = 1$
R.H.S.
$= 2^5 + 5(2)^4.(-1) + 10(2)^3.(-1)^2 + 10(2)^2.(-1)^3 + 5(2).(-1)^4 + (-1)^5$
$= 32 - 80 + 80 - 40 + 10 - 1 = 1$
i.e. L.H.S. $= 1 =$ R.H.S.

The student may note that the expansion is self-terminating: if we had attempted in error to write down a further term

$$\frac{5(5 - 1)(5 - 2)(5 - 3)(5 - 4)(5 - 5)}{1.2.3.4.5.6}a^{5-6}x^6,$$

i.e. $0a^{-1}x^6$, the term vanishes. This happens only when n is a positive integer.

10.2 Expand $(2x - 3y)^6$ by the binomial theorem. Check the resulting identity using the values $x = \frac{1}{2}$ and $y = \frac{1}{3}$.

$$(a + x)^n = a^n + \frac{n}{1}a^{n-1}.x^1 + \frac{n(n - 1)}{1.2}a^{n-2}x^2 + \dots x^n$$

Let $a = 2x$, $x = -3y$, $n = 6$.

$$(2x + [-3y])^6 = (2x)^6 + 6(2x)^5(-3y) + \frac{6.5}{1.2}(2x)^4(-3y)^2$$

$$+ \frac{6.5.4}{1.2.3}(2x)^3(-3y)^3 + \frac{6.5.4.3}{1.2.3.4}(2x)^2(-3y)^4$$

$$+ \frac{6.5.4.3.2}{1.2.3.4.5}(2x)(-3y)^5 + \frac{6.5.4.3.2.1}{1.2.3.4.5.6}(-3y)^6$$

$$= 64x^6 - 6.32.3x^5y + 15.16.9x^4y^2 - 20.8.27x^3y^3$$
$$+ 15.4.81x^2y^4 - 6.2.243xy^5 + 729y^6$$

$$= 64x^6 - 576x^5y + 2160x^4y^2 - 4320x^3y^3 + 4860x^2y^4$$
$$- 2916xy^5 + 729y^6$$

i.e. $(2x - 3y) \equiv 64x^6 - 576x^5y + 2160x^4y^2 - 4320x^3y^3 + 4860x^2y^4$
$$- 2916xy^5 + 729y^6$$

When $x = \frac{1}{2}$, $y = \frac{1}{3}$,
   L.H.S. $= (1 - 1)^6 = 0$

R.H.S. $= \dfrac{64}{2^6} - \dfrac{576}{2^5 \cdot 3} + \dfrac{2160}{2^4 \cdot 3^2} - \dfrac{4320}{2^3 \cdot 3^3} + \dfrac{4860}{2^2 \cdot 3^4} - \dfrac{2916}{2 \cdot 3^5} + \dfrac{729}{3^6}$

$\qquad = 1 - 6 + 15 - 20 + 15 - 6 + 1$
$\qquad = 0$

i.e.   L.H.S. $= 0 =$ R.H.S.

10.3 (*a*) Expand $(a + 3b)^5$ by the binomial theorem.
(*b*) Test the expansion when $a = 1$ and $b = \frac{1}{3}$.
(*c*) Apply the expansion to evaluate $(1 \cdot 03)^5$ correct to five decimal places, and also exactly.

(*a*) $(a + x)^n = a^n + \dfrac{n}{1}a^{n-1}x^1 + \dfrac{n(n-1)}{1 \cdot 2}a^{n-2}x^2 + \ldots + x^n$

Let $x = 3b$ and $n = 5$.

$(a + 3b)^5 = a^5 + 5a^4(3b) + \dfrac{5 \cdot 4}{1 \cdot 2}a^3(3b)^2 + \dfrac{5 \cdot 4 \cdot 3}{1 \cdot 2 \cdot 3}a^2(3b)^3$

$\qquad\qquad + \dfrac{5 \cdot 4 \cdot 3 \cdot 2}{1 \cdot 2 \cdot 3 \cdot 4}a(3b)^4 + (3b)^5$

i.e.
$(a + 3b)^5 \equiv a^5 + 15a^4b + 90a^3b^2 + 270a^2b^3 + 405ab^4 + 243b^5$ (i)

(*b*) When $a = 1$ and $b = \frac{1}{3}$,

L.H.S. $= \left(1 + \dfrac{3}{3}\right)^5 = 2^5 = 32$

R.H.S. $= 1^5 + \dfrac{15 \cdot 1^4}{3} + \dfrac{90 \cdot 1^3}{3^2} + \dfrac{270 \cdot 1^2}{3^3} + \dfrac{405}{3^4} + \dfrac{243}{3^5}$

$\qquad = 1 + 5 + 10 + 10 + 5 + 1 = 32$
$\therefore$ L.H.S. $= 32 =$ R.H.S.

(*c*) Let $a = 1$ and $3b = 0 \cdot 03$, i.e. $b = 0 \cdot 01$. Substituting in (i),

$(1 + 0 \cdot 03)^5 = 1 + 15(0 \cdot 01) + 90(0 \cdot 01)^2 + 270(0 \cdot 01)^3 + 405(0 \cdot 01)^4 + \ldots$
$\qquad = 1 + 0 \cdot 15 + 0 \cdot 009 + 0 \cdot 000\,27 + 0 \cdot 000\,004\,05 + \ldots$ (ii)
$\qquad = 1 \cdot 159\,274$ to six decimal places
$\qquad = 1 \cdot 159\,27$ correct to five decimal places.

To obtain the exact value we rewrite line (ii) with the omitted term.

$(1 \cdot 03)^5 = 1 + 0 \cdot 15 + 0 \cdot 009 + 0 \cdot 000\,27 + 0 \cdot 000\,004\,05 + 0 \cdot 000\,000\,024\,3$
$\qquad = 1 \cdot 159\,274\,074\,3$

which can be verified by long multiplication.

The expansion of $(1 + x)^n$ where n is a positive integer

Since $(a + x)^n = a^n + \dfrac{n}{1}a^{n-1}x^1 + \dfrac{n(n-1)}{1.2}a^{n-2}x^2 + \ldots + x^n$

let $a = 1$, then

$$(1 + x)^n = 1 + nx + \frac{n(n-1)}{1.2}x^2 + \ldots + x^n$$

Example
10.4 (*a*) Expand $(1 + 2u)^5$ by the binomial theorem.
(*b*) Deduce the expansion of $(1 + 2u)^5 - (1 - 2u)^5$, and test the result.
(*c*) Hence evaluate $(1 \cdot 02)^5 - (0 \cdot 98)^5$ correct to five decimal places, and also exactly.

(*a*) $(1 + x)^n = 1 + nx + \dfrac{n(n-1)}{1.2}x^2 + \ldots + x^n$

Replacing x by $2u$ and n by 5,

$$(1 + 2u)^5 = 1 + 5(2u) + \frac{5.4}{1.2}(2u)^2 + \frac{5.4.3}{1.2.3}(2u)^3 + \frac{5.4.3.2}{1.2.3.4}(2u)^4$$
$$+ (2u)^5$$
$$= 1 + 10u + 40u^2 + 80u^3 + 80u^4 + 32u^5$$

(*b*) Replacing u by $-u$ in the above result,

$$(1 - 2u)^5 = 1 - 10u + 40u^2 - 80u^3 + 80u^5 - 32u^5$$

Subtracting

$$(1 + 2u)^5 - (1 - 2u)^5 = 20u + 160u^3 + 64u^5$$

Check

When $u = \dfrac{1}{2}$, L.H.S. $= (1 + 1)^5 - (1 - 1)^5 = 2^5 = 32$

$$\text{R.H.S.} = \frac{20}{2} + \frac{160}{2^3} + \frac{64}{2^5} = 10 + 20 + 2 = 32$$

$\therefore$ L.H.S. $= 32 =$ R.H.S.

(*c*) Replacing u by $0 \cdot 01$,

$(1 + 0 \cdot 02)^5 - (1 - 0 \cdot 02)^5 = 20(0 \cdot 01) + 160(0 \cdot 01)^3 + 64(0 \cdot 01)^5$
i.e. $(1 \cdot 02)^5 - (0 \cdot 98)^5 = 0 \cdot 2 + 0 \cdot 000\ 160 + 0 \cdot 000\ 000\ 006\ 4$
$= 0 \cdot 200\ 160\ 006\ 4$ exactly
$= 0 \cdot 200\ 16$ correct to five decimal places.

By long multiplication,

$$(1 \cdot 02)^5 = 1 \cdot 104\ 080\ 803\ 2$$
$$(0 \cdot 98)^5 = 0 \cdot 903\ 920\ 796\ 8$$
$$(1 \cdot 02)^5 - (0 \cdot 98)^5 = 0 \cdot 200\ 160\ 006\ 4$$

The factorial notation

It is convenient to have a notation for a repeated product such as $1 . 2 . 3 . 4 . 5 . 6$, which is called 'factorial 6' and is written as 6! or $\lfloor 6$. Thus, if r is a positive integer, then

$$r! = 1 . 2 . 3 . 4 \ldots (r-1) . r$$

Determination of a particular term in the expansion of $(a + x)^n$

$$(a + x)^n = \underbrace{a^n}_{\text{1st term}} + \underbrace{\frac{n}{1!}a^{n-1}x^1}_{\text{2nd term}} + \underbrace{\frac{n(n-1)}{2!}a^{n-2}x^2}_{\text{3rd term}}$$

$$+ \underbrace{\frac{n(n-1)(n-2)}{3!}a^{n-3}x^3}_{\text{4th term}} + \ldots + x^n$$

Suppose that we write down the 4th, 5th and 6th term as the $(3+1)^{\text{th}}$, $(4+1)^{\text{th}}$ and $(5+1)^{\text{th}}$ term.

The $(3+1)^{\text{th}}$ term is

$$\frac{n(n-1)(n-2)}{3!}a^{n-3}x^3 = \frac{n(n-1)(n-3+1)}{3!}a^{n-3}x^3$$

The $(4+1)^{\text{th}}$ term is

$$\frac{n(n-1)(n-2)(n-3)}{4!}a^{n-4}x^4 = \frac{n(n-1)\ldots(n-4+1)}{4!}a^{n-4}x^4$$

The $(5+1)^{\text{th}}$ term is

$$\frac{n(n-1)(n-2)(n-3)(n-4)}{5!}a^{n-5}x^5 = \frac{n(n-1)\ldots(n-5+1)}{5!}a^{n-5}x^5$$

We infer that, following the pattern of the above results, the $(r+1)^{\text{th}}$ term of the expansion of $(a+x)^n$, where r is a positive integer, is

$$\frac{n(n-1)\ldots(n-r+1)}{r!}a^{n-r}x^r$$

A difficulty arises when $r = 0$, although the first term of the binomial expansion is otherwise obvious. We may, however, rearrange the above expression as follows.

$$\frac{n(n-1)\ldots(n-r+1)}{r!} = \frac{n(n-1)\ldots(n-r+1)}{r!} \cdot \frac{(n-r)\ldots2.1}{(n-r)\ldots2.1}$$

$$= \frac{n!}{r!(n-r)!}$$

i.e. the $(r+1)^{\text{th}}$ term is $\dfrac{n!}{r!(n-r)!}a^{n-r}x^r$

When $r = 0$, the first term is $\dfrac{n!}{0!\,n!} \cdot a^n x^0 = \dfrac{1}{0!} \cdot a^n$. If we define $0!$ to be 1, then the first term is a^n, i.e. the $(r+1)$th term of the expansion of $(a + x)^n$, where n is a positive integer, is

$$\frac{n!}{r!(n - r)!} a^{n-r} x^r$$

for $r = 0, 1, 2, \ldots n$.

Examples

10.5 Find the ninth term of $(2x + 1)^{12}$.

Here $a = 2x$, $x = 1$, $n = 12$, $r + 1 = 9$, i.e. $r = 8$,

$\therefore$ since the $(r + 1)^{\text{th}}$ term of $(a + x)^n$ is $\dfrac{n!}{r!(n - r)!}\, a^{n-r} x^r$,

the $(8 + 1)^{\text{th}}$ term of $(2x + 1)^{12}$ is $\dfrac{12!}{8!(12 - 8)!} (2x)^{12-8} 1^8$

$$= \frac{12 \cdot 11 \cdot 10 \cdot 9 \cdot 8!}{8! \cdot 4 \cdot 3 \cdot 2 \cdot 1}(2x)^4 \cdot 1 = 495 \cdot 16x^4$$

$$= 7920x^4$$

10.6 Find the coefficients of z^3 and z^{11} in the expansion of $\left(z^3 + \dfrac{1}{z}\right)^9$.

Since the $(r + 1)^{\text{th}}$ term of $(a + x)^n$ is $\dfrac{n!}{r!(n - r)!}\, a^{n-r} x^r$,

the $(r + 1)^{\text{th}}$ term of $\left(z^3 + \dfrac{1}{z}\right)^9$ is $\dfrac{9!}{r!(9 - r)!}(z^3)^{9-r}(\tfrac{1}{z})^r$

$$= \frac{9!}{r!(9 - r)!}z^{27 - 3r} \cdot \frac{1}{z^r}$$

$$= \frac{9!}{r!(9 - r)!}z^{27 - 4r}$$

Putting $27 - 4r = 3$, then $24 = 4r$ and $r = 6$,

i.e. the $(6 + 1)^{\text{th}}$ term of $\left(z^3 + \dfrac{1}{z}\right)^9$ is $\dfrac{9!}{6!(9 - 6)!}z^{27 - 24}$

$$= \frac{9 \cdot 8 \cdot 7 \cdot 6!}{6! 1 \cdot 2 \cdot 3} \cdot z^3 = 84z^3$$

The required coefficient is 84.

Putting $27 - 4r = 11$, then $16 = 4r$ and $r = 4$,

i.e. the $(4 + 1)^{\text{th}}$ term of $\left(z^3 + \dfrac{1}{z}\right)^9$ is $\dfrac{9!}{4!(9-4)!} \cdot z^{27-16}$

$$= \frac{9 \cdot 8 \cdot 7 \cdot 6 \cdot 5!}{1 \cdot 2 \cdot 3 \cdot 4 \cdot 5!} \cdot z^{11} = 126z^{11}$$

The required coefficient is 126.

Proof of the binomial theorem for a positive integral index

We will assume that the binomial theorem is true for a positive integral index n and then show that, if so, it must be true for an index $n + 1$.

If $(a + x)^n = a^n + \dfrac{n!}{1!(n-1)!}a^{n-1}x^1 + \dfrac{n!}{2!(n-2)!}a^{n-2}x^2 + \ldots x^n$

then
$(a + x)^{n+1} = (a + x)(a + x)^n$

$$= (a + x)\left[a^n + \frac{n!}{1!(n-1)!}a^{n-1}x^1 + \frac{n!}{2!(n-2)!}a^{n-2}x^2 + \ldots x^n\right]$$

The term in x^r in the expansion of $(a + x)^{n+1}$ will be obtained by multiplying

$$\frac{n!}{r!(n-r)!} \cdot a^{n-r}x^r \quad \text{by } a$$

and $\quad \dfrac{n!}{(r-1)!(n-r+1)!} \cdot a^{n-r+1}x^{r-1} \quad \text{by } x,$

and will be

$$\frac{n!}{r!(n-r)!} \cdot a^{n-r+1}x^r + \frac{n!}{(r-1)!(n-r+1)!} \cdot a^{n-r+1}x^r$$

i.e. $\quad \dfrac{n!}{r!(n-r)!}\left[1 + \dfrac{r}{n-r+1}\right]a^{n+1-r}x^r$

i.e. $\quad \dfrac{n!}{r!(n-r)!}\left[\dfrac{n+1}{n-r+1}\right]a^{n+1-r}x^r$

i.e. $\quad \dfrac{(n+1)!}{r![(n+1)-r]!} \cdot a^{(n+1)-r}x^r.$

Putting $r = 0, 1, 2, \ldots n + 1$, we obtain the expansion of $(a + x)^{n+1}$ as follows.

$$(a + x)^{n+1} = a^{n+1} + \frac{n+1!}{1![(n+1)-1]!} \cdot a^{(n+1)-1}x^1$$

$$+ \frac{n+1!}{2![(n+1)-2]!} \cdot a^{(n+1)+2}x^2 + \ldots x^{n+1}$$

proving that *if* the binomial theorem is true for a positive integral index n, *then* it is true for an index $n + 1$.

But $(a + x)^2 = a^2 + 2ax + x^2 = a^2 + \dfrac{2!}{1!\,(2-1)!} \cdot ax + x^2$, and the binomial theorem is true for $n = 2$, and hence for $n = 2 + 1$, i.e. $n = 3$, and, by repeating this argument, for $n = 4, 5, 6 \dots$ etc., in fact for all positive integral indices.

EXERCISE 10.1

1 Use the binomial theorem to obtain the identity
$$(2x + 3)^5 \equiv 32x^5 + 240x^4 + 720x^3 + 1080x^2 + 810x + 243$$
Show that when $x = -1$ then L.H.S. $= -1 =$ R.H.S.

2 Use the binomial theorem to obtain the identity
$$\left(x - \frac{1}{2y}\right)^6 \equiv x^6 - \frac{3x^5}{y} + \frac{15x^4}{4y^2} - \frac{5x^3}{2y^3} + \frac{15x^2}{16y^4} - \frac{3x}{16y^5} + \frac{1}{64y^6}$$
Show that when $x = 1$ and $y = \frac{1}{2}$ then L.H.S. $= 0 =$ R.H.S.

3 Using the binomial theorem   (*a*) evaluate $(1 \cdot 0007)^6$ correct to five decimal places, (*b*) expand $\left(x - \dfrac{4}{x}\right)^5$, each term to be reduced to its simplest form,   (*c*) find, in its simplest form, the sixth term of $\left(3x + \dfrac{1}{x}\right)^9$.

UEI

4 (*a*) Solve $6^{3-4x} \cdot 4^{x+5} = 8$.
(*b*) Using the binomial theorem   (i) write down the expansion of $\left(3x - \dfrac{1}{x}\right)^8$ as far as the term independent of x, (ii) find the value of $(0 \cdot 996)^4$ correct to five decimal places,   (iii) write down the sixth term in the expansion of $\left(x + \dfrac{1}{2x}\right)^{10}$.

UEI

5 By means of the binomial theorem, or otherwise, find the coefficient of b^3 in the expansion of $\left(3 + \dfrac{b}{2}\right)^5$.

NCTEC part question

6 (*a*) Expand and simplify $(x - \sqrt{5})^4 + (x + \sqrt{5})^4$.   (*b*) Find the middle term in the expansion of $\left(\dfrac{2x}{3} - \dfrac{3}{2x}\right)^6$. (*c*) A metal sphere of diameter 25 mm is heated so that the diameter registers an increase of 0·005 mm. Find to three significant figures the increase (i) in surface area, (ii) in volume.

MEDWAY

7 (*a*) If $\left(1 + \dfrac{c}{2}\right)^5 + \left(1 - \dfrac{c}{2}\right)^5 = 2 \cdot 002$, where c is a small positive quantity, powers of which above the second can be neglected, find the value of c.
(*b*) Use the expansion of $(1 + x)^n$ to find the values of (i) $(1 \cdot 05)^4$, (ii) $(0 \cdot 998)^3$, correct to five significant figures.

ULCI

8 Write down and simplify the binomial expansion of $(a + x)^6$.

ULCI part question

9 Give the first four terms of the expansion of $(a - x)^n$. Use this expansion to evaluate $(0.99)^4$ accurately to five decimal places.

ULCI part question

10 Write down the expansion of $(a + x)^4$. By replacing x by $b + c$, prove that

$$(a + b + c)^4 \equiv a^4 + b^4 + c^4 + 4(a^3b + ab^3 + b^3c + c^3a + ca^3)$$
$$+ 6(a^2b^2 + b^2c^2 + c^2a^2) + 12(a^2bc + b^2ca + c^2ab)$$

Show that when $a = 2$, $b = 1$, $c = -1$, then L.H.S. $= 16 =$ R.H.S.

MID-CHESHIRE

The binomial theorem for negative and fractional indices

The expansion of $(a + x)^n$ may be used when n is negative and/or fractional, but in this case *the series does not terminate: it is an infinite series.* Further, the expansion is *only true for values of* x *which lie between* $+a$ and $-a$, apart from special cases when $x = +a$ and $x = -a$. The following example will serve to illustrate these points.

Examples

10.7 (*a*) Expand $\dfrac{1}{1 - x}$ to five terms by the binomial theorem.

(*b*) Compare the approximate values of each side of the result when $x = \frac{1}{100}$.

(*c*) Discuss the case when $x = 100$.

(*a*) $\dfrac{1}{1 - x} = (1 - x)^{-1} = 1 + \dfrac{-1}{1}(-x)^1 + \dfrac{-1 \cdot -2}{1 \cdot 2}(-x)^2$

$+ \dfrac{-1 \cdot -2 \cdot -3}{1 \cdot 2 \cdot 3}(-x)^3 + \dfrac{-1 \cdot -2 \cdot -3 \cdot -4}{1 \cdot 2 \cdot 3 \cdot 4}(-x^4) + \ldots$

i.e. $\dfrac{1}{1 - x} = 1 + x + x^2 + x^3 + x^4 + \ldots$ . . . .(i)

Note that it is essential to indicate that the series does not terminate. The above series has been previously obtained by long division, and is an infinite geometric progression.

(*b*) When $x = \dfrac{1}{100}$, referring to (i),

$$\text{L.H.S.} = \frac{1}{1 - \frac{1}{100}} = \frac{1}{\frac{99}{100}} = \frac{100}{99}$$

$$= 1.010\ 101 \text{ to six decimal places.}$$

$$\text{R.H.S.} = 1 + \frac{1}{100} + \left(\frac{1}{100}\right)^2 + \left(\frac{1}{100}\right)^3 + \ldots$$

$$= 1 + 0{\cdot}01 + 0{\cdot}0001 + 0{\cdot}000\ 01 + \ldots$$
$$= 1{\cdot}010\ 101 \text{ to six decimal places.}$$

(c) When $x = 100$,

$$\text{L.H.S.} = \frac{1}{1 - 100} = \frac{1}{-99}$$

$$\text{R.H.S.} = 1 + 100 + (100)^2 + (100)^3 + (100)^4 + \ldots$$

We see that the R.H.S. is essentially positive and that, by taking a sufficient number of terms of the series, the sum can be made to exceed any assigned number, however large. The series is said to be divergent. It does not represent the value of $\dfrac{1}{1-x}$ when $x = 100$, namely $-\dfrac{1}{99}$. We have illustrated a particular case of the fact that (i) is invalid for values of x which do not lie between $+1$ and -1.

10.8 Write out the expansion of $(1 - x)^{-\frac{1}{2}}$ to 4 terms. By using the substitution $x = \frac{1}{50}$, calculate $\sqrt{2}$ correct to five decimal places.

$$(1 - x)^{-\frac{1}{2}} = 1 + \frac{-\frac{1}{2}}{1}(-x)^1 + \frac{-\frac{1}{2} \cdot -\frac{3}{2}}{1 \cdot 2}(-x)^2 + \frac{-\frac{1}{2} \cdot -\frac{3}{2} \cdot -\frac{5}{2}}{1 \cdot 2 \cdot 3}(-x)^3 + \ldots$$

$$= 1 + \frac{x}{2} + \frac{3}{8}x^2 + \frac{5}{16}x^3 + \ldots$$

Putting $x = \dfrac{1}{50}$,

$$\left(1 - \frac{1}{50}\right)^{-\frac{1}{2}} = 1 + \frac{1}{2}\left(\frac{1}{50}\right) + \frac{3}{8}\left(\frac{1}{50}\right)^2 + \frac{5}{16}\left(\frac{1}{50}\right)^3 + \ldots$$

i.e. $$\left(\frac{50}{49}\right)^{\frac{1}{2}} = 1 + \frac{1}{100} + \frac{3}{20\ 000} + \frac{5}{2\ 000\ 000} + \ldots$$

i.e. $$\left(\frac{2 \cdot 5^2}{1 \cdot 7^2}\right)^{\frac{1}{2}} = 1 + 0{\cdot}01 + 0{\cdot}000\ 15 + 0{\cdot}000\ 002\ 5 + \ldots$$

$$\left(\frac{2}{1}\right)^{\frac{1}{2}} \cdot \frac{5}{7} = 1{\cdot}010\ 152\ 5 + \ldots$$

$$\therefore \quad \sqrt{2} \simeq \tfrac{7}{5}(1{\cdot}010\ 152\ 5) = 1{\cdot}414\ 213\ 50$$
$$= 1{\cdot}414\ 21 \text{ correct to five decimal places.}$$

Check
$$(1{\cdot}414\ 21)^2 = 1{\cdot}999\ 989\ 924\ 1$$
$$= 1{\cdot}999\ 99 \text{ correct to five decimal places.}$$

The student will note that by continuing to evaluate more and more terms of the above series we can evaluate $\sqrt{2}$ to as many decimal places as we please, getting closer and closer approximations, but we can never obtain an exact value of $\sqrt{2}$. This is characteristic of an irrational number.

10.9 Find an approximate formula for $\dfrac{(4-x)^{\frac{1}{2}}}{(1+x)^3}$, if x is so small that x^3 and higher powers of x may be neglected. Test the approximation when $x = 0.001$. Verify the result also by finding the reciprocal of the expansion of $\dfrac{(1+x)^3}{(4-x)^{\frac{1}{2}}}$

MID-CHESHIRE

$$\frac{(4-x)^{\frac{1}{2}}}{(1+x)^3} = (4-x)^{\frac{1}{2}}(1+x)^{-3}$$

$$(4-x)^{\frac{1}{2}} \simeq \left[4^{\frac{1}{2}} + \frac{\frac{1}{2}}{1} . 4^{-\frac{1}{2}}(-x)^1 + \frac{\frac{1}{2} . -\frac{1}{2}}{1 . 2} . 4^{-\frac{3}{2}}(-x)^2 \right]$$

$$\left(\text{Note } 4^{-\frac{1}{2}} = \frac{1}{4^{\frac{1}{2}}} = \frac{1}{2} \quad \text{and} \quad 4^{-\frac{3}{2}} = \frac{1}{4^{\frac{3}{2}}} = \frac{1}{2^3} = \frac{1}{8} \right)$$

$$= 2 - \frac{x}{4} - \frac{x^2}{64}$$

$$(1+x)^{-3} \simeq 1 - \frac{3}{1} . x + \frac{-3 . -4}{1 . 2} . x^2 = 1 - 3x + 6x^2$$

$$\therefore \quad (4-x)^{\frac{1}{2}}(1+x)^{-3} \simeq \left(2 - \frac{x}{4} - \frac{x^2}{64} \right)(1 - 3x + 6x^2)$$

Multiplication may now be carried out, and terms in x^3 and higher powers of x are ignored.

$$\therefore \quad (4-x)^{\frac{1}{2}}(1+x)^{-3} \simeq 2 - \frac{x}{4} \quad - \frac{x^2}{64}$$

$$- 6x \quad + \frac{3x^2}{4}$$

$$+ 12x^2$$

$$= 2 - 6\tfrac{1}{4}x + 12\tfrac{47}{64}x^2$$

Hence $\dfrac{(4-x)^{\frac{1}{2}}}{(1+x)^3} \simeq 2 - 6.25x + 12.73x^2$, if x is sufficiently small.

When $x = 0.001$,

L.H.S. $= \dfrac{(3.999)^{\frac{1}{2}}}{(1.001)^3} = \dfrac{1.9997}{1.0030} = 1.9940$ correct to four decimal places.

R.H.S. $= 2 - 6 \cdot 125 \times 10^{-3} + 12 \cdot 73 \times 10^{-6}$
$= 2 \cdot 000\ 012\ 73 - 0 \cdot 006\ 125$
$= 1 \cdot 9940$ correct to four decimal places.

$$\frac{(1+x)^3}{(4-x)^{\frac{1}{2}}} = (1+x)^3(4-x)^{-\frac{1}{2}}$$

$$\simeq (1+3x+3x^2)[4^{-\frac{1}{2}} + \left(\frac{-\frac{1}{2}}{1}\right) . 4^{-\frac{3}{2}}(-x)$$

$$+ \frac{(-\frac{1}{2})(-\frac{3}{2})}{1.2} . 4^{-\frac{5}{2}}(-x^2)]$$

$$= (1+3x+3x^2)\left(\frac{1}{2} + \frac{x}{16} + \frac{3x^2}{256}\right)$$

$$\simeq \frac{1}{2} + \frac{3x}{2} + \frac{3x^2}{2}$$

$$+ \frac{x}{16} + \frac{3x^2}{16}$$

$$+ \frac{3x^2}{256}$$

$$= \frac{1}{2} + \frac{25}{16}x + 1\frac{179}{256}x^2 \text{ by multiplication}$$

$$\therefore \frac{(4-x)^{\frac{1}{2}}}{(1+x)^3} = \frac{1}{\frac{1}{2} + \frac{25}{16}x + 1\frac{179}{256}x^2} = 2 - 6\frac{1}{4}x + 12\frac{47}{64}x^2 \text{ by long division.}$$

EXERCISE 10.2

Write down the expansions in questions 1–10.

1 $(x-3)^5$

2 $(3x+2y)^4$

3 $\left(x - \frac{2}{x}\right)^6$

4 $(1-xy)^7$

5 $(a^2 - 2c)^4$

6 $(2x - 3y)^3$

7 $\left(1 - \frac{1}{x}\right)^{10}$

8 $(\frac{1}{2} + a)^8$

9 $(x + \Delta x)^{-2}$ to three terms

10 $\dfrac{1}{(m+n)^4}$ to three terms [i.e. $(m+n)^{-4}$]

Evaluate the following, by using the binomial theorem.

11 $(1 \cdot 002)^6$ to three decimal places

12 $\dfrac{1}{(0 \cdot 999)^3}$ to three decimal places

13 $(1 \cdot 000\ 012)^3$ to six decimal places

Find the following terms.

14 Fifth term of $\left(3a - \dfrac{b}{2}\right)^7$

15 Fourth term of $(2x - 3y)^7$

The following miscellaneous examples have a bearing on the work of this chapter.

16 Show that, as far as the term in x^2,

$$\frac{1 - x}{(1 + x)^2} = 1 - 3x + 5x^2$$

NUNEATON

17 (a) Using the first four terms of the binomial expansion for $(1 + x)^{10}$, obtain an approximate value for $1\cdot02^{10}$, giving the answer correct to three decimal places.

   (b) The fifth and eleventh terms of an A.P. are 20 and 11 respectively. Find the A.P.

NCTEC

18 (a) Give the first four terms of the binomial expansion of $(x + 2)^{11}$.

   (b) A strip of ground 200 m long has the following widths measured at regular intervals of 25 m commencing at one end: 20, 32, 28, 21, 19, 15, 18, 25 and 29 m. Use Simpson's rule to determine its area.

ULCI

19 (a) The seventh term of an A.P. is 13·5 and the sixteenth is 27. How many terms are needed to give a sum of 175·5?

   (b) The fifth term of a G.P. is 128 and the eleventh is 2. Find the sum to infinity.

   (c) Use the binomial theorem to expand $(1 - 2x)^4$ and $(2 + x)^4$, and show that $(1 - \frac{1}{3}x)^{-3} - (1 + 3x)^{\frac{1}{3}}$ is approximately equal to $\frac{5}{3}x^2 - \frac{3}{2}\frac{5}{7}x^3$.

NUNEATON

20 Using the binomial theorem, expand $(1 + 4n)^4$.

21 Using the binomial theorem, find the first five terms of the expansions of $(1 + 2x)^7$, (b) $(1 - 2x)^{\frac{1}{2}}$.

   Show that if x is small $(1 + x)^{\frac{1}{2}} \cdot (1 - x)^{-\frac{1}{2}}$ is approximately $1 + x + \frac{1}{2}x^2$.

NUNEATON

22 (a) A geometric progression has 2 for its third term and $\frac{1}{64}$ for its tenth term. Find the first term and the common ratio, the sum of the first eight terms and the sum to infinity.

   (b) Show that, as far as the term in x^2,

$$\frac{(1 + x)^{-\frac{3}{2}} + (1 - x)^{\frac{3}{2}}}{1 + x} = 2 - 5x + \frac{29}{4}x^2$$

NUNEATON

23 (a) Write down the full expansions of $(1 + x)^6$ and $(1 - x)^6$.

   (b) Write down and simplify the expansions of $(1 + 3x)^{\frac{1}{3}}$ and $(1 - 2x)^{-\frac{1}{2}}$ as far as the term in x^2, and show that, if x is so small that x^3 and higher powers are neglected,

$$(1 + 3x)^{\frac{1}{3}}(1 - 2x)^{-\frac{1}{2}} = 1 + 2x + \frac{3}{2}x^2$$

NUNEATON

24 Using the binomial theorem, show that

(a) the expansion of $(x + 2)^{-3}$ is $\dfrac{1}{x^3} - \dfrac{6}{x^4} + \dfrac{24}{x^5} - \dfrac{80}{x^6} + \dots$

and of $(2 + x)^{-3}$ is $\frac{1}{8} - \frac{3}{16}x + \frac{3}{16}x^2 - \frac{5}{32}x^3 + \dots$

(b) the seventh term of $\left(3m - \dfrac{2}{n}\right)^{10}$ is $\dfrac{1\,088\,640\,m^4}{n^6}$,

(c) $(1 \cdot 0005)^4 = 1 \cdot 0020$ to four decimal places.

25 (a) Write down and simplify the binomial expansion of $(a + x)^8$.

(b) The first, second and fourth terms in the binomial expansion of $(1 + x)^8$ form a geometric progression. Find the value of x.

ULCI

Chapter Eleven
The Trigonometrical Ratios

The trigonometrical ratios of an acute angle θ
Consider a right-angled triangle OPC, in which OP is 1 unit of length and $\angle POC = \theta$ (Fig. 11.1).

$$\cos \theta = \frac{OC}{OP} = \frac{OC}{1}, \quad \text{or} \quad OC = 1 \cos \theta$$

$$\sin \theta = \frac{CP}{OP} = \frac{CP}{1}, \quad \text{or} \quad CP = 1 \sin \theta$$

We thus obtain an extremely important diagram (Fig. 11.2):

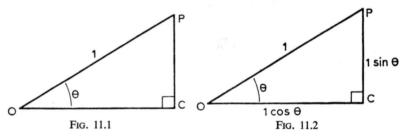

FIG. 11.1   FIG. 11.2

(Note that $\tan \theta = \dfrac{CP}{OC} = \dfrac{\sin \theta}{\cos \theta}$ and $OP^2 = OC^2 + CP^2$, i.e. $1 = \cos^2\theta + \sin^2 \theta$.)

Or, completing the rectangle OCPS in which OS = CP (Fig. 11.3):

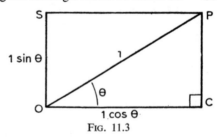

FIG. 11.3

This diagram suggests a method of defining the trigonometrical functions $\sin \phi$, $\cos \phi$ and $\tan \phi$ of the general real angle ϕ, which may be of any magnitude.

The cosine, sine and tangent of any real angle ϕ
Consider a circle of unit radius with centre O, the origin of coordinates. Let P be *any* point on the circumference of this circle and P_0 its initial

position on the axis OX. Then P is the point (cos ϕ, sin ϕ), i.e. *cos ϕ is the coordinate of P measured along the x-axis and sin ϕ is the coordinate of P measured along the y-axis.* Tan ϕ is defined as $\dfrac{\sin \phi}{\cos \phi}$.

If $\angle$P$_0$OP is an angle of magnitude θ, described in the anticlockwise sense from OP$_0$ to OP, then $\phi = +\theta$ (Fig. 11.4). If $\angle$P$_0$OP is an angle of magnitude θ, described in the clockwise sense from OP$_0$ to OP, then $\phi = -\theta$ (Fig. 11.5).

Diagram for
$\phi = +\theta$

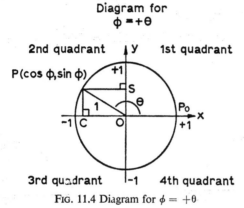

FIG. 11.4 Diagram for $\phi = +\theta$

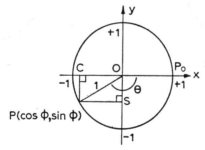

FIG. 11.5 Diagram for $\phi = -\theta$

If PC is perpendicular to Ox, then the point C represents cos ϕ on the x-axis. If PS is perpendicular to Oy, then the point S represents sin ϕ on the y-axis. Thus, with the diagrams drawn as shown,

(a) when S is above 0, sin ϕ is *positive* between 0 and $+1$,
(b) when S is below 0, sin ϕ is *negative* between 0 and -1,
(c) when C is to the right of 0, cos ϕ is *positive* between 0 and $+1$,
(d) when C is to the left of 0, cos ϕ is *negative* between 0 and -1.

Note that the length OC always represents the *magnitude* or *numerical value* of cos ϕ, and that the length OS always represents the *magnitude* or *numerical value* of sin ϕ.

Examples
11.1 Find (*a*) cos 0°, sin 0°, tan 0°; (*b*) cos 90°, sin 90°, tan 90°; (*c*) cos 180°, sin 180°, tan 180°; (*d*) cos 270°, sin 270°, tan 270°

(*a*) Referring to Fig. 11.6,

$$\cos 0° = +1, \quad \sin 0° = 0, \quad \tan 0° = \frac{\sin 0°}{\cos 0°} = \frac{0}{+1} = 0.$$

(*b*) Referring to Fig. 11.7,

$$\cos 90° = 0, \quad \sin 90° = +1, \quad \tan 90° = \frac{\sin 90°}{\cos 90°} = \frac{+1}{0} = +\infty \text{ (conventionally).}$$

(*c*) Referring to Fig. 11.8,

$$\cos 180° = -1, \quad \sin 180° = 0, \quad \tan 180° = \frac{\sin 180°}{\cos 180°} = \frac{0}{-1} = 0.$$

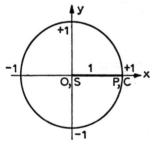

FIG. 11.6

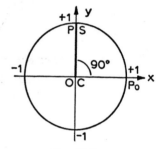

FIG. 11.7

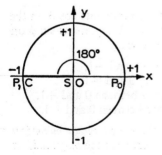

FIG. 11.8

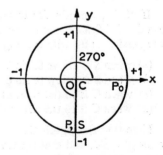

FIG. 11.9

(*d*) Referring to Fig. 11.9,

$$\cos 270° = 0, \quad \sin 270° = -1, \quad \tan 270° = \frac{\sin 270°}{\cos 270°} = \frac{-1}{0} = -\infty$$

(conventionally)

It may be noted that we have not included the sign of the angle, writing cos 270°, for example, when cos(+270°) is the more proper form. This is the usual custom, but it is important that the student should realise that cos 270° has no meaning except as an abbreviation for cos(+270°).

11.2 Find by scale drawing (*a*) sin 55°, cos 55°, sin(−305°), cos(−305°); (*b*) sin 135°, cos 135°, sin(−225°), cos(−225°); (*c*) sin 230°, cos 230°, sin(−130°), cos(−130°); (*d*) sin 320°, cos 320°, sin(−40°), cos(−40°).
Check by calculation using trigonometrical tables.

The student is urged to draw the following diagrams to the largest convenient scale.

sin 55° = sin(−305°) = +0·82   (Fig. 11.10)
By tables, sin(55°) = sin(−305°) = +0·8192
cos 55° = cos(−305°) = +0·58
By tables, cos(55°) = cos(−305°) = +0·5736

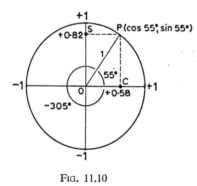

FIG. 11.10

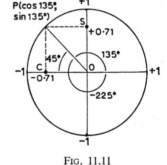

FIG. 11.11

sin(135°) = sin(−225°) = +0·71   (Fig. 11.11)
By tables, length OS = 1 sin 45° = +0·7071
sin(135°) = sin(−225°) = +0·7071
cos(135°) = cos(−225°) = −0·71
By tables, length OC = 1 cos 45° = 0·7071
Point C represents −0·7071
cos(135°) = cos(−225°) = −0·7071

$\sin(230°) = \sin(-130°) = -0.76$ (Fig. 11.12)
By tables, length OS $= 1 \sin 50° = 0.7660$
$\sin(230°) = \sin(-130°) = -0.7660$
$\cos(230°) = \cos(-130°) = -0.64$
By tables, length OC $= 1 \cos 50° = 0.6428$
$\cos(230°) = \sin(-130°) = -0.6428$

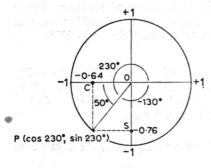

FIG. 11.12

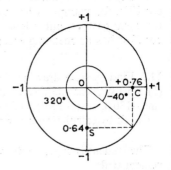

FIG. 11.13

$\sin(320°) = \sin(-40) = -0.64$ (Fig. 11.13)
By tables, length OS $= 1 \sin 40° = 0.6428$
$\sin(320°) = \sin(-40°) = -0.6428$
$\cos(320°) = \cos(-40°) = 0.76$
By tables, length OC $= 1 \cos 40° = 0.7660$
$\cos(320°) = \cos(-40°) = +0.7660$

The functions sin ϕ and cos ϕ as periodic functions

Since the coordinates of P are unchanged if $\angle P_0OP$ is changed by *complete revolutions of* 360° (Fig. 11.13A) in the positive or negative sense,

$$\sin \theta = \sin(\theta + n \times 360°)$$
$$\cos \theta = \cos(\theta + n \times 360°)$$

where n is a positive or negative integer (whole number).

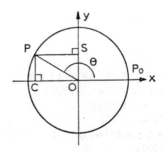

FIG. 11.13A

For example,

$$\sin 60° = \sin(60° + 360°) = \sin(60° + 720°) = \sin(60° - 360°)$$
$$= \sin(60° - 720°)$$

i.e. $\sin 60° = \sin 420° = \sin 780° = \sin(-300°) = \sin(-660°)$

Similarly,

$$\cos 60° = \cos 420° = \cos 780° = \cos(-300°) = \cos(-660°)$$
$$\sin(-30°) = \sin(-30° - 360°) = \sin(-30° + 360°) = \sin(-30° + 720°)$$

i.e. $\sin(-30°) = \sin(-390°) = \sin(330°) = \sin(690°)$

Sin ϕ and cos ϕ are said to be periodic functions of period 360° or 2π radians. The student will note how, on the accompanying graphs of $y = \sin \phi$ and $y = \cos \phi$ (Figs 11.14 and 11.15), the values of the functions repeat at angular intervals of 360°.

The graph of $y = \sin \phi$

In Fig. 11.14 the point A represents $\angle P_0OP$ on the ϕ scale. The vertical line through A and the horizontal line through P and S meet at A_1, a point on the curve. The curve extends indefinitely in the direction of ϕ increasing and ϕ decreasing, but the function $\sin \phi$ is bounded between $+1$ and -1 for real values of ϕ.

Since, in the case illustrated in Fig. 11.14, OS $= 1 \sin 60° = 0·8660$, the horizontal line through P represents the line $y = 0·8660$. Points such as A_1, A_2, etc. at the intersection of $y = 0·8660$ and $y = \sin \phi$ are solutions of the equation $\sin \phi = 0·8660$.

Thus $\phi = 60°$ and $\phi = 120°$ are the solutions of the equation $\sin \phi = 0·8660$ which lie between 0° and 360°. There is an infinite number of other solutions, all of which differ from 60° or 120° by integral multiples of 360°.

The graph of $y = \cos \phi$

In Fig. 11.15 it is convenient to first rotate OP_0 into the vertical position, since the point C represents cos ϕ. The point B represents $\angle P_0OP$ on the ϕ scale. The vertical line through B and the horizontal line through P and C meet at B_1, a point on the curve. The curve extends indefinitely in the direction of ϕ increasing and ϕ decreasing but the function cos ϕ is bounded between $+1$ and -1 for real values of ϕ.

Since, in the case illustrated in Fig. 11.15, OC $= 1 \cos 60° = 0·5$, the horizontal line through P represents the line $y = 0·5$. Points such as B_1, B_2, etc. at the intersection of $y = 0·5$ and $y = \cos \phi$ are solutions of the equation $\cos \phi = 0·5$.

Thus $\phi = 60°$ and $\phi = 300°$ are the solutions of the equation $\cos \phi = 0·5$ which lie between 0° and 360°. There is an infinite number of other solutions, all of which differ from 60° or 300° by integral multiples of 360°.

The student is recommended to redraw Figs. 11.14 and 11.15 to a large scale, using as many different values of the angle P_0OP as possible.

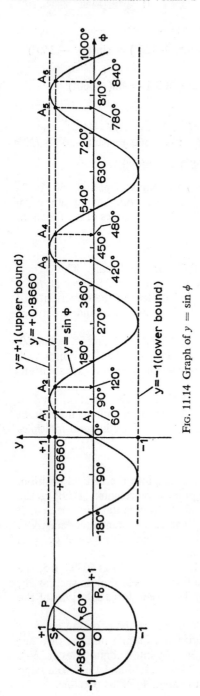

FIG. 11.14 Graph of $y = \sin \phi$

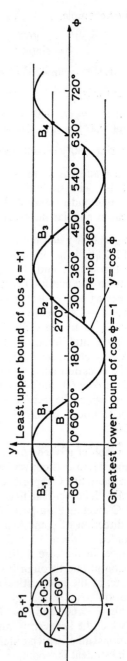

FIG. 11.15 Graph of $y = \cos \phi$

For example,

$$\sin 60° = \sin(60° + 360°) = \sin(60° + 720°) = \sin(60° - 360°)$$
$$= \sin(60° - 720°)$$

i.e. $\sin 60° = \sin 420° = \sin 780° = \sin(-300°) = \sin(-660°)$

Similarly,

$$\cos 60° = \cos 420° = \cos 780° = \cos(-300°) = \cos(-660°)$$
$$\sin(-30°) = \sin(-30° - 360°) = \sin(-30° + 360°) = \sin(-30° + 720°)$$

i.e. $\sin(-30°) = \sin(-390°) = \sin(330°) = \sin(690°)$

Sin ϕ and cos ϕ are said to be periodic functions of period 360° or 2π radians. The student will note how, on the accompanying graphs of $y = \sin \phi$ and $y = \cos \phi$ (Figs 11.14 and 11.15), the values of the functions repeat at angular intervals of 360°.

The graph of $y = \sin \phi$
In Fig. 11.14 the point A represents $\angle P_0OP$ on the ϕ scale. The vertical line through A and the horizontal line through P and S meet at A_1, a point on the curve. The curve extends indefinitely in the direction of ϕ increasing and ϕ decreasing, but the function sin ϕ is bounded between $+1$ and -1 for real values of ϕ.

Since, in the case illustrated in Fig. 11.14, $OS = 1 \sin 60° = 0.8660$, the horizontal line through P represents the line $y = 0.8660$. Points such as A_1, A_2, etc. at the intersection of $y = 0.8660$ and $y = \sin \phi$ are solutions of the equation sin $\phi = 0.8660$.

Thus $\phi = 60°$ and $\phi = 120°$ are the solutions of the equation sin $\phi = 0.8660$ which lie between 0° and 360°. There is an infinite number of other solutions, all of which differ from 60° or 120° by integral multiples of 360°.

The graph of $y = \cos \phi$
In Fig. 11.15 it is convenient to first rotate OP_0 into the vertical position, since the point C represents cos ϕ. The point B represents $\angle P_0OP$ on the ϕ scale. The vertical line through B and the horizontal line through P and C meet at B_1, a point on the curve. The curve extends indefinitely in the direction of ϕ increasing and ϕ decreasing but the function cos ϕ is bounded between $+1$ and -1 for real values of ϕ.

Since, in the case illustrated in Fig. 11.15, $OC = 1 \cos 60° = 0.5$, the horizontal line through P represents the line $y = 0.5$. Points such as B_1, B_2, etc. at the intersection of $y = 0.5$ and $y = \cos \phi$ are solutions of the equation cos $\phi = 0.5$.

Thus $\phi = 60°$ and $\phi = 300°$ are the solutions of the equation cos $\phi = 0.5$ which lie between 0° and 360°. There is an infinite number of other solutions, all of which differ from 60° or 300° by integral multiples of 360°.

The student is recommended to redraw Figs. 11.14 and 11.15 to a large scale, using as many different values of the angle P_0OP as possible.

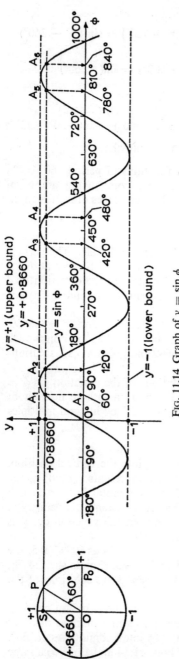

Fig. 11.14 Graph of $y = \sin \phi$

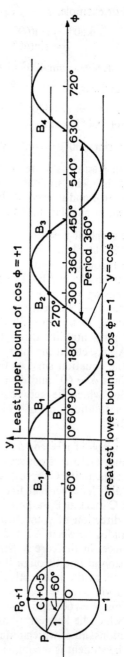

Fig. 11.15 Graph of $y = \cos \phi$

Note on the graphical solution of the equations of the type sin ϕ = $F(\phi)$ or cos ϕ = $F(\phi)$ where $F(\phi)$ is a polynomial in ϕ

If the graph of $F(\phi)$ is plotted on the same base as sin ϕ (or cos ϕ), it must be remembered that ϕ will usually be a pure number and not a number of degrees, i.e. ϕ must be expressed in radians. Since $360° = 2\pi$ radians, $180° = \pi$ radians $\simeq 3\cdot14$ radians. Thus in Figs. 11.14 and 11.15, considered previously, $180°$ on the ϕ scale would be replaced by $3\cdot14$, and other angles would be replaced in the same proportion.

For the purpose of illustration the equation $\cos \phi = 1 - \dfrac{\phi}{\pi}$ has been considered (Fig. 11.16).

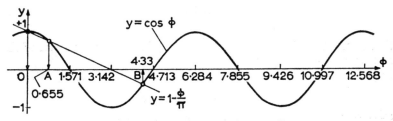

Fig. 11.16 Graph of $\cos \phi = 1 - \phi/\pi$

The graph of the line $y = 1 - \dfrac{\phi}{\pi}$ is obtained by noting that when $\phi = 0$,

$y = 1$, and when $y = 0$, $0 = 1 - \dfrac{\phi}{\pi}$, i.e. $\dfrac{\phi}{\pi} = 1$ and $\phi = \pi \simeq 3\cdot142$:

i.e. (0, 1) and (3·142, 0) are points on the line.

The points 0, A and B represent the values of ϕ which are the solutions of $\cos \phi = 1 - \dfrac{\phi}{\pi}$.

To express the ratios of the angle $-\theta$ in terms of the ratios of the acute angle $+\theta$

Since $OC = OC'$, and C and C' coincide, $\cos(-\theta) = \cos(+\theta)$, Fig. 11.17.
Since $OS = OS'$, but S' represents a negative number,

$$\sin(-\theta) = -\sin(+\theta)$$

$$\text{and } \tan(-\theta) = \frac{\sin(-\theta)}{\cos(-\theta)} = \frac{-\sin(+\theta)}{\cos(+\theta)}$$

$$\text{i.e. } \tan(-\theta) = -\tan(+\theta).$$

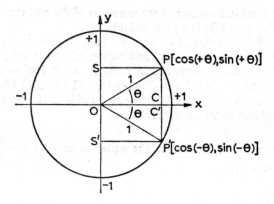

Fig. 11.17

To express the ratios of the angle (90° + θ) in terms of the ratios of the acute angle +θ

Since OS′ = OC, and C and S′ represent positive numbers, Fig. 11.18,

$$\sin(90° + \theta) = \cos(+\theta).$$

Since OC′ = CP, but C′ represents a negative number,

$$\cos(90° + \theta) = -\sin(+\theta)$$

and $\tan(90° + \theta) = \dfrac{\sin(90° + \theta)}{\cos(90° + \theta)} = \dfrac{\cos(+\theta)}{-\sin(+\theta)} = -\dfrac{1}{\tan(+\theta)}$

$$= -\cot(+\theta)$$

where cot is an abbreviation for cotangent, and $\cot \phi = \dfrac{1}{\tan \phi}$.

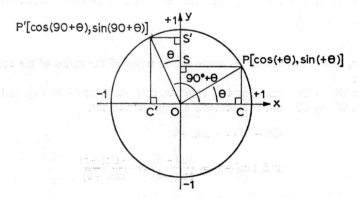

Fig. 11.18

EXERCISE 11.1

By drawing suitable diagrams, obtain the following results. (cot θ = 1/tan θ.)

(a) sin(90° − θ) = cos(+θ), cos(90° − θ) = sin(+θ), tan(90° − θ) = cot(+θ)

(b) sin(180° − θ) = sin(+θ), cos(180° − θ) = −cos(+θ), tan(180° − θ) = −tan(+θ)

(c) sin(180° + θ) = −sin(+θ), cos(180° + θ) = −cos(+θ), tan(180° + θ) = tan(+θ)

(d) sin(270° − θ) = −cos(+θ), cos(270° − θ) = −sin(+θ), tan(270° − θ) = cot(+θ)

(e) sin(270° + θ) = −cos(+θ), cos(270° + θ) = +sin(+θ), tan(270° + θ) = −cot(+θ)

Although these results have been demonstrated when θ is an acute angle, it can be verified by the addition theorems which are considered later in this book that the results are true for all magnitudes of θ.

Solution of the equations $R \sin \phi = D$ and $R \cos \phi = D$ for real values of θ where R and D are constants

Writing the equations as $\sin \phi = \dfrac{D}{R}$ and $\cos \phi = \dfrac{D}{R}$, since, for real values of ϕ, sin ϕ and cos ϕ cannot exceed 1, it is necessary that $\dfrac{D}{R}$ is not greater than 1. This being so, the equations may be solved as illustrated in the following examples.

Examples

11.3 Solve the equation 5 sin ϕ = 4 (*a*) for values of ϕ between 0° and 360°, (*b*) for all values of ϕ.

(*a*) Since 5 sin ϕ = 4, sin ϕ = $\frac{4}{5}$ = +0·8.

From the quadrant diagram (Fig. 11.19), when S corresponds to +0·8

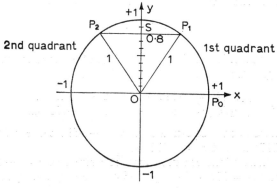

Fig. 11.19

there are two corresponding positions of P, i.e. P_1 and P_2, and two angles whose sine is 0·8, i.e. $\angle P_0OP_1$ and $\angle P_0OP_2$. From the trigonometrical tables, the acute angle whose sine is 0·8 is 53° 8′.

$$\text{Thus} \quad \angle P_0OP_1 = +53° 8′$$
$$\angle P_0OP_2 = +180° - 53° 8′ = +126° 52′$$

i.e. the solutions of the equation $5 \sin \theta = 4$ between 0° and 360° are $\phi = +53° 8′$ and $\phi = +126° 52′$.

(b) Any angle obtained by rotation from OP_0 to OP_1 or OP_2, making any number of positive or negative revolutions in the process will satisfy the equation, since S will always correspond to $+0·8$, i.e. the solutions of the equation $5 \sin \phi = 4$ may be written as $\phi = +53° 8′ + n_1 \times 360°$ and $\phi = +126° 52′ + n_2 \times 360°$, where n_1 and n_2 may be any positive or negative integers or zero.

11.4 Solve the equation $6 \sin \phi = -3·636$ (a) for values of ϕ between 0° and 360°, (b) for all values of ϕ.

(a) Since $6 \sin \phi = -3·636$, $\sin \phi = -0·606$. When S corresponds to $-0·606$, there are two corresponding positions of P, i.e. P_1 and P_2 in the

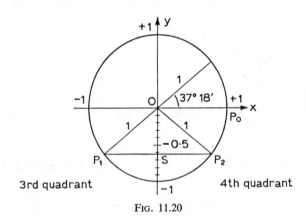

Fig. 11.20

third and fourth quadrants (Fig. 11.20). From the trigonometrical tables, the acute angle whose sine is $+0·606$ is 37° 18′.

$$\text{Thus} \quad \angle P_0OP_1 = +180° + 37° 18′ = +217° 18′$$
$$\text{and} \quad \angle P_0OP_2 = +360° - 37° 18′ = +322° 42′$$

i.e. the solutions of the equation $6 \sin \phi = -3·636$ between 0° and 360° are $\phi = +217° 18′$ and $\phi = +322° 42′$.

(b) The solutions of the equation may be written as $\phi = +217° 18′ + n_1 \times 360°$ and $\phi = 322° 42′ + n_2 \times 360°$, where n_1 and n_2 may be any positive or negative integers or zero.

11.5 Solve the equation $4 \cdot 5 \cos \phi = 1 \cdot 35$ (a) for values of ϕ between $0°$ and $360°$,   (b) for all values of ϕ.

(a) Since $4 \cdot 5 \cos \phi = 1 \cdot 35$, $\cos \phi = \dfrac{1 \cdot 35}{4 \cdot 5} = 0 \cdot 3$. When C corresponds to $+0 \cdot 3$, there are two corresponding positions of P, i.e. P_1 and P_2 in the first and fourth quadrants (Fig. 11.21). From the trigonometrical tables, the acute angle whose cosine is $+0 \cdot 3$ is $72° 32'$.

$$\text{Thus} \quad \angle P_0 O P_1 = +72° 32'$$
$$\angle P_0 O P_2 = +360° - 72° 32' = +287° 28'$$

i.e. the solutions of the equation $4 \cdot 5 \cos \phi = 1 \cdot 35$ between $0°$ and $360°$ are $\phi = 72° 32'$ and $\phi = 287° 28'$.

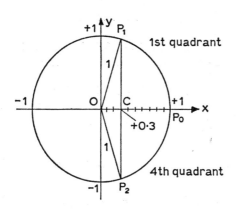

FIG. 11.21

(b) The solutions of the equation $4 \cdot 5 \cos \phi = 1 \cdot 35$ may be written as $\phi = 72° 32' + n_1 \times 360°$ or $\phi = 287° 28' + n_2 \times 360°$, where n_1 and n_2 may be any positive or negative integers or zero.

11.6 Solve the equation $11 \cos \phi = -1 \cdot 76$ (a) for values of ϕ between $0°$ and $360°$,   (b) for all values of ϕ.

(a) Since $11 \cos \phi = -1 \cdot 76$, $\cos \phi = -0 \cdot 16$. When C corresponds to $-0 \cdot 16$, there are two corresponding positions of P, i.e. P_1 and P_2 in the second and third quadrants (Fig. 11.22). From the trigonometrical tables, the acute angle whose cosine is $+0 \cdot 16$ is $80° 48'$.

$$\therefore \quad \angle P_0 O P_1 = +180° - 80° 48' = +99° 12'$$
$$\angle P_0 O P_2 = +180° + 80° 48' = +260° 48'$$

i.e. the solutions of the equation $11 \cos \phi = -1 \cdot 76$ between $0°$ and $360°$ are $\phi = 99° 12'$ and $\phi = 260° 48'$.

(b) The solutions of the equation $11 \cos \phi = -1 \cdot 76$ may be written as

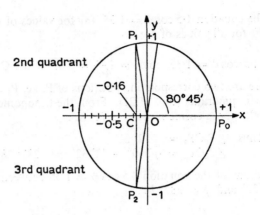

Fig. 11.22

$\phi = 99° 12' + n_1 \times 360°$ and $\phi = 260° 48' + n_2 \times 360°$, where n_1 and n_2 may be any positive or negative integers or zero.

Solution of the equation $R \sin p \, \phi = D$ where R, p and D are constants
In this case, if ϕ is required between 0° and 360°, it is necessary to obtain the values of $p\phi$ between 0° and p. 360°.

Examples
11.7 Solve the equation $12 \sin 3\phi = 1{\cdot}3212$ for values of ϕ between 0° and 180° (Fig. 11.23).

Since $12 \sin 3\phi = 1{\cdot}3212$, $\sin 3\phi = +0{\cdot}1101$. From the tables, the corresponding acute angle is 6° 19′,

$$\text{i.e.} \quad \angle P_0OP_1 = +6° 19'$$
$$\angle P_0OP_2 = +180° - 6° 19' = +173° 41'$$

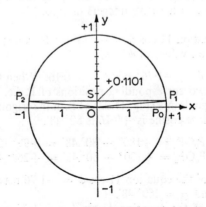

Fig. 11.23

If ϕ is between $0°$ and $180°$ then 3ϕ is between $3 \times 0°$ and $3 \times 180°$, i.e. $0°$ and $540°$.

Hence $3\phi = 6°\ 19',\ 173°\ 41',\ 6°\ 19' + 360°$ and $173°\ 41' + 360°$

i.e. $3\phi = 6°\ 19',\ 173°\ 41',\ 366°\ 19'$ and $533°\ 41'$
i.e. $\phi = 2°\ 6',\ 57°\ 54',\ 122°\ 6'$ and $177°\ 54'$

11.8 Solve the equation $5\cos(4\phi + 30°) = -3{\cdot}7575$ for values of ϕ between $-90°$ and $+90°$ (Fig. 11.24).

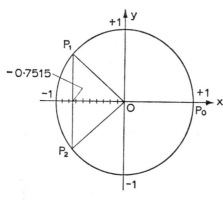

FIG. 11.24

Since $5\cos(4\phi + 30°) = -3{\cdot}7575$, $\cos(4\phi + 30°) = -0{\cdot}7515$. From the trigonometrical tables, the acute angle whose cosine is $+0{\cdot}7515$ is $41°\ 17'$,

i.e. $\angle P_0OP_1 = +180° - 41°\ 17' = 138°\ 43'$
$\angle P_0OP_2 = +180° + 41°\ 17' = 221°\ 17'$

If ϕ is between $-90°$ and $+90°$ then 4ϕ is between $-360°$ and $+360°$, and $4\phi + 30°$ is between $-330°$ and $+390°$.

i.e. $4\phi + 30° = +138°\ 43',\ +138°\ 43' - 360°,\ +221°\ 17'$ or $+221°\ 17' - 360°$
i.e. $4\phi + 30° = +138°\ 43',\ -221°\ 17',\ +221°\ 17'$ or $-138°\ 43'$
i.e. $4\phi = +108°\ 43',\ -251°\ 17',\ +191°\ 17'$ or $-168°\ 43'$
i.e. $\phi = +27°\ 11',\ -62°\ 49',\ +47°\ 49'$ or $-42°\ 11'$

Solution of the equation $a \cdot \sin^2\phi + b \cdot \sin\phi + d = 0$
Since the equation $a \cdot \sin^2\phi + b \cdot \sin\phi + d = 0$ is a quadratic equation in $\sin\phi$, it may be solved by the usual methods of solving a quadratic equation. It is not necessary, but may be helpful, to make the substitution $s = \sin\phi$ in the equation, as shown in the following example.

Examples

11.9 Solve the equation $3 \sin^2\phi - 7 \sin\phi + 2 = 0$ for values of ϕ from $0°$ to $360°$.

In the equation $3 \sin^2\phi - 7 \sin\phi + 2 = 0$, let $s = \sin\phi$.

Then $\qquad\qquad 3s^2 - 7s + 2 = 0$

i.e. $\qquad\qquad (3s - 1)(s - 2) = 0$
i.e. $\qquad\qquad 3s - 1 = 0$, or $s - 2 = 0$, i.e. $s = \frac{1}{3}$ or $s = 2$
i.e. $\qquad\qquad \sin\phi = \frac{1}{3}$ or $\sin\phi = 2$

We reject the equation $\sin\phi = 2$ since, if ϕ is real, $\sin\phi$ cannot exceed 1 in magnitude.

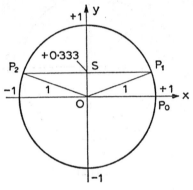

FIG. 11.25

If $\sin\phi = +\frac{1}{3} = +0.3333$, the corresponding acute angle, from the tables, is $19° 28'$ (Fig. 11.25)

$$\angle P_0OP_1 = +19° 28'$$
$$\angle P_0OP_2 = +180° - 19° 28' = +160° 32'$$

i.e. $\quad \phi = 19° 28'$ or $160° 32'$

11.10 Solve these equations $6 \sin^2\phi + 5 \cos\phi - 7 = 0$ for the values of ϕ between $0°$ and $360°$.

The equation is not a quadratic in $\sin\phi$ but, by using the identity $\sin^2\phi + \cos^2\phi = 1$, i.e. $\sin^2\phi = 1 - \cos^2\phi$, it may be transformed into a quadratic in $\cos\phi$.

$$6 \sin^2\phi + 5 \cos\phi - 7 = 0$$
$$\therefore \quad 6(1 - \cos^2\phi) + 5 . \cos\phi - 7 = 0$$
$$\therefore \quad 6 - 6 \cos^2\phi + 5 \cos\phi - 7 = 0$$

i.e. $\qquad\qquad -6 \cos^2\phi + 5 \cos\phi - 1 = 0$

Multiplying both sides of the equation by -1

$$6 \cos^2\phi - 5 \cos \phi + 1 = 0$$
i.e. $\quad (3 \cos \phi - 1)(2 \cos \phi - 1) = 0$
i.e. $\quad (3 \cos \phi - 1)(2 \cos \phi - 1) = 0$
i.e. $\quad \cos \phi = \frac{1}{3} \quad$ or $\quad \cos \phi = \frac{1}{2}$
i.e. $\quad \cos \phi = 0 \cdot 3333 \quad$ or $\quad \cos \phi = 0 \cdot 5$

When $\cos \phi = +0 \cdot 3333$, the corresponding acute angle, from the tables, is $70° \, 32'$ (see Fig. 11.26),

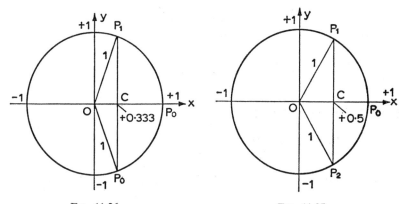

FIG. 11.26   FIG. 11.27

i.e. $\quad \angle P_0OP_1 = +70° \, 32'$
$\quad\quad\quad \angle P_0OP_2 = +360° - 70° \, 32' = +289° \, 28'$

When $\cos \phi = +0 \cdot 5$ the corresponding acute angle, from the tables, is $60°$ (see Fig. 11.27),

i.e. $\quad \angle P_0OP_1 = +60°$
$\quad\quad\quad \angle P_0OP_2 = +360° - 60° = +300°$

i.e. the solutions of the equation $6 \sin^2\phi + 5 \cos \phi - 7 = 0$ between $0°$ and $360°$ are $\phi = 60°, \, 70° \, 32', \, 289° \, 28'$ and $300°$.

The algebraic signs of the cosine, sine and tangent of an angle in each of the four quadrants

Let ϕ be the general angle $\theta + k \cdot 360°$, where k is any positive or negative integer or zero.

Consider the points S and C, representing $\sin \phi$ and $\cos \phi$, in each quadrant. The point P has co-ordinates $(\sin \phi, \cos \phi)$.

In the 1st quadrant (Fig. 11.28):

$\quad\quad \sin(\phi)$ is positive,
$\quad\quad \cos(\phi)$ is positive

$\quad\quad \tan(\phi)$ is positive, since $\tan(\phi) = \dfrac{\sin(\phi)}{\cos(\phi)} . \left(\dfrac{'+'}{'+'} = \, '+'. \right)$

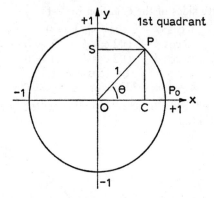

FIG. 11.28

In the 2nd quadrant (Fig. 11.29):

$\sin(\phi)$ is positive,

$\cos(\phi)$ is negative (between 0 and -1),

$\tan(\phi)$ is negative, since $\tan(\phi) = \dfrac{\sin(\phi)}{\cos(\phi)}.$ $\left(\dfrac{'+'}{'-'} = '-'. \right)$

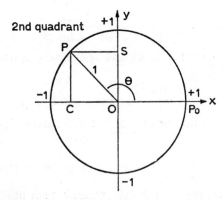

FIG. 11.29

In the 3rd quadrant (Fig. 11.30):

$\sin(\phi)$ is negative,

$\cos(\phi)$ is negative,

$\tan(\phi)$ is positive, since $\tan(\phi) = \dfrac{\sin(\phi)}{\cos(\phi)}.$ $\left(\dfrac{'-'}{'-'} = '+'. \right)$

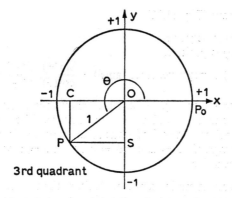

FIG. 11.30

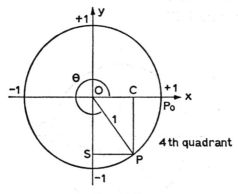

FIG. 11.31

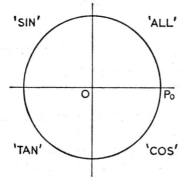

FIG. 11.32

In the 4th quadrant (Fig. 11.31):

> $\sin(\phi)$ is negative,
> $\cos(\phi)$ is positive,
> $\tan(\phi)$ is negative, since $\tan(\phi) = \dfrac{\sin(\phi)}{\cos(\phi)}. \left(\dfrac{'-'}{'+'} = '-'.\right)$

This is summarised in Fig. 11.32, which shows the functions which are positive in each quadrant. All others are negative. The diagram may be used as follows to formulate rules for the solution of simple trigonometrical equations of the type $\sin \phi = a$, $\cos \phi = b$ or $\tan \phi = c$, where a, b and c are positive or negative.

(a) Find the corresponding acute angle from the tables.
(b) Select the appropriate quadrants where the given function will have its appropriate sign.
(c) Position OP in these quadrants at the acute angle found in (a) to the line OP_0 or P_0O produced.
(d) Assess the angle P_0OP.

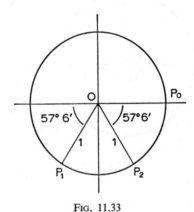

FIG. 11.33

Examples
11.11 Solve the equation $\sin \phi = -0{\cdot}8396$ for values of ϕ from $0°$ to $360°$.

(a) The corresponding acute angle is $57° 6'$.
(b) The sine is negative in the third and fourth quadrants using 'ALL, SIN, TAN, COS'.
(c) The acute angle of $57° 6'$ is placed in the third and fourth quadrants as shown in Fig. 11.33.

(d) $\angle P_0OP_1 = +180° + 57° 6' = +237° 6'$
             $\angle P_0OP_2 = +360° - 57° 6' = +302° 54'$

i.e. $\phi = +237° 6'$ or $+302° 54'$

11.12 Solve the equation $\tan \phi = -9\cdot677$ for values of ϕ from $0°$ to $360°$.

(*a*) The corresponding acute angle is $84° 6'$.
(*b*) The tangent is negative in the second and fourth quadrants using 'ALL, SIN, TAN, COS'.
(*c*) The acute angle of $84° 6'$ is placed in the second and fourth quadrants as shown in Fig. 11.34.

(*d*)
$$\angle P_0OP = +180° - 84° 6' = +95° 54'$$
$$\angle P_0OP_2 = +360° - 84° 6' = +275° 54'$$

i.e.
$$\phi = 95° 54' \quad \text{or} \quad +275° 54'$$

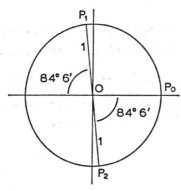

FIG. 11.34

Construction of the graph of $y = \tan \phi$
The tangent of an angle in the four quadrants is represented to scale by the point T, the intersection of OP produced or OP produced with the tangent to the unit circle at P_0 (Figs. 11.35 to 11.38).
In all cases

$$\frac{P_0T}{OP_0} = \frac{P_0T}{1} = \frac{CP}{OC} = \frac{\text{magnitude of the sine of the angle}}{\text{magnitude of the cosine of the angle}}$$

$$= \text{magnitude of the tangent of the angle}$$

Thus P_0T always represents the magnitude of the tangent of the angle: hence the point T always represents the tangent of the angle to scale, since, as the diagrams show, the correct sign is obtained in each quadrant. The construction fails, however, when OP is parallel to the tangent at P_0.
In Fig. 11.35, as ϕ approaches $+90°$, P_0T becomes very large, i.e. as $\phi \rightarrow +90°$, $\tan \phi \rightarrow +\infty$.
In Fig. 11.36, as ϕ approaches $+90°$, P_0T becomes very large, i.e. as $\phi \rightarrow +90°$ (from above), $\tan \phi \rightarrow -\infty$.

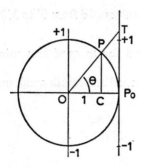

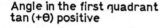

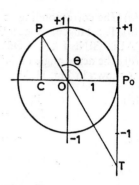

Fig. 11.35   In the 1st quadrant tan $(+\theta)$ positive

Fig. 11.36   In the 2nd quadrant tan $(+\theta)$ negative

In Fig. 11.37, as ϕ approaches $+270°$, P_0T becomes very large, i.e. as $\phi \to +270°$, $\tan \phi \to +\infty$.

In Fig. 11.38, as ϕ approaches $+270°$, P_0T becomes very large, i.e. as $\phi \to +270°$ (from above), $\tan \phi \to -\infty$.

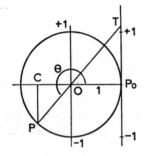

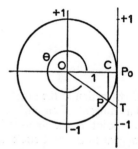

Fig. 11.37   In the 3rd quadrant tan $(+\theta)$ positive

Fig. 11.38   In the 4th quadrant tan $(+\theta)$ negative

The graph (Fig. 11.39) shows that $\tan \phi$ is a periodic function of ϕ, the period being $180°$. At $90°$ and angular intervals of $180°$ there is a discontinuity in the graph, but within each $180°$ interval or period $\tan \phi$ takes all positive and negative values, being zero at the centre of each period. Within each period, as ϕ increases, $\tan \phi$ increases from $-\infty$ to $+\infty$ (we say conventionally). We see that $\tan(-45°) = -\tan 45°$ and more generally $\tan(-\theta) = -\tan(+\theta)$.

Determination of tan ϕ using the trigonometrical tables

Since the period of $\tan \phi$ is $180°$, we may add or subtract any integral multiple of $180°$ from the angle ϕ and use the relation $\tan(-\theta) = -\tan(+\theta)$.

Examples

(a) $\tan 63°$ $= 1{\cdot}9626$ (direct from tables).
(b) $\tan(-63°)$ $= -\tan 63° = -1{\cdot}9626$.
(c) $\tan(125°)$ $= \tan(125° - 180°) = \tan(-55°) = -\tan 55° = -1{\cdot}4281$.
(d) $\tan(226°)$ $= \tan(226° - 180°) = \tan 46° = 1{\cdot}0355$.
(e) $\tan(342°)$ $= \tan(342°) = \tan(-18°) = -\tan 18° = -0{\cdot}3249$.
(f) $\tan 1000°$ $= \tan(1000° - 3.360°) = \tan(1000° - 1080°) = \tan(-80°)$
$= -\tan 80° = -5{\cdot}6713$.
(g) $\tan(-600°)$ $= \tan(-600° + 3.180°) = \tan(-600° + 540°) = \tan(-60°)$
$= -\tan 60° = -1{\cdot}7321$.

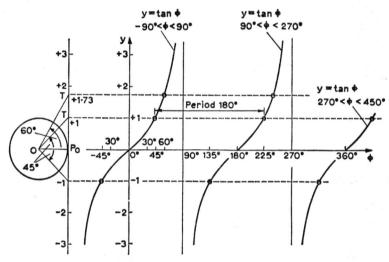

FIG. 11.39 Graph of $y = \tan \phi$

A slight variation on this method is to (a) position OP correctly on the quadrant diagram, (b) find the acute angle made by OP and OP_0 (or P_0O produced), (c) find the tangent of the acute angle, and (d) affix the correct sign using the ALL, SIN, TAN, COS mnemonic.
For example, to find $\tan 250°$:

(a) See Fig. 11.40,
(b) the acute angle is 70°,
(c) the tangent of 70° is $2{\cdot}7475$,
(d) the tangent is positive in the third quadrant, i.e. $\tan 250° = +2{\cdot}7475$.

The secant, cosecant and cotangent of any real angle ϕ

$\dfrac{1}{\text{cosine } \phi}$ is called the secant of the angle ϕ and is written $\sec \phi$.

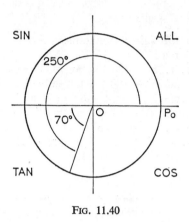

FIG. 11.40

$\dfrac{1}{\text{sine } \phi}$ is called the cosecant of the angle ϕ and is written cosec ϕ.

$\dfrac{1}{\text{tangent } \phi}$ is called the cotangent of the angle ϕ and is written cot ϕ.

Example
11.13 If $\sin \theta = +\frac{3}{5}$, find $\cos \theta$, $\tan \theta$, $\text{cosec } \theta$, $\sec \theta$ and $\cot \theta$ without using tables when (*a*) θ is between $0°$ and $+90°$, (*b*) θ is between $90°$ and $+180°$.

Let P and P′ be the possible positions of θ (Fig. 11.41). Since $OP^2 = OC^2 + CP^2$, i.e. $1 = OC^2 + (\frac{3}{5})^2$

$$\therefore \quad OC = \sqrt{\left(1 - \frac{9}{25}\right)} = \frac{4}{5} \quad \text{and} \quad OC' = \frac{4}{5}$$

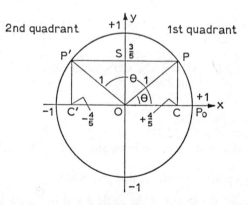

FIG. 11.41

(*a*) When θ is in the first quadrant,

$$\sin \theta = +\frac{3}{5}, \ \cos \theta = +\frac{4}{5}, \ \tan \theta = \frac{\sin \theta}{\cos \theta} = +\frac{3}{5} \cdot \frac{5}{4} = +\frac{3}{4}$$

$$\therefore \quad \operatorname{cosec} \theta = +\frac{5}{3}, \ \sec \theta = +\frac{5}{4}, \ \cot \theta = +\frac{4}{3}$$

(*b*) When θ is in the second quadrant,

$$\sin \theta = +\frac{3}{5}, \ \cos \theta = -\frac{4}{5}, \ \tan \theta = \frac{\sin \theta}{\cos \theta} = \frac{+\frac{3}{5}}{-\frac{4}{5}} = -\frac{3}{4}$$

$$\therefore \quad \operatorname{cosec} \theta = +\frac{5}{3}, \ \sec \theta = -\frac{5}{4}, \ \cot \theta = -\frac{4}{3}$$

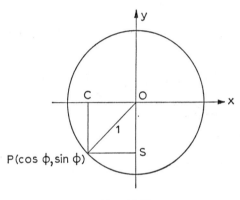

FIG. 11.42

The fundamental identities $\sin^2 \phi + \cos^2 \phi = 1$, $1 + \tan^2 \phi = \sec^2 \phi$ and $1 + \cot^2 \phi = \operatorname{cosec}^2 \phi$ for any real angle ϕ

For any position of P on the unit circle in the four quadrants (Fig. 11.42),
$OP^2 = OC^2 + CP^2$

i.e. $OP^2 = OC^2 + OS^2$

i.e. $1 = $ (magnitude of $\cos \phi)^2 + $ (magnitude of $\sin \phi)^2$.

Hence $1 = \cos^2\phi + \sin^2\phi$

since $\cos^2\phi$ and $\sin^2\phi$ are always positive when $\cos \phi$ and $\sin \phi$ are positive or negative.

Since $\cos^2\phi + \sin^2\phi = 1$,

dividing both sides of the identity by $\cos^2\phi$,

$$\frac{\cos^2\phi}{\cos^2\phi} + \frac{\sin^2\phi}{\cos^2\phi} = \frac{1}{\cos^2\phi}$$

i.e. $1 + \tan^2\phi = \sec^2\phi$

or,   since $\cos^2\phi + \sin^2\phi = 1$,

dividing both sides of the identity by $\sin^2\phi$,

$$\frac{\cos^2\phi}{\sin^2\phi} + \frac{\sin^2\phi}{\sin^2\phi} = \frac{1}{\sin^2\phi}$$

Now $\tan\phi = \dfrac{\sin\phi}{\cos\phi}$ and $\cot\phi = \dfrac{\cos\phi}{\sin\phi}$

$$\therefore \quad 1 + \cot^2\phi = \mathrm{cosec}^2\phi$$

Examples

11.14 Prove that $\dfrac{\sec^2 A - 2}{2 - \mathrm{cosec}^2 A} = \tan^2 A$

$$\text{L.H.S.} = \frac{\sec^2 A - 2}{2 - \mathrm{cosec}^2 A} = \frac{1 + \tan^2 A - 2}{2 - (1 + \cot^2 A)} = \frac{\tan^2 A - 1}{1 - \cot^2 A} = \frac{\tan^2 A - 1}{1 - \dfrac{1}{\tan^2 A}}$$

$$= \frac{(\tan^2 A - 1)\,.\,\tan^2 A}{\left(1 - \dfrac{1}{\tan^2 A}\right)\,.\,\tan^2 A} = \frac{(\tan^2 A - 1)\,.\,\tan^2 A}{(\tan^2 A - 1)} = \tan^2 A$$

$$= \text{R.H.S.}$$

11.15 Prove that $\dfrac{\tan\theta + \tan\theta}{\cot\theta + \cot\phi} = \tan\theta\,.\,\tan\phi$

$$\text{L.H.S.} = \frac{\tan\theta + \tan\phi}{\dfrac{1}{\tan\theta} + \dfrac{1}{\tan\phi}} = \frac{\tan\theta + \tan\phi}{\dfrac{\tan\phi + \tan\theta}{\tan\theta\,.\,\tan\phi}} = \tan\theta\,.\,\tan\phi$$

$$= \text{R.H.S.}$$

EXERCISE 11.2

1 Write down the values of the sine, cosine and tangent of:

(a) 102°	(f) 6·222 radians	(k) 149° 33′
(b) 242° 32′	(g) −51°	(l) 201° 13′
(c) 315° 20′	(h) −300°	(m) 343° 8′
(d) π radians	(i) −256°	
(e) $\dfrac{\pi}{2}$ radians	(j) 2000°	

2 Find the value of:

(a) cosec 154°	(c) cot 321°	(e) sec 300°
(b) sec 235°	(d) cosec 251°	(f) cot 163°

3 Given that an angle is in the third quadrant, and its sine $= -\frac{2}{5}$, find its cosine and tangent.

4 If $\cos \theta = \dfrac{a^2 - b^2}{2lm}$, find all the possible values of θ up to 360°, when $a = 6, b = 7, l = 9, m = 10$.

5 Given that $2 \sin x = 0.96$, find all values of x up to 360°.

6 If $\sin \theta = -\frac{5}{13}$, find all possible values of $\sec \theta$ and $\cot \theta$ without evaluating θ.

7 If $\sec \theta = 5$ (θ being an acute angle), find $\sin \theta$ and $\cot \theta$.

8 If $\sin A = 0.5$ (A being between 90° and 180°), find $\sec A + \tan A$ (without evaluating A).

9 If $\sec A + \tan A = 5$ (A being an acute angle), find $\cos A$.

10 The cosine of an acute angle is m: find its cosecant and also its tangent.

11 Evaluate $ae^{-kt} \sin (nt + g)$, when $a = 6, t = 0.004, k = 40, n = 1000$, $e = 2.718, g = 1.235$, where $(nt + g)$ is in radians.

12 Evaluate $\sin \dfrac{3\pi}{2} \cos \dfrac{\pi}{2} - \cos \dfrac{3\pi}{2} \sin \dfrac{\pi}{2}$ (the angle is in radians).

13 Evaluate $\sin \dfrac{\pi}{3} \cos \dfrac{\pi}{4} + \cos \dfrac{\pi}{3} \sin \dfrac{\pi}{4}$ (the angle is in radians).

14 Evaluate $\cos \dfrac{\pi}{4} \cos \dfrac{\pi}{6} - \sin \dfrac{\pi}{4} \sin \dfrac{\pi}{6}$ (the angle is in radians).

15 Evaluate $\dfrac{2}{\sec \dfrac{\pi}{5} \cos \dfrac{\pi}{5}}$ (the angle is in radians).

Prove the following.

16 $\dfrac{1 - \tan^2\theta}{1 + \tan^2\theta} = \cos^2\theta - \sin^2\theta$.

17 $(1 + \tan^2 A) \cos^2 A = 1$.

18 $\dfrac{2 \tan B}{1 + \tan^2 B} = 2 \sin B . \cos B$.

19 $\cos^2 x - \sin^2 x = 1 - 2 \sin^2 x$.

20 $1 - 2 \sin^2\theta = 2 \cos^2\theta - 1$.

21 $(\sec \theta + \tan \theta)(\sec \theta - \tan \theta) = 1$.

22 $(\sec \theta - \cos \theta) = \sqrt{\{(\tan \theta + \sin \theta)(\tan \theta - \sin \theta)\}}$.

23 $\tan \phi + \cot \phi = \sec \phi . \operatorname{cosec} \phi$.

24 $\dfrac{1 - \sin A}{1 + \sin A} = (\sec A - \tan A)^2$.

25 Graph each of the functions $2 \cdot 5 \sin x°$ and $(2 - \cos x°)$ for values of x from $-30°$ to $+140°$, using the same reference axes for both graphs, and 1 mm to represent $1°$. By means of the graph, solve the equation

$$\cos x° + 2 \cdot 5 \sin x° = 2 \qquad \text{NCTEC}$$

26 On the same axes and to the same scales, graph the functions $3 \cdot 5 \sin x$ and $(2 \cdot 5 - \cos x)$, taking values of x from $0°$ to $180°$. Then, using your diagrams, solve the equation

$$3 \cdot 5 \sin x + \cos x = 2 \cdot 5 \qquad \text{UEI}$$

27 Solve the following equations (a) for values of ϕ between $0°$ and $360°$, (b) for all values of ϕ: (i) $3 \cos \phi = 1 \cdot 9242$, (ii) $12 \cos \phi = -1 \cdot 2540$.

28 Solve the equation $8 \cos 3\phi = 6 \cdot 9072$ for values of ϕ between $0°$ and $180°$.

29 Solve the equation $10 \sin(4\phi - 20°) = 4 \cdot 617$ for values of ϕ between $-90°$ and $+90°$.

30 Solve the equation $8 \sin^2\theta - 2 \sin \theta - 1 = 0$ for values of θ between $0°$ and $360°$.

31 Solve the equation $50 \cos^2\theta + 15 \cos \theta - 2 = 0$ for values of θ between $0°$ and $360°$.

32 Solve the equation $1 - 2 \sin^2\theta = \cos \theta$ for values of θ between $0°$ and $360°$ inclusive.

33 Solve the following equations for values of x between $0°$ and $360°$: (a) $\tan x = 71 \cdot 615$, (b) $\tan 2x = 71 \cdot 615$.

Chapter Twelve
The Addition Formulae

It is sometimes expedient to express the trigonometrical ratios of a compound angle in terms of the ratios of its component angles; for instance, $\sin(A + B)$ may be conveniently expressed in terms of $\sin A$, $\cos A$, $\cos B$ and $\sin B$ in the course of a calculation. A common mistake is to identify $\sin(A + B)$ with $\sin A + \sin B$. Why does this error arise? Previously $a(b + c) \equiv ab + ac$, where a, b and c are numbers, has been used. Remembering that $\sin(A + B)$ does not mean multiply a number 'sin' by the sum of A and B, we can see that we have as yet to derive the trigonometrical ratios of the compound angle $(A + B)$ in terms of the ratios of the angle A and of the angle B.

Suppose that A, B and $A + B$ represent acute angles, we may sketch the following diagrams, Figs. 12.1 to 12.3.

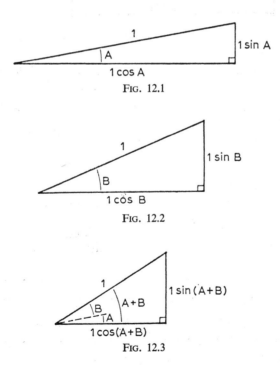

FIG. 12.1

FIG. 12.2

FIG. 12.3

Figure 12.3 may now be adapted to relate the ratios of the angle $A + B$ to those of the angles A and B (Fig. 12.4).

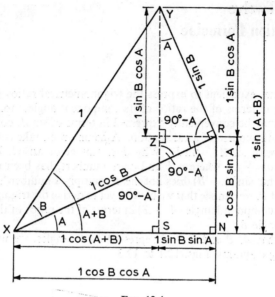

Fig. 12.4

Since XY = 1 unit, XS = 1 cos(A + B), YS = 1 sin(A + B)
XR = 1 cos B, YR = 1 sin B.

From right-angled triangle YZR since YR = 1 sin B, YZ = (1 sin B)cos A and ZR = (1 sin B)sin A.

From the right-angled triangle XRN, since XR = 1 cos B, YZ = (1 sin B)cos A and ZR = (1 sin B)sin A.

Marking these on the diagram, we may read off horizontally along XN

$$1 \cos(A + B) + 1 \sin A \sin B = 1 \cos A \cos B$$

i.e. $$\cos(A + B) = \cos A \cos B - \sin A \sin B. \quad . \quad . \quad . \quad . \quad (i)$$

Replacing B by −B, *assuming that this is valid*,

$$\cos(A - B) = \cos A \cos(-B) - \sin A \sin(-B)$$

but $$\cos(-B) = \cos B \text{ and } \sin(-B) = -\sin B$$

∴ $$\cos(A - B) = \cos A \cos B + \sin A \sin B. \quad . \quad . \quad . \quad . \quad (ii)$$

Similarly, we may read off vertically that

$$1 \sin(A + B) = 1 \sin A \cos B + 1 \cos A \sin B$$

i.e. $$\sin(A + B) = \sin A \cos B + \cos A \sin B. \quad . \quad . \quad . \quad . \quad (iii)$$

Replacing B by −B, *assuming that this is valid*,

$$\sin(A - B) = \sin A \cos(-B) + \cos A \sin(-B)$$

i.e. $$\sin(A - B) = \sin A \cos B - \cos A \sin B \quad . \quad . \quad . \quad . \quad (iv)$$

The identities (i), (ii), (iii) and (iv) are known as the *addition theorems*. They are not independent. From any one of them the other three may be deduced.

General proof of the addition theorems

Consider lines OP and OQ, each of unit length, making angles B and A respectively with the x-axis. The coordinates of P and Q will be $(1 \cos B, 1 \sin B)$ and $(1 \cos A, 1 \sin A)$ respectively. Note that P and Q may be in *any* quadrant: in Fig. 12.5 they are drawn in the first quadrant merely for convenience.

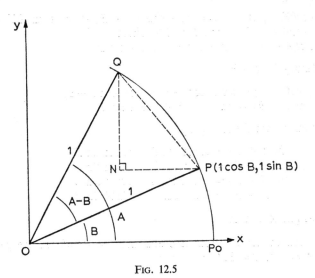

FIG. 12.5

From $\triangle$QNP $QP^2 = QN^2 + NP^2$
i.e. $QP^2 = (\cos B - \cos A)^2 + (\sin B - \sin A)^2$

and this result is true no matter which quadrants Q and P inhabit.

Thus $QP^2 = \cos^2 B - 2 \cos B \cos A + \cos^2 A + \sin^2 B$
          $- 2 \sin A \sin B + \sin^2 A$
But $\cos^2 B + \sin^2 B = 1$ and $\cos^2 A + \sin^2 A = 1$
$\therefore$ $QP^2 = 2 - 2 \cos A \cos B - 2 \sin A . \sin B$ . . .   (i)

Applying the cosine formula to $\triangle$OQP

$$QP^2 = 1^2 + 1^2 - 2 . 1 . 1 \cos \angle QOP$$

and $\cos \angle QOP = \cos(A - B)$ no matter which quadrants Q and P inhabit.

i.e. $QP^2 = 2 - 2 \cos(A - B)$ . . . .   (ii)

Equating the expressions for QP² from (i) and (ii),

$$2 - 2\cos(A - B) = 2 - 2\cos A \cos B - 2\sin A \sin B$$

Cancelling the 2 and dividing both sides by −2,

$$\cos(A - B) = \cos A \cos B + \sin A \sin B$$

Replacing B by −B,

$$\cos(A + B) = \cos A \cos(-B) + \sin A \sin(-B)$$

i.e. $\cos(A + B) = \cos A \cos B - \sin A \sin B$

Replacing B by 90° − B,

$$\cos(90° + A - B) = \cos A \cos(90° - B) - \sin A \sin(90° - B)$$

i.e. $-\sin(A - B) = \cos A \sin B - \sin A \cos B$

i.e. $\sin(A - B) = \sin A \cos B - \cos A \sin B$

Replacing B by −B,

$$\sin(A + B) = \sin A \cos(-B) - \cos A \sin(-B)$$

i.e. $\sin(A + B) = \sin A \cos B + \cos A \sin B$

We thus obtain the following:

$$\sin(A + B) = \sin A \cos B + \cos A \sin B$$
$$\sin(A - B) = \sin A \cos B - \cos A \sin B$$
$$\cos(A + B) = \cos A \cos B - \sin A \sin B$$
$$\cos(A - B) = \cos A \cos B + \sin A \sin B$$

for all angles A and B.

The truth of the addition theorems may be tested for the particular values $A = 60°$, $B = 30°$ (Fig. 12.6).

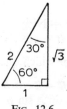

FIG. 12.6

$$\sin(60° + 30°) = \sin 90° = 1,$$

$$\sin 60° \cos 30° + \cos 60° \sin 30 = \frac{\sqrt{3}}{2} \cdot \frac{\sqrt{3}}{2} + \frac{1}{2} \cdot \frac{1}{2}$$

$$= \frac{3}{4} + \frac{1}{4} = 1$$

Similarly,

$$\sin(60° - 30°) = \sin 30° = \frac{1}{2},$$

$$\sin 60° \cos 30° - \cos 60° \sin 30° = \frac{3}{4} - \frac{1}{4} = \frac{1}{2}$$

$$\cos(60° + 30°) = \cos 90° = 0,$$

$$\cos 60° \cos 30° - \sin 60° \sin 30° = \frac{1}{2} \cdot \frac{\sqrt{3}}{2} - \frac{\sqrt{3}}{2} \cdot \frac{1}{2} = 0$$

Similarly,

$$\cos(60° - 30°) = \cos 30° = \frac{\sqrt{3}}{2},$$

$$\cos 60° \cos 30° + \sin 60° \sin 30° = \frac{\sqrt{3}}{4} + \frac{\sqrt{3}}{4} = \frac{\sqrt{3}}{2}$$

To express tan $(A + B)$ and tan $(A - B)$ in terms of tan A and tan B
For acute angles, $\tan(A + B) = \dfrac{\sin(A + B)}{\cos(A + B)}$

$$= \frac{\sin A \cos B + \cos A \sin B}{\cos A \cos B - \sin A \sin B}$$

Now, divide numerator and denominator each by cos A cos B.

$$\tan(A + B) = \frac{\dfrac{\sin A \cos B}{\cos A \cos B} + \dfrac{\cos A \sin B}{\cos A \cos B}}{\dfrac{\cos A \cos B}{\cos A \cos B} - \dfrac{\sin A \sin B}{\cos A \cos B}}$$

$$= \frac{\dfrac{\sin A}{\cos A} + \dfrac{\sin B}{\cos B}}{1 - \dfrac{\sin A}{\cos A} \cdot \dfrac{\sin B}{\cos B}}$$

$$\therefore \quad \tan(A + B) = \frac{\tan A + \tan B}{1 - \tan A \tan B}$$

and $\tan(A - B) = \dfrac{\tan A - \tan B}{1 + \tan A \tan B}$

The latter will easily be obtained from the first by the change of sign, i.e. by replacing B by $-B$, and using $\tan(-B) = -\tan B$. These formulae are valid for all angles which are not an odd multiple of 90°.

Examples
12.1 If sin $A = 0.5$ and cos $B = 0.3$, find the value of sin$(A + B)$ and of cos$(A - B)$ without evaluating the acute angles A and B.

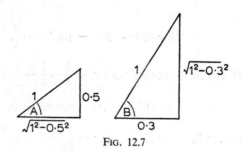

Fig. 12.7

Since $\sin(A + B) = \sin A \cos B + \cos A \sin B$
and $\cos(A - B) = \cos A \cos B + \sin A \sin B$

we must first find the values of $\cos A$ and $\sin B$.

Drawing a $\triangle$ in which A represents the angle, and with hypotenuse and perpendicular 1 and 0·5 as shown, Fig. 12.7, we see that the length of the base is $\sqrt{(1^2 - 0\cdot5^2)} = 0\cdot866$

$$\therefore \quad \cos A = 0\cdot866$$

Similarly,

$$\sin B = \sqrt{(1^2 - 0\cdot3^2)} = 0\cdot9539$$
$$\therefore \quad \sin(A + B) = (0\cdot5 \times 0\cdot3) + (0\cdot866 \times 0\cdot9539)$$
$$= 0\cdot15 + 0\cdot8260 = 0\cdot9760$$
and
$$\cos(A - B) = (0\cdot866 \times 0\cdot3) + (0\cdot5 \times 0\cdot9539)$$
$$= 0\cdot2598 + 0\cdot4770 = 0\cdot7368$$

12.2 Given that $\tan A = 0\cdot7$ and $\tan(A - B) = 0\cdot3$, find $\tan B$.

$$\tan(A - B) = \frac{\tan A - \tan B}{1 + \tan A \tan B}$$

$$\therefore \quad 0\cdot3 = \frac{0\cdot7 - \tan B}{1 + 0\cdot7 \tan B}$$

$$\therefore \quad (0\cdot3)(1 + 0\cdot7 \tan B) = 0\cdot7 - \tan B$$
$$\therefore \quad 0\cdot3 + 0\cdot21 \tan B = 0\cdot7 - \tan B$$
$$\therefore \quad \tan B + 0\cdot21 \tan B = 0\cdot7 - 0\cdot3$$
$$\therefore \quad 1\cdot21 \tan B = 0\cdot4$$
$$\therefore \quad \tan B = \frac{0\cdot4}{1\cdot21}$$
$$\therefore \quad \tan B = 0\cdot3306$$

A very useful application of these identities is found in electrical and mechanical problems, where it is sometimes necessary to change an expression of such a form as $a \sin pt + b \cos pt$ into the form $R \sin(pt + c)$, where a, b and p are constants, and c is an angle.

Since

$$R \sin(pt + c) = R(\sin pt \cos c + \cos pt \sin c)$$
$$= R \sin pt \cos c + R \cos pt \sin c$$

if $a \sin pt + b \cos pt$ is to be identical with $R(\sin pt + c)$, then $R \sin pt \cos c + R \cos pt \sin c$ must be identical with

$$a \sin pt + b \cos pt$$

∴ the coefficients of $\sin pt$ in both expressions must be equal, and similarly the coefficients of $\cos pt$.

∴ $R \cos c = a$

and $R \sin c = b$

By division, $\dfrac{R \sin c}{R \cos c} = \tan c = \dfrac{b}{a}$

By squaring and adding,

$$R^2(\cos^2 c + \sin^2 c) = a^2 + b^2$$
$$\therefore \quad R = \surd(a^2 + b^2), \text{ since } \cos^2 c + \sin^2 c = 1,$$
$$\therefore \quad a \sin pt + b \cos pt = R \sin(pt + c)$$

where $R = \surd(a^2 + b^2)$ and $\tan c = b/a$, i.e. $c = \arctan b/a$.

If b/a is positive, c may be taken as an acute angle in the first quadrant. If b/a is negative, c is taken in the fourth quadrant, as in Example 12.2.

The method is illustrated for pt and $\arctan b/a$ acute angles in the following diagram (Fig. 12.8).

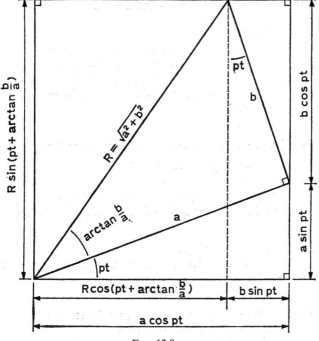

Fɪɢ. 12.8

Comparing lengths vertically,

$$a \sin pt + b \cos pt = R \sin(pt + \arctan b/a)$$

Also, comparing lengths horizontally,

$a \cos pt - b \sin pt = R \cos(pt + \arctan b/a)$, where $R = \sqrt{(a^2 + b^2)}$

Maximum and minimum values of $a \sin pt$, $b \cos pt$ as t varies

Since $a \sin pt + b \cos pt = \sqrt{(a^2 + b^2)} \sin(pt + \arctan b/a)$, the maximum value will be attained when $\sin(pt + \arctan b/a)$ reaches its maximum value of $+1$, and will be then $\sqrt{(a^2 + b^2)}$. The minimum value will be attained when $\sin(pt + \arctan b/a)$ reaches its minimum value of -1, and will be then $-\sqrt{(a^2 + b^2)}$.

Examples

12.3 Change $9\cdot2 \sin 3t + 8\cdot4 \cos 3t$ into the form $R \sin(3t + c)$. State the maximum and minimum values of $9\cdot2 \sin 3t + 8\cdot4 \cos 3t$, and the corresponding smallest positive values of t.

Here
$$\begin{aligned} R &= \sqrt{(9\cdot2^2 + 8\cdot4^2)} \\ &= \sqrt{(155\cdot2)} \\ &= 12\cdot46 \end{aligned}$$

Also, $\tan c = b/a$, where $b = 8\cdot4$ and $a = 9\cdot2$
$$= \frac{8\cdot4}{9\cdot2} = 0\cdot9131$$

$\therefore$ $c = 42° 24'$, or (as is more usual) $0\cdot74$ radian.

$\therefore$ $9\cdot2 \sin 3t + 8\cdot4 \cos 3t = 12\cdot46 \sin(3t + 0\cdot74)$ and has a maximum value of $12\cdot46$ when $3t + 0\cdot74 = \pi/2 \simeq 1\cdot57$; i.e. $3t = 0\cdot83$, $t = 0\cdot28$. The minimum value of $-12\cdot46$ occurs when $3t + 0\cdot74 = 3\pi/2 \simeq 4\cdot71$; i.e. $3t = 3\cdot97$, $t = 1\cdot32$.

12.4 Express $5 \sin 2\pi ft - 3 \cos 2\pi ft$ in the form $A \sin(2\pi ft + g)$.

Here $A = \sqrt{\{5^2 + (-3)^2\}} = 5\cdot831$
and $\tan g = -\frac{3}{5}$
$\therefore$ $g = -30° 58' = -0\cdot5405$ radian (using smallest value)
$\therefore$ $5 \sin 2\pi ft - 3 \cos 2\pi ft = 5\cdot831 \sin(2\pi ft - 0\cdot5405)$.

Ratios of multiple and sub-multiple angles

Perhaps the most useful applications of the previous addition formulae occur when $A = B$.

Then

$$\sin(A + B) = \sin A \cos B + \cos A \sin B \text{ becomes}$$
$$\sin 2A = \sin A \cos A + \cos A \sin A$$
$\therefore$ $$\mathbf{\sin 2A = 2 \sin A \cos A}$$

Similarly,

$$\cos(A + B) = \cos A \cos B - \sin A \sin B \quad \text{becomes}$$
$$\cos 2A = \cos A \cos A - \sin A \sin A$$
$$\therefore \quad \mathbf{\cos 2A = \cos^2 A - \sin^2 A}$$

Also,

$$\tan(A + B) = \frac{\tan A + \tan B}{1 - \tan A \tan B} \quad \text{becomes}$$

$$\tan 2A = \frac{\tan A + \tan A}{1 - \tan A \tan A}$$

$$\therefore \quad \mathbf{\tan 2A = \frac{2 \tan A}{1 - \tan^2 A}}$$

An important development of $\cos 2A$ is met with when its value, $\cos^2 A - \sin^2 A$, is coupled with

$$\cos^2 A + \sin^2 A = 1$$

For
$$\cos 2A = \cos^2 A - \sin^2 A$$
and
$$1 = \cos^2 A + \sin^2 A$$

which, by addition, give

$$\cos 2A + 1 = 2 \cos^2 A$$
$$\therefore \quad \cos 2A = 2 \cos^2 A - 1$$

Also, since $2 \cos^2 A = 1 + \cos 2A$

$$\cos^2 A = \tfrac{1}{2}(1 + \cos 2A)^*$$

and, by subtraction,

$$\cos 2A - 1 = -2 \sin^2 A$$
$$\therefore \quad \cos 2A = 1 - 2 \sin^2 A$$

Also, since
$$2 \sin^2 A = 1 - \cos 2A,$$

$$\sin^2 A = \tfrac{1}{2}(1 - \cos 2A)^*$$

If $2A$ be written as θ,

then
$$A = \theta/2$$
$$\therefore \quad \sin \theta = 2 \sin \theta/2 \cos \theta/2$$
$$\cos \theta = \cos^2 \theta/2 - \sin^2 \theta/2$$
$$\tan \theta = \frac{2 \tan \theta/2}{1 - \tan^2 \theta/2}$$

by substitution in the previous formulae.

---

* The starred formulae are important in later work on integration.

To express sin θ, cos θ and tan θ in terms of *t*, where *t* = tan θ/2

Since $\tan \theta = \dfrac{2 \tan \theta/2}{1 - \tan^2\theta/2}$, putting $\tan \theta/2 = t$,

$$\tan \theta = \frac{2t}{1 - t^2}$$

By constructing the appropriate right-angled triangle, the trigonometrical ratios of θ can be expressed in terms of *t*.

$$AB^2 = (1 - t^2)^2 + (2t)^2 = 1 - 2t^2 + t^4 + 4t^2$$
$$= 1 + 2t^2 + t^4 = (1 + t^2)^2$$
$$AB = 1 + t^2 \quad \text{(see Fig. 12.9)}$$

$$\therefore \quad \sin \theta = \frac{2t}{1 + t^2}$$

$$\cos \theta = \frac{1 - t^2}{1 + t^2}$$

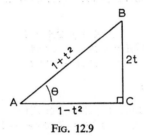

FIG. 12.9

Alternatively, $\quad \sin \theta = 2 \sin \theta/2 \cos \theta/2 = \dfrac{2 \sin \theta/2}{\cos \theta/2} \cdot \cos^2 \theta/2$

$$= \frac{2 \tan \theta/2}{\sec^2\theta/2}$$

$$= \frac{2 \tan \theta/2}{1 + \tan^2\theta/2} = \frac{2t}{1 + t^2}$$

$$\cos \theta = \cos^2\theta/2 - \sin^2\theta/2 = \frac{\dfrac{\cos^2\theta/2}{\cos^2\theta/2} - \dfrac{\sin^2\theta/2}{\cos^2\theta/2}}{\dfrac{1}{\cos^2\theta/2}}$$

$$= \frac{1 - \tan^2\theta/2}{\sec^2\theta/2}$$

$$= \frac{1 - \tan^2\theta/2}{1 + \tan^2\theta/2} = \frac{1 - t^2}{1 + t^2}$$

By inverting these results,

$$\cot \theta = \frac{1 - t^2}{2t}, \quad \sec \theta = \frac{1 + t^2}{1 - t^2} \quad \text{and} \quad \operatorname{cosec} \theta = \frac{1 + t^2}{2t}$$

Using these transformations, a trigonometrical equation (or, in later work, a trigonometrical integrand) may be turned into an algebraic form.

Examples

12.5 Solve the equation $2 \sin \theta + \cos \theta = 1.866$ for values of θ between $0°$ and $180°$.

Since $2 \sin \theta + \cos \theta = 1.866$, using triangle ABC (Fig. 12.10)

$$2 \frac{2t}{1 + t^2} + \frac{1 - t^2}{1 + t^2} = 1.866$$
$$\therefore \quad 4t + 1 - t^2 = 1.866(1 + t^2)$$
$$\therefore \quad 0 = 2.866t^2 - 4t + 0.866$$
$$\therefore \quad t = \frac{+4 \pm \sqrt{(16 - 4 \times 2.866 \times 0.866)}}{2 \times 2.866}$$
$$= 1.128 \quad \text{or} \quad 0.268$$
$$\therefore \quad \tan \theta/2 = 1.128 \quad \text{or} \quad \tan \theta/2 = 0.268$$
$$\therefore \quad \theta/2 = 48° 27' \quad \text{or} \quad \theta/2 = 15°$$
$$\therefore \quad \theta = 96° 54' \quad \text{or} \quad \theta = 30°$$

N.B., if θ is between $0°$ and $180°$, $\theta/2$ is between $0°$ and $90°$.

Check

$$2 \sin(96° 54') + \cos(96° 54') = 2(0.9928) - 0.1201 = 1.8655$$
$$2 \sin 30° + \cos 30° = 2(0.5) + 0.866 = 1.866$$

The solutions are correct to three decimal places.

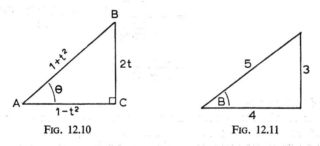

FIG. 12.10   FIG. 12.11

12.6 If $\cos B = 0.8$, find $\sin 2B$, $\cos 2B$ and $\tan 2B$ without evaluating the acute angle B.

Since $\cos B = \frac{4}{5}$ $\qquad \sin B = \frac{3}{5}$ (see Fig. 12.11)

$$\therefore \quad \sin 2B = 2 \sin B \cos B$$
$$= 2 \times \tfrac{3}{5} \times \tfrac{4}{5} = \tfrac{24}{25}$$
$$= 0.96$$
$$\cos 2B = \cos^2 B - \sin^2 B$$
$$= \tfrac{16}{25} - \tfrac{9}{25} = \tfrac{7}{25}$$
$$= 0.28$$
$$\tan 2B = \frac{\sin 2B}{\cos 2B} = \frac{0.96}{0.28}$$
$$= 3.428$$

12.7 Prove that $\dfrac{\sin 2A}{1 + \cos 2A} = \tan A$

$$\frac{\sin 2A}{1 + \cos 2A} = \frac{2 \sin A \cos A}{1 + (2 \cos^2 A - 1)}$$

$$= \frac{2 \sin A \cos A}{2 \cos^2 A}$$

$$= \frac{\sin A}{\cos A} = \tan A.$$

12.8 Prove that $\dfrac{\cos 2\theta}{1 + \sin 2\theta} = \dfrac{1 - \tan \theta}{1 + \tan \theta}.$

$$\frac{\cos 2\theta}{1 + \sin 2\theta} = \frac{\cos^2\theta - \sin^2\theta}{1 + 2 \sin \theta \cos \theta}$$

$$= \frac{(\cos \theta + \sin \theta)(\cos \theta - \sin \theta)}{\sin^2\theta + \cos^2\theta + 2 \sin \theta \cos \theta}$$

$$= \frac{(\cos \theta + \sin \theta)(\cos \theta - \sin \theta)}{(\sin \theta + \cos \theta)(\sin \theta + \cos \theta)}$$

$$= \frac{\cos \theta - \sin \theta}{\cos \theta + \sin \theta}$$

$$= \frac{1 - \dfrac{\sin \theta}{\cos \theta}}{1 + \dfrac{\sin \theta}{\cos \theta}} \quad \text{(by dividing each term by } \cos \theta\text{)}$$

$$= \frac{1 - \tan \theta}{1 + \tan \theta}.$$

The Product formulae

By adding or subtracting the addition theorems, we may obtain formulae which are known as the *product formulae*, since they express the product of trigonometrical ratios in terms of the sum or difference of trigonometrical ratios.

$$\sin(A + B) = \sin A \cos B + \cos A \sin B$$
$$\sin(A - B) = \sin A \cos B - \cos A \sin B$$

Adding, $\sin(A + B) + \sin(A - B) = 2 \sin A \cos B$ . . (i)

Subtracting, $\sin(A + B) - \sin(A - B) = 2 \cos A \sin B$ . . (ii)

$$\cos(A + B) = \cos A \cos B - \sin A \sin B$$
$$\cos(A - B) = \cos A \cos B + \sin A \sin B$$

Adding, $\cos(A - B) + \cos(A + B) = 2 \cos A \cos B$ . . (iii)

Subtracting, $\cos(A - B) - \cos(A + B) = -2 \sin A \sin B$ . . (iv)

(i), (ii), (iii) and (iv) are known as the product formulae, and are useful in later work on integration.

Example

12.9 Express the following products as sums or differences: (a) $\sin 6x \cos 4x$, (b) $\cos 8\theta \cos 4\theta$, (c) $\sin 10\phi \sin 8\phi$.

(a)
$$2 \sin A \cos B = \sin(A + B) + \sin(A - B)$$
$$\therefore \quad 2 \sin 6x \cos 4x = \sin(6x + 4x) + \sin(6x - 4x)$$
$$\therefore \quad \sin 6x \cos 4x = \tfrac{1}{2}(\sin 10x + \sin 2x)$$

(b)
$$2 \cos A \cos B = \cos(A + B) + \cos(A - B)$$
$$\therefore \quad 2 \cos 8\theta \cos 4\theta = \cos(8\theta + 4\theta) + \cos(8\theta - 4\theta)$$
$$\therefore \quad \cos 8\theta \cos 4\theta = \tfrac{1}{2}(\cos 12\theta + \cos 4\theta)$$

(c)
$$-2 \sin A \sin B = \cos(A + B) - \cos(A - B)$$
$$\therefore \quad -2 \sin 10\phi \sin 8\phi = \cos(10\phi + 8\phi) - \cos(10\phi - 8\phi)$$
$$\therefore \quad \sin 10\phi \sin 8\phi = -\tfrac{1}{2}(\cos 18\phi - \cos 2\phi)$$

The sum and difference rules

If in the product formulae we let $A + B = C$, and $A - B = D$, then, adding, we find that $2A = C + D$, or $A = \dfrac{C + D}{2}$, and, subtracting, we find that $2B = C - D$, or $B = \dfrac{C - D}{2}$.

Since
$$\sin(A + B) + \sin(A - B) = 2 \sin A \cos B$$
$$\sin(A + B) - \sin(A - B) = 2 \cos A \cos B$$
$$\cos(A + B) + \cos(A - B) = 2 \cos A \cos B$$
$$\cos(A + B) - \cos(A - B) = 2 \sin A \sin B$$

replacing $A + B$ by C, $A - B$ by D, A by $\dfrac{C + D}{2}$, and B by $\dfrac{C - D}{2}$, we have

$$\sin C + \sin D = 2 \sin \frac{C + D}{2} \cos \frac{C - D}{2}. \quad . \quad . \quad \text{(i)}$$

$$\sin C - \sin D = 2 \cos \frac{C + D}{2} \sin \frac{C - D}{2}. \quad . \quad . \quad \text{(ii)}$$

$$\cos C + \cos D = 2 \cos \frac{C + D}{2} \cos \frac{C - D}{2} \quad . \quad . \quad \text{(iii)}$$

$$\cos C - \cos D = -2 \sin \frac{C + D}{2} \sin \frac{C - D}{2} \quad . \quad . \quad \text{(iv)}$$

(i), (ii), (iii) and (iv) are known as the sum and difference rules. They are usually remembered by mentally repeating

'sin' + 'sin' = '2 sin $\tfrac{1}{2}$ sum . cos $\tfrac{1}{2}$ diff'
'sin' − 'sin' = '2 cos $\tfrac{1}{2}$ sum . sin $\tfrac{1}{2}$ diff'
'cos' + 'cos' = '2 cos $\tfrac{1}{2}$ sum . cos $\tfrac{1}{2}$ diff'
'cos' − 'cos' = '−2 sin $\tfrac{1}{2}$ sum . sin $\tfrac{1}{2}$ diff'.

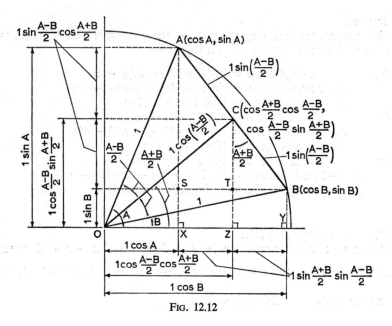

Fig. 12.12

A demonstration of these formulae in the particular case where all the angles are acute is given in Fig. 12.12.

Since $AC = CB$, $XZ = ZY$ and $CT = \frac{1}{2}AS$, OZ is the mean of OX and OY. CZ is the mean of AX and BY.

Comparing lengths along OY,

$$1 \cos A + 2 \sin \frac{A-B}{2} \sin \frac{A+B}{2} = 1 \cos B$$

or

$$\cos A - \cos B = -2 \sin \frac{A-B}{2} \sin \frac{A+B}{2}$$

$$\frac{1 \cos A + 1 \cos B}{2} = 1 \cos \frac{A-B}{2} \cos \frac{A+B}{2}$$

or

$$\cos A + \cos B = 2 \cos \frac{A+B}{2} \cos \frac{A-B}{2}$$

Comparing lengths perpendicular to OY,

$$\sin A - \sin B = 2 \cos \frac{A+B}{2} \sin \frac{A-B}{2}$$

$$\frac{1 \sin A + 1 \sin B}{2} = 1 \cos \frac{A-B}{2} \sin \frac{A+B}{2}$$

i.e.

$$\sin A + \sin B = 2 \sin \frac{A+B}{2} \cos \frac{A-B}{2}$$

Examples
12.10 Express the following sums or differences as products: (*a*) cos 10θ − cos 6θ, (*b*) cos 6*x* + cos 4*x*, (*c*) sin 5*y* + sin 3*y*, (*d*) sin(*x* + δ*x*) − sin *x*.

(*a*) $\cos 10\theta - \cos 6\theta = -2 \sin \dfrac{10\theta + 6\theta}{2} \sin \dfrac{10\theta - 6\theta}{2} = -2 \sin 8\theta \sin 2\theta$

(*b*) $\cos 6x + \cos 4x = 2 \cos \dfrac{6x + 4x}{2} \cos \dfrac{6x - 4x}{2} = 2 \cos 5x \cos x$

(*c*) $\sin 5y + \sin 3y = 2 \sin \dfrac{5y + 3y}{2} \cos \dfrac{5y - 3y}{2} = 2 \sin 4y \cos y$

(*d*) $\sin(x + \delta x) - \sin x = 2 \cos \dfrac{x + \delta x + x}{2} \sin \dfrac{x + \delta x - x}{2}$

$$= 2 \cos\left(x + \frac{\delta x}{2}\right)\left(\sin \frac{\delta x}{2}\right)$$

12.11 Solve the equation sin 5*x* + sin *x* = 0 for values of *x* between 0° and 90° inclusive.

If sin 5*x* + sin *x* = 0, then $2 \sin \dfrac{5x + x}{2} \cos \dfrac{5x - x}{2} = 0$

i.e.   2 sin 3*x* cos 2*x* = 0
∴   either   sin 3*x* = 0   or   cos 2*x* = 0.
If *x* is between 0° and 90° inclusive, then 2*x* is between 0° and 180° inclusive and 3*x* is between 0° and 270° inclusive.
∴   3*x* = 0° or 180° and 2*x* = 90° i.e. *x* = 0°, 45° or 60°.
Check
When *x* = 0°,
        L.H.S. = sin 0 + sin 0 = 0 = R.H.S.

when *x* = 45°
        $\text{L.H.S.} = \sin 225° + \sin 45° = -\dfrac{1}{\sqrt{2}} + \dfrac{1}{\sqrt{2}} = 0 = \text{R.H.S.}$

when *x* = 60°,
        $\text{L.H.S.} = \sin 300° + \sin 60° = -\dfrac{\sqrt{3}}{2} + \dfrac{\sqrt{3}}{2} = 0 = \text{R.H.S.}$

Further trigonometrical equations

Examples
12.12 Given 2·5 cos 2θ + 3 cos θ = 0·25, find all positive values of θ up to 360°.

Here there are involved two different functions, cos 2θ and cos θ. But, since $\cos 2\theta = 2\cos^2\theta - 1$, we can reduce the equation to a quadratic in $\cos\theta$, and so solve it.

$$2\cdot5 \cos 2\theta + 3 \cos \theta = 0\cdot25$$
$$2\cdot5(2 \cos^2\theta - 1) + 3 \cos \theta = 0\cdot25$$
$$5 \cos^2\theta - 2\cdot5 + 3 \cos \theta = 0\cdot25$$
$$\therefore \quad 5 \cos^2\theta + 3 \cos \theta - 2\cdot75 = 0$$

$$\therefore \quad \cos \theta = \frac{-3 \pm \sqrt{(9 + 55)}}{10}$$

$$= \frac{-3 \pm 8}{10}$$

$$= -\tfrac{11}{10} \quad \text{or} \quad \tfrac{5}{10}$$

$$= -\tfrac{11}{10} \quad \text{or} \quad \tfrac{1}{2}$$

But it is the values of θ which are required, and not merely those of $\cos\theta$. Therefore, we must find all values of θ up to 360°, such that $\cos\theta = -\tfrac{11}{10}$ or $\tfrac{1}{2}$. Since no angle has a cosine value greater than 1 or less than -1, $\cos\theta = -\tfrac{11}{10}$ is inadmissible.

If $\cos\theta = \tfrac{1}{2}$, $\theta = 60°$, or the corresponding angle in any quadrant where the cosine value is positive.

This is found only in the fourth quadrant, where $\cos(360° - \theta) = \cos\theta$.

$$\therefore \quad \theta = 60° \quad \text{or} \quad 300°.$$

12.13 Solve $\cos 2\theta = 1 - 6 \cos^2\theta$ for all positive values of θ up to 360°.

Substituting $2\cos^2\theta - 1$ for $\cos 2\theta$,

$$2 \cos^2\theta - 1 = 1 - 6 \cos^2\theta$$
$$\therefore \quad 8 \cos^2\theta = 2$$
$$\therefore \quad \cos^2\theta = \tfrac{1}{4}$$
$$\therefore \quad \cos \theta = \pm\tfrac{1}{2}$$

If $\cos\theta = +\tfrac{1}{2}$, $\theta = 60°$ or 300°.
If $\cos\theta = -\tfrac{1}{2}$, $\theta = 120°$ or 240°
$$\therefore \quad \theta = 60°, 120°, 240° \text{ or } 300°$$

Solution of the equation $a \sin\theta + b \cos\theta = c$
The method is illustrated in the following example.

12.14 (a) Express $2 \sin\theta + \cos\theta$ in the form $R \sin(\theta + \alpha)$.
(b) Solve the equation $2 \sin\theta + \cos\theta = 1\cdot866$ (previously solved by the substitution $t = \tan\theta/2$) for values of θ between 0° and 180°.

(a) Let $2 \sin\theta + \cos\theta = R \sin(\theta + \alpha)$
$$= R \sin\theta \cos\alpha + R \cos\theta \sin\alpha$$
Then
$$2 = R \cos\alpha \quad \text{(i)}$$
$$1 = R \sin\alpha \quad \text{(ii)}$$
$$\therefore \quad \tfrac{1}{2} = \frac{R \sin\alpha}{R \cos\alpha} = \tan\alpha$$
$$\therefore \quad \alpha = 26° 34'$$

also	$2^2 + 1^2 = R^2\cos^2\alpha + R^2\sin^2\alpha$
i.e.	$5 = R^2$ and $R = 2\cdot236$
i.e.	$2 \sin \theta - \cos \theta = 2\cdot236 \sin(\theta + 26° 34')$

(b) Using the above result, the equation becomes

	$2\cdot236 \sin(\theta + 26° 34') = 1\cdot866$
i.e.	$\sin(\theta + 26° 34') = 0\cdot8345$
i.e.	$\theta + 26° 34' = 56° 34'$ or $123° 26'$
i.e.	$\theta = 30° 0'$ or $96° 52'$

It will be noted that in the previous working R and α are eliminated in turn by dividing the equations to find α, and squaring and adding the equations to find R. A slight variation is as follows

$$R \cos \alpha = 2 \quad \therefore \quad \cos \alpha = \frac{2}{R} \quad \text{(iii)}$$

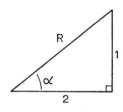

FIG. 12.13

also $\qquad R \sin \alpha = 1$ and $\sin \alpha = \dfrac{1}{R}$ (iv)

If a right-angled triangle is drawn to conform to (iii) and (iv), then $\tan \alpha = \frac{1}{2}$ and $\alpha = 26° 34'$, as before. Also $R^2 = 1^2 + 2^2 = 5$ and $R = 2\cdot236$, as before.

EXERCISE 12.1

In questions 1–10, the angle must not be evaluated.

1 If $\sin A = 0\cdot52$ and $\cos B = 0\cdot78$, find the values of $\sin(A + B)$ and $\cos(A + B)$.

2 If $\sin A = 0\cdot72$ and $\cos B = 0\cdot91$, find $\cos(A - B)$.

3 If $\sin B = 0\cdot23$ and $\cos A = 0\cdot309$, find $\sin(A - B)$.

4 If $\cos B = 0\cdot32$ and $\sin A = 0\cdot71$, find $\sin(A + B)$ and $\tan(A + B)$.

5 If $\tan A = 1\cdot2$ and $\tan B = 0\cdot4$, find $\tan(A + B)$ and $\tan(A - B)$.

6 If $\sin A = \frac{60}{61}$ and $\cos B = \frac{40}{41}$, find $\cos(A - B)$.

7 If $\tan A = 1\cdot782$ and $\tan(A - B) = 0\cdot7054$, find $\tan B$.

8 Given that $\sin A = \frac{3}{5}$, find $\sin 2A$, $\cos 2A$ and $\tan 2A$.

9 If $\cos B = 0\cdot66$, find $\sin 2B$ and $\cos 2B$.

10 Find $\sin A$ and $\sin A/2$, if $\sin 2A = 0\cdot97$.

11 Express $3·6 \cos 6t + 2·7 \sin 6t$ in the form $M \sin(6t + c)$.

12 Express $30 \sin(2\pi nt) + 25 \cos(2\pi nt)$ in the form $A \sin(2\pi nt + g)$.

13 Express $5 \sin 4t - 3 \cos 4t$ in the form $A \sin(4t + g)$, using the smallest possible value of g in radians.

Prove the following identities for all values of the given angle.

14 $\cos(A + B) + \cos(A - B) = 2 \cos A \cos B$

15 $\sin(A + B) - \sin(A - B) = 2 \cos A \sin B$

16 $\dfrac{\sin(A + B)}{\cos A \cos B} = \tan A + \tan B$

17 $\cot A + \tan B = \dfrac{\cos(A - B)}{\sin A \cos B}$

18 $\cos(A + 45°) + \sin(A - 45°) = 0$

19 $\dfrac{1 - \cos x}{1 + \cos x} = \tan^2\dfrac{x}{2}$

20 $\dfrac{2 \operatorname{cosec} 2\theta}{\operatorname{cosec} \theta} = \sec \theta$

21 $\dfrac{2 \tan A}{1 + \tan^2 A} = \sin 2A$

22 $\dfrac{1 + \tan^2\theta}{1 - \tan^2\theta} = \sec 2\theta$

23 $\dfrac{\sin A}{1 - \cos A} = \cot \dfrac{A}{2}$

24 $\dfrac{1 + \sec \theta}{\sec \theta} = 2 \cos^2\dfrac{\theta}{2}$

25 $\dfrac{\sin^2 2A}{2 \cos^2 A} = 1 - \cos 2A$

26 $\cos^4 B - \sin^4 B = \cos 2B$

27 $\dfrac{\sin 4A}{\sin 2A} = 2 \cos 2A$

28 $\dfrac{\cos(A - 45°)}{\cos(A + 45°)} = \dfrac{1 + \sin 2A}{\cos 2A}$

29 $\sin 3A = 3 \sin A - 4 \sin^3 A$
(Start with $\sin 3A = \sin(2A + A)$.)

30 $\cos 3A = 4 \cos^3 A - 3 \cos A$

31 $\tan 3A = \dfrac{3 \tan A - \tan^3 A}{1 - 3 \tan^2 A}$

Solve the following equations for all values of the angle from $0°$ to $360°$.

32 $6 \sin \theta = \tan \theta$

33 $2 \tan^2\theta - 3 \tan \theta + 1 = 0$

34 $\cos \theta + \tan \theta = \sec \theta$

35 $5 \tan^2 x - \sec^2 x = 11$

36 $4 \cos \theta = 3 \tan \theta$
37 $4 \sin^2\theta - 3 \cos \theta = 1\cdot5$
38 $4 \sin^2\theta = \cos^2\theta$
39 $3 \cos^2 A + 5 \sin^2 A = 4$
40 $4 \sin^2 x + 5 \cos^2 x = 4\cdot25$
41 $\cos \theta - \sin \theta = 0\cdot8$
42 $4 \cos \theta = 3 \sec \theta$
43 $\tan x \operatorname{cosec} x = 5$
44 $\cot^2 x + \operatorname{cosec}^2 x = 3$
45 $\sin \theta + \sin 2\theta = 0$
46 $\sin 2\theta = \sin \theta$
47 $\cos \theta = \cos 2\theta$
48 $\cos 2\theta - \cos \theta = 2$

Express the following products as sums or differences.
49 (*a*) $\sin 10y \cos 6y$ (*b*) $2 \sin 6\theta \cos 2\theta$ (*c*) $7 \sin 10y \cos 4y$
50 (*a*) $\cos 10y \sin 6y$ (*b*) $2 \cos 5z \sin 3z$ (*c*) $5 \cos 9\theta \sin 5\theta$
51 (*a*) $\sin 8\phi \sin 4\phi$ (*b*) $2 \sin 7\theta \sin \theta$ (*c*) $3 \sin 11\phi \sin 7\phi$
52 (*a*) $\cos 8x \cos 6x$ (*b*) $2 \cos 3\theta \cos \theta$ (*c*) $8 \cos 5z \cos 3z$

Express the following sums or differences as products.
53 (*a*) $\sin 4\theta + \sin 2\theta$ (*b*) $\sin 5\phi + \sin 3\phi$ (*c*) $\sin 8y + \sin 6y$
54 (*a*) $\sin 4\theta - \sin 2\theta$ (*b*) $\sin 5\phi - \sin 3\phi$ (*c*) $\sin 8y - \sin 6y$
55 (*a*) $\cos 4\theta + \cos 2\theta$ (*b*) $\cos 5\theta + \cos 3\theta$ (*c*) $\cos 8y + \cos 6y$
56 (*a*) $\cos 4\theta - \cos 2\theta$ (*b*) $\cos 5\phi - \cos 3\phi$ (*c*) $\cos 8y - \cos 6y$

Using the results of examples 53–56, or otherwise, prove the following identities.

57 $\dfrac{\sin 4\theta - \sin 2\theta}{\cos 4\theta + \cos 2\theta} = \tan \theta$ **58** $\dfrac{\sin 5\phi + \sin 3\phi}{\sin 5\phi - \sin 3\phi} = \tan 4\phi \cot \phi$

59 $\dfrac{(\sin 8y + \sin 6y)(\cos 8y + \cos 6y)}{(\sin 8y - \sin 6y)(\cos 8y - \cos 6y)} = -\cot^2 y$

60 Verify that both sides of the identity in
(*a*) question 57 have the value 1 when $\theta = 45°$,
(*b*) question 58 have the value -3 when $\phi = 30°$,
(*c*) question 59 have the value -3 when $y = 30°$.

EXERCISE 12.2 (miscellaneous)

1 Figure 12.14 shows a double inverted-V guideway with truncated V's. All faces can be assumed to be inclined at 60° to the horizontal. Rollers of various exactly known diameters are available, also outside micrometers and a micrometer depth gauge. Work out a method for checking the correct spacing of the V's.

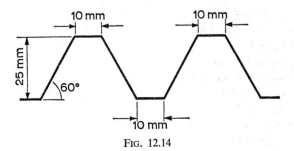

FIG. 12.14

2 A rectangular packing-case $2 \cdot 7$ m $\times$ 2 m is jammed in an alley between two vertical walls 3 m apart (Fig. 12.15).

Write down an equation for θ, and hence find θ and the height of the corner A above the floor.   COVENTRY

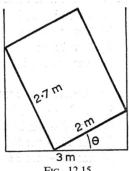

FIG. 12.15

3 (a) Write down the expansion of (i) $\sin(A + B)$, (ii) $\cos(A + B)$.

Hence show that $\dfrac{\sin 2A}{1 - \cos 2A} = \cot A$.

(b) Solve the equation $\sin 2\theta = \sin \theta$ for values of θ from $0°$ to $360°$.

(c) State the formula for the area of a triangle in terms of the sides.

The lengths of the sides of a triangle are 200 mm, 150 mm and 100 mm. Find the area of the triangle and the value of the largest angle (to the nearest degree).   HANDSWORTH

4 Find all the roots of the following equations which lie between $0°$ and $360°$:   (a) $\sin^2\theta - 2\cos^2\theta + 1 = 0$,   (b) $1 \cdot 5 \sin \theta + 2 \cdot 8 \cos \theta = 1 \cdot 5$.
(c) Solve the triangle ABC, where AB $= 67$ mm, BC $= 60$ mm and CA $= 40$ mm.   RUGBY

5 Figure 12.16 is the dimensional plan of a roof, one end splayed and the other end square. All slopes are $50°$ to the horizontal.

Calculate (a) the height of the roof,   (b) the roof area,   (c) the length of the ridge EF,   (d) the true length of the hip-rafter FC.

COVENTRY

6 If $5\cos^2 x - \sin^2 x + \cos x - 1 = 0$, find the value of x, giving all possible values between $0°$ and $360°$.

Write down the formula you would use to solve a triangle given (*a*) two sides and an included angle, (*b*) two sides and one angle (not included).

In the triangle ABC, CA $= 84$ mm, AB $= 102$ mm, $\angle$CAB $= 37°$. Calculate BC, $\angle$BCA, $\angle$ABC, and the area of the triangle.

WORCESTER

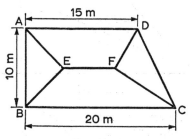

Fɪɢ. 12.16

7 (*a*) Write the equations (i) $3x - y + 5 = 0$, (ii) $2x + 3y - 1 = 0$ in the general form of the straight-line equation $y = mx + c$.

(*b*) Using the formula for $\tan(A - B)$, calculate the tangent of the angle between the two straight lines specified in part (*a*).

(*c*) Determine the rectangular coordinates of the point of intersection of these two lines.

(*d*) Give the polar coordinates of this point.   NCTEC

8 A triangle ABC is such that A $= 53°$, $a = 10·5$ m and $c = 12$ m. Show that two solutions for the triangle are possible, and give these solutions.

9 (*a*) Express $4·7 \sin 3\theta + 2·9 \cos 3\theta$ in the form $R \sin(3\theta + \alpha)$, and so find the two smallest positive values of θ for which $4·7 \sin 3\theta + 2·9 \cos 3\theta = 3·25$.

(*b*) Solve the equation $4 \sin \theta + 3 \cos 2\theta = 1·8$, giving all solutions between $0°$ and $360°$.   SW ESSEX

10 (*a*) State and prove the cosine formula for the solution of triangles.

(*b*) In a loosely jointed framework of steel rods in the form of a quadrilateral ABCD, the lengths of the rods are AB $= 4·93$ m, BC $= 5·76$ m, CD $= 3·89$ m and DA $= 6·14$ m. The framework is stiffened by a diagonal rod AC which is $8·23$ m long. Calculate the length of the other diagonal BD.   SW ESSEX

11 Using the cosine rule, or otherwise, find the length of AB in Fig. 12.17 to three significant figures, and its angle to the horizontal.

COVENTRY

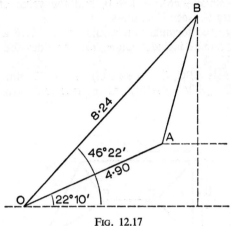

FIG. 12.17

12 (*a*) A chord AB of a circle, radius *r*, subtends an angle $\theta°$ at the centre O (see Fig. 12.18). Show that the chord bisects the sector AOB if $\sin \theta = \frac{1}{2}\theta$. Show that the angle AOB in degrees is 109° approximately.

(*b*) The currents in a three-phase system are proportional to $\sin \theta$, $\sin\left(\theta + \frac{2\pi}{3}\right)$, $\sin\left(\theta + \frac{4\pi}{3}\right)$, where θ depends on the position of the rotor. Prove that the sum of the currents is zero.

NUNEATON

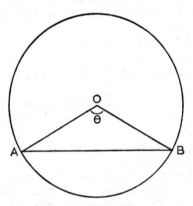

FIG. 12.18

13 The roof shown in plan in Fig. 12.19 has uniform pitch of 48° to the horizontal.

Calculate (*a*) the length of the common rafter, (*b*) the length of the hip rafter FB, (*c*) the length of ridge EF, (*d*) the total roof area.

COVENTRY

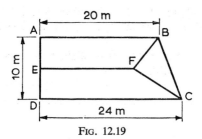

FIG. 12.19

14 A pin gauge of known length g units is employed to measure the diameter d units of a large hole into which it will enter freely. Whilst kept within a plane at right angles to the axis of the hole, the gauge is rocked from side to side, and its free end is found to have an amplitude of motion of $2a$ units (see Fig. 12.20).

Find an expression for d in terms of g and a. Also deduce a simple expression for a correction to be added to g to give d when a is small in comparison with g and d.

15 Express $35 \cos \theta - 12 \sin \theta$ in the form $R \cos(\theta + \alpha)$, stating the values of R and α.   NUNEATON

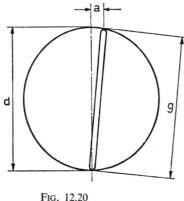

FIG. 12.20

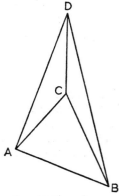

FIG. 12.21

16 ABC (Fig. 12.21) is the triangular base, and D the vertex of a pyramid. ABC is a horizontal plane and D is vertically above C. $\angle DAC = 48°$, $\angle CAB = 83°$, $\angle CBA = 53°$, and AB = 1·5 m. Calculate (a) the volume of the pyramid, and (b) the angle between planes DAB and ABC.   WEST RIDING

17 (a) Solve the following equations giving, in each case, all the solutions between $0°$ and $360°$: (i) $\sin x + \tan x = 0$, (ii) $\tan x - 2 \cot x - 1 = 0$.

(b) Simplify $\cos 60° + \sin 120° - \tan 275° - \sin 300°$.

(c) Calculate the height h in Fig. 12.22.   COVENTRY

18 (a) If $\tan \theta = -\frac{1}{3}$, and $\tan \phi = \frac{5}{12}$, where θ and ϕ are less than 180°, find the values of $\sin 2\theta$, $\cos 2\phi$ and $\cos(\theta + \phi)$ without using tables.

   (b) Solve the equation $4 \cos 2\theta - 2 \sin \theta - 1 = 0$ for values of θ between 0° and 360°. CHELTENHAM

19 In the triangle ABC, angle $A = 30°$, side $b = 100$ mm and side $c = 180$ mm. Calculate the angles B and C, the length of a, and the area of the triangle. ULCI

20 (a) Prove the identity $\dfrac{1 - \cos x}{1 + \cos x} = \tan^2 \dfrac{x}{2}$.

   (b) Solve the equation $3 \sin \theta = \tan \theta$ for values of θ between 0° and 360°.

   (c) Evaluate the following: (i) $\sin 150°$, (ii) $\cos 162°$, (iii) $\cos 240°$, (iv) $\tan 263°$, (v) $\tan 1064°$. HANDSWORTH

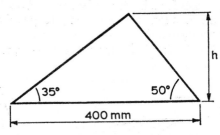

FIG. 12.22

21 A belt passes tightly around two pulleys without crossing. The pulleys have diameters of 500 mm and 250 mm respectively, and their centres are 1·2 m apart. Calculate: (a) the angle a straight portion of the belt makes with the line of centres, (b) the length of the two straight portions of the belt, (c) the total length of the belt.

WEST RIDING

22 Find the length of the open belt connecting wheels of radii 600 mm and 150 mm placed with their centres 1·8 m apart and in a common plane.

   If the larger wheel is rotating at 200 rev/min, find the speed of any point P on the rim of the smaller wheel in metres per second. (Take $\pi = 3·142$.) SUNDERLAND

23 A roof truss consists of five bars. Three of them, AB, BC, CA, form a triangle, right-angled at C and with AB horizontal. The bar CD is vertical, and DE makes an angle of 45° with the horizontal, D and E being points on AB and BC respectively. If angle $CAB = 50°$ and $AB = 5$ m, find the lengths of DB, CD and DE. SUNDERLAND

24 The diagram (Fig. 12.23) shows the side view of a window AB which is free to rotate about a horizontal axis at C.

The window is retained at an angle of 38° with the vertical by a string AD fastened to a small hook at D, which is 450 mm vertically below C. If AC = 225 mm, find (a) the length of string AD, (b) the angle the string makes with the vertical.

25 (a) A flagstaff PQ is due N of an observer A, and the angle of elevation of the top P is 18°. After walking 75 m due W to a point B, the observer finds that the flagstaff bears in a direction N 63° E. Find the height of the flagstaff.

   (b) Find the resultant of 3 forces of 12 newtons, 16 newtons and 20 newtons acting on a single point in directions due E, NE and due N respectively.   HANDSWORTH

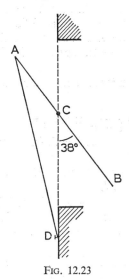

Fig. 12.23

26 (a) If $\sin(x + 60°) = a \sin x + b \cos x$, what are the values of a and b?
   (b) If $\tan A = \frac{3}{4}$, and A is an acute angle, find, without using tables, the values of $\sin 2A$, $\cos 2A$ and $\tan 2A$.

27 (a) Solve the following equations for values of θ between 0° and 360°:
   (i) $\sin^2\theta - 3 \cos^2\theta = 4 \sin \theta$, (ii) $\sin 3\theta = \sin \theta$.
   (b) If $\tan A = 0.5$, and $\tan(A + B) = 0.7$, find the value of $\tan B$.
       CHELTENHAM

28 In a triangle ABC, $a = 5$ m, $b = 3$ m and angle $C = 50°$. Calculate:
   (a) the remaining sides and angles, and the area of the triangle,
   (b) the radius of the circumcircle.   CHELTENHAM

29 (a) Draw the graph of $y = 5(\sin 3\theta + 90°)$ between 0° and 180°. State the values of θ for which $y = 0$, and the values of θ for which $y = 2.5$.
   (b) Solve the equation $2 \sec^2\theta = \frac{1}{2} \tan \theta + 2$ for values of θ between 0° and 360°.   HANDSWORTH

30 (a) Given that $\sin A = \frac{5}{13}$ and $\cos B = \frac{7}{25}$, where A and B are acute angles, evaluate without the use of tables   (i) $\sin(A + B)$, (ii) $\tan(A + B)$.

(b) Solve the equation $\sin(\theta - 40°) = 2\cos(\theta + 25°)$, giving the values of θ between $0°$ and $360°$.   RUGBY

31 (a) If $\sin A = \frac{1}{3}$, and $\cos B = \frac{1}{4}$, show, without using tables, that
$$\sin(A + B) = \tfrac{1}{12}(1 + 2\sqrt{30}).$$

(b) Use tables to find the values of $\sec 152°$ and $\tan(-490°)$.

                                                                SUNDERLAND

32 Two lighthouses A and B are situated 5 international nautical miles apart on a stretch of coastline running due north and south, A being to the north of B. At a certain time the bearings of A and B from a ship are observed to be N 60° W and S 75° W respectively. Half an hour later the lighthouse A is due west of the ship, and B is on a bearing S 25° W. Calculate the speed of the ship in knots.

                                                                SUNDERLAND

33 (a) Write down the identities for $\sin 2\theta$ and $\cos 2\theta$ in terms of $\sin \theta$ and $\cos \theta$.

(b) Prove that $\tan \theta = \dfrac{2t}{1 - t^2}$, if $\tan \dfrac{\theta}{2} = t$.

(c) If $R \sin \theta = 10$, and $R \cos \theta = 5$, find R and θ when R is positive, if θ is less than $90°$.   WEST RIDING

34 Find the angles θ between $0°$ and $360°$ which satisfy the equation $4{\cdot}2 \sin \theta + 5{\cdot}5 \cos \theta = 2{\cdot}7$.   COVENTRY

35 (a) If $\tan \alpha = \frac{5}{12}$ and $\tan \beta = -\frac{4}{3}$, and α and β are less than $180°$, find the values of $\sin 2\alpha$, $\cos 2\beta$ and $\sin(\alpha + \beta)$, without using tables.

(b) Solve the equation $2 \sin \theta = 4 \cos 2\theta - 1$ for values of θ between $0°$ and $360°$.   CHELTENHAM

36 A tapered hole is bored through an 80 mm thick plate, its axis being at right angles to the two faces of the plates.

Balls, respectively 40 mm and 30 mm in diameter, placed in the hole give depth-gauge readings as shown in Fig. 12.24.

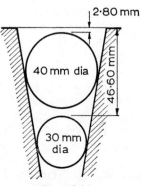

Fig. 12.24

What is the included angle of the taper, and what are the diameters of the hole at the two ends?

37 (a) Write down a formula for cos($A + B$) in terms of the trigonometrical ratios of A and B, and deduce an expression for cos $2A$ in terms of sin A.

    If cos $2A = 0.4$, and A is acute, find the value of sin A, without using trigonometrical tables.

(b) If the sides of a triangle ABC are 130 mm, 100 mm and 70 mm, find its area.

    The shortest side of a geometrically similar triangle PQR is 100 mm. Deduce the area of the triangle PQR.   SUNDERLAND

38 PQ is the jib of a crane, and is supported by a chain RQ, as shown in Fig. 12.25. If the jib is rotated upwards about P through an angle of 20°, calculate by how much the length RQ has been shortened.

NUNEATON

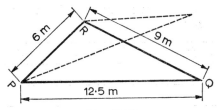

FIG. 12.25

39 (a) If $R \cos \theta = +2.6$, and $R \sin \theta = -3.86$, find the values of R and θ.

(b) If $V_{0°} = 5_{40°} + 3_{150°} + 7_{200°}$, find the values of V and θ.

NUNEATON

40 (a) Prove that

$$\tan\left(\frac{\pi}{4} + A\right) + \tan\left(\frac{\pi}{4} - A\right) = 2 \tan 2A$$

(b) Calculate all the values of θ between 0° and 360° which satisfy the equation $2\theta + 3 \sin \theta = 0$.

(c) The value of an alternating current at time t seconds is given by

$$i = 25 \sin\left(100\pi t - \frac{\pi}{3}\right),$$ the angle being in radians.

    State (i) its period, (ii) its amplitude, (iii) its frequency, and calculate i when $t = 0.005$ seconds.   NUNEATON

41 Two pipes, each 750 mm diameter, are placed side by side in contact on a wagon, and secured by a chain passing over the pipes and hooked to the floor of the wagon at two points 2 m apart. The chain is in a vertical plane perpendicular to the axes of the pipes. The line of contact of the cylinders is equidistant from the hooks. What is the length of chain required?   EMEU

42 Calculate the length AE in the gusset plate given in Fig. 12.26. Find also the area of the plate.   SURREY

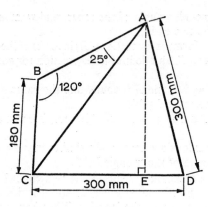

FIG. 12.26

43 Copy out and complete the following table, giving your answers in fractional and surd form.

Angle A	sin A	cos A	tan A
30°			
45°			
60°			

Find all the values of the angle A between 0° and 360° for which sec $A = -2/\sqrt{3}$.

Prove, without tables, that

(a) sin 420° cos 390° − cos(−300°) sin(330°) = 1,   and

(b) cos 210° sin 150° + sin 330° sin 240° = 0   EMEU

44 (a) Write down formulae for sin$(\theta + \phi)$ and cos$(\theta + \phi)$. Deduce a formula for tan$(\theta + \phi)$ in terms of tan θ and tan ϕ.

If tan$(A + B) = 2$, and tan $A = 1$, show, without using tables, that tan $B = \frac{1}{3}$, and evaluate tan$(180° + 2B)$.

(b) Find the smallest angle of the triangle ABC, where $a = 75$ mm, $b = 60$ mm and $c = 55$ mm.   SUNDERLAND

45 (a) Given that sin $A = \frac{12}{13}$ and cos $B = \frac{4}{5}$, find, without using tables,

(i) sin$(A + B)$, (ii) cos$(A − B)$.

(b) Solve the equation $\sqrt{2}$ cos x − sin $x = 1$, giving values of x between 0° and 360°.   RUGBY

46 (a) In the framework shown in Fig. 12.27, AB is parallel to DC. Find the length of AD and its inclination to AB.

(b) Determine the area of a triangle in which sides of 5 m and 3 m include an angle of 102°.   COVENTRY

47 (a) A surveyor finds that, instead of being able to work in a straight line from a station A to another station C, he has to go first 5 km to B, at an angle of 65° to AC. From B, the direct line to C measures 7·5 km. Calculate the direct distance from A to C.

(b) Forces of 25 newtons and 18 newtons act at a point, at 43° to one another. Calculate the magnitude of the resultant force, and its direction relative to the 25 newton force.   DUDLEY

48 (a) Given that A is a second quadrant angle whose sine is $\frac{5}{13}$, find, without using tables, the values of sin 2A, cos 2A and tan 2A.

(b) Obtain the values of R and α, so that 24 sin θ — 7 cos θ can be expressed in the form $R \sin(\theta - \alpha)$.

Hence solve the equation 24 sin θ — 7 cos θ = 20, giving the possible values of θ between 0° and 360°.   COVENTRY

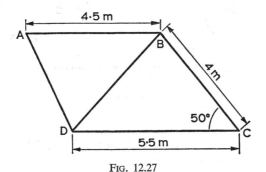

FIG. 12.27

49 (a) If tan $\alpha = \frac{7}{24}$, show that 100 cos($\theta + \alpha$) = 96 cos θ — 28 sin θ. Use the result to solve the equation 96 cos θ — 28 sin θ = 47 for values of θ between —180° and 180°.

(b) The displacement x at time t is given by

$$x = 4 \sin\left(4t - \frac{\pi}{4}\right)$$

Find the first value of t when $x = 3$. State the period and amplitude of x.   NUNEATON

50 (a) The sides of a triangle are 140 mm, 100 mm and 60 mm. Find the greatest angle of the triangle.

(b) Prove that

(cosec A — sin A)(sec A — cos A)(tan A + cot A) = 1   CANNOCK

51 ABC is a triangle in which $\angle$BAC is 105°, the side $b = 73\cdot2$ mm and $c = 87\cdot5$ mm. Solve the triangle completely and find its area.

                                                                WEST RIDING

52 (a) If $x = 3 \sin(5t + 1)$, the angular measure being in radians, find the first two values of t which make $x = -2$. State the period and amplitude of x.

(b) In determining the maximum velocity of a piston of a certain engine, the crank angle θ is given by the equation $\cos\theta + 0\cdot16 \cos 2\theta = 0$. Solve this for values of θ between $-180°$ and $180°$.

(c) Given that $\tan A = \frac{12}{35}$ and $\cos B = \frac{40}{41}$ evaluate (without tables) $\sin 2A$ and $\cos(A - B)$.   NUNEATON

53 A metal plate has dimensions as shown in Fig. 12.28. Find (a) the length of the diagonal BD, (b) the area of the plate.   RUGBY

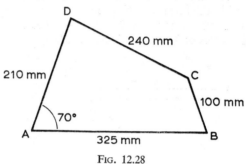

FIG. 12.28

54 (a) Prove that
$$\sin^2 A \cos^2 B - \cos^2 A \sin^2 B = \sin^2 A - \sin^2 B$$
(b) Solve the simultaneous equations
$$\frac{1}{x} + \frac{1}{y} + 1 = \frac{2}{x} + \frac{3}{y} = 1\tfrac{3}{4}$$

(c) A board the shape of an equilateral triangle is placed in the corner of a room, with one edge resting on the floor, and one against each wall. Find the inclination of the board to the floor.   RUGBY

55 (a) The sides OA and OB of a parallelogram OBCA represent forces of 3·7 newtons and 7·4 newtons, and the diagonal OC represents their resultant of 8·2 newtons. Calculate $\angle$AOB.

(b) Express $150 \sin 2\pi ft - 75 \cos 2\pi ft$ in the form $A \sin(2\pi ft - \alpha)$, giving the values of A and α (in radians).

(c) If $\tan A = a/b$, find the values of $\sin A$, $\cos A$, $\sin 2A$ and $\cos 2A$ in terms of a and b.

(d) Solve $5 \sin \theta + 3 \sin 2\theta = 0$ for values of θ between $0°$ and $360°$.   NUNEATON

56 (a) Find the angles between $0°$ and $360°$ which satisfy the equation $4 \sin \theta + 5 \cos \theta = 2$.

(b) In a triangle ABC, $A = 54°$, $b = 126$ mm and $c = 162$ mm. Find the other side and angles. Also calculate the area of the triangle.   RUGBY

57 (a) Prove that $\dfrac{\cot A + \tan B}{\cot B + \tan A} = \cot A \tan B$.

(*b*) An open belt passes over two pulleys whose diameters are 200 mm and 100 mm respectively, and whose centres are 400 mm apart. Calculate the length of belt necessary.   CANNOCK

58 The length of the crank OA of a reciprocating engine is 150 mm and the length of the connecting-rod AB is 600 mm. OA rotates about the fixed point O, and B slides along the fixed arm OB (see Fig. 12.29). Calculate (*a*) OB when the angle AOB is 67°, and (*b*) the angle AOB when OB is 500 mm.

What is the greatest angle that BA makes with BO during the motion?   NUNEATON

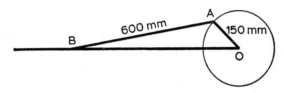

FIG. 12.29

59 A triangle of metal ABC is fitted into a cylinder with its plane at right angles to the axis of the cylinder, such that the points are just touching the inside wall. If the known sides are $a = 50$ mm and $b = 75$ mm, and angle $C = 30°$, find the side c and angles A and B, and the inside diameter of the cylinder.   WORCESTER

60 (*a*) If $\sin A = \frac{4}{5}$ and $\sin B = \frac{5}{13}$, and A and B are acute angles, write down, without using tables, the values of (i) $\cos(A - B)$, (ii) $\tan(A - B)$.

(*b*) Express $i = 10 \cos(180\pi t - 0.8)$ in the form $i = a \cos 180\pi t + b \sin 180\pi t$, giving the values of a and b correct to one decimal place.

(*c*) In (*b*), i represents an alternating current, i_{max} the maximum current, i_0 the value of i when $t = 0$, and f the frequency. Write down the values of i_{max}, i_0 and f.

(*d*) Solve the equation $\cos \theta + \frac{1}{4} \cos 2\theta = 0$, giving the values of θ lying between $\pm 90°$.   NUNEATON

Chapter Thirteen
Differentiation

The notion of a function
A function is a relation between real numbers. If $y = x^2$, then y is said to be a single-valued function of x, since to each value of x there corresponds one, and only one, value of y. x is called the independent variable, since its value does not depend upon the value of any other variable. y is called the dependent variable, since its value depends upon the value given to x.

Thus we can choose the value 2 for x, but then $y = 2^2 = 4$ is the corresponding value of y, which is seen to depend upon the value 2 given to x. The idea of correspondence is inherent in the notion of a function.

It is not necessary that an analytical formula such as $y = x^2$ exists in order to define a function. Suppose that, in a given interval of time, five cars pass a point on a road, and that the cars are numbered $n = 1, 2, 3, 4, 5$ successively as they pass the point, whilst the registration numbers N are noted.

n	1	2	3	4	5
N	1101	46	327	425	3251

Then N is a function of n, since to each value of n there corresponds a value of N. We shall, however, be concerned in the next three chapters with functional relationships where an analytical formula exists, where the *representing curve is continuous* and where *to each value of x, in the region of interest, there corresponds one and only one value of y*.

Definition of a function
y is a function of x if to each value of x, in the region of interest, there corresponds one and only one value of y.

Functional notation
The notation $f(x)$ may be used to denote a function of x. $f(x)$ should be regarded as an abbreviation for 'a function of x'. It does not, of course, mean the product of f and (x). The value of $f(x)$ when $x = a$ is denoted by $f(a)$.

Advantage of the functional notation
If $f(x) = x^2$ we may draw up a table of corresponding values of $f(x)$ and x for some values of x. Thus $f(-3) = (-3)^2 = 9$.

x	-3	-2	-1	0	1	2	3
$f(x)$	9	4	1	0	1	4	9

These results may be exhibited on a graph (Fig. 13.1). $f(x)$ is thus an alternative symbol to y, which has been used with previous graphs. Indeed, y is a more compact symbol than $f(x)$ in itself. What is the advantage of the symbol $f(x)$?

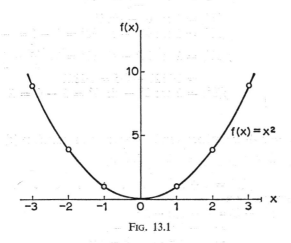

FIG. 13.1

When $f(x) = x^2$, $f(-3) = 9$, $f(-2) = 4$, $f(-1) = 1$, $f(0) = 0$, $f(1) = 1$, etc: a series of statements are obtained in which the value of the independent variable and the corresponding value of the dependent variable are exhibited in compact form. If we avoid the functional notation and state the relationship in the form $y = x^2$, then the previous statements could be written thus: when $x = -3$, $y = 9$; when $x = -2$, $y = 4$; and so on. This is obviously less compact. An excellent notation, which is unfortunately not widely used in elementary work, is $y(x) = x^2$, $y(-3) = 9$, $y(-2) = 4$, etc.

Examples
13.1 (*a*) If $f(x) = x^3 - 1$, calculate $f(2)$, $f(0)$ and $f(-1)$.

(*b*) If $y(x) = x + \dfrac{1}{x}$, calculate $y(2)$, $y(0)$ and $y(-2)$.

(*c*) If $f(\theta°) = 2 \cos \theta° - \sin \theta°$, calculate $f(90°)$, $f(30°)$ and $f(0°)$.

(*a*) If
then
$$f(x) = x^3 - 1$$
$$f(2) = (2)^3 - 1 = 8 - 1 = 7$$
$$f(0) = (0)^3 - 1 = 0 - 1 = -1$$
$$f(-1) = (-1)^3 - 1 = -1 - 1 = -2.$$

(b) If $\qquad y(x) = x + \dfrac{1}{x}$, replacing *every* x by 2,

then $\qquad\qquad y(2) = 2 + \frac{1}{2} = 2\frac{1}{2}$

$\qquad\qquad\qquad y(0) = 0 + \dfrac{1}{0}$ (Now $1 \div 0$ is meaningless. There is no value of y corresponding to $x = 0$: we say that y is undefined when $x = 0$.)

$\qquad\qquad\quad y(-2) = -2 + \dfrac{1}{-2} = -2\frac{1}{2}$

(c) If $\qquad\quad f(\theta°) = 2 \cos \theta° - \sin \theta°$

then $\qquad\quad f(90°) = 2 \cos 90° - \sin 90° = 0 - 1 = -1$

$\qquad\quad f(30°) = 2 \cos 30° - \sin 30° = 2 . \dfrac{\sqrt{3}}{2} - \dfrac{1}{2}$

$\qquad\qquad\qquad = 1{\cdot}7321 - 0{\cdot}5 = 1{\cdot}2321$

$\qquad\quad f(0°) = 2 \cos 0° - \sin 0° = 2 - 0 = 2$

13.2 If $f(x) = a + bx^2$, where a and b are constants, calculate (a) $f(x + h)$, (b) $f(x + h) - f(x)$ and (c) $\dfrac{f(x + h) - f(x)}{h}$.

(a) Since $\qquad f(x) = a + bx^2$ . . . . (i)

replacing every x by $x + h$,

$\qquad\quad f(x + h) = a + b(x + h)^2$
$\qquad\qquad\qquad = a + b(x^2 + 2xh + h^2)$
$\qquad\qquad\qquad = a + bx^2 + 2bxh + bh^2$ . . . . (ii)

(b) Subtracting (i) from (ii)

$\qquad\qquad f(x + h) - f(x) = 2bxh + bh^2$

(c) $\dfrac{f(x + h) - f(x)}{h} = \dfrac{2bxh + bh^2}{h} = 2bx + bh$

Use of the notations $F(x)$, $\phi(x)$, $A(x)$, etc

A function of x may also be denoted by $F(x)$, $\phi(x)$, $A(x)$, etc. Different functions of x used in the same context may be distinguished from one another in this way.

$f(x) \;\;= x^3 - 1 \qquad$ may be read 'a function of x, namely $x^3 - 1$'.

$F(x) \;\;= x^2 + \dfrac{2}{x} \qquad$ may be read 'a function of x, namely $x^2 + 2/x$'.

$\phi(x) \;\;= \sin x \qquad$ may be read 'a function of x, namely $\sin x$'.

$A(x) \;\;= \pi x^2 \qquad$ may be read 'a function of x, namely πx^2'.

It does not alter the functional relationship if we use a different symbol for the independent variable: thus $F(x) = x^4$ and $F(\theta) = \theta^4$ define the same functional relationship. We are merely choosing to use a different letter to represent the independent variable.

Examples
13.3 If $f(x) = a + bx$, and $F(x) = 2a + bx^2$, where a and b are constants, (a) calculate a and b, given that $f(1) = 1$ and $F(2) = 4$, (b) show that $[f(x)]^2 = F(x)$ and (c) calculate $f(2)$ and $F(-1)$.

(a) If $f(x) = a + bx$, then $f(1) = a + b \cdot 1$.

$$\text{Since } f(1) = 1, \text{ then } 1 = a + b \quad \ldots \ldots \quad \text{(i)}$$

If $F(x) = 2a + bx^2$, then $F(2) = 2a + b \cdot 2^2$.

$$\text{Since } F(2) = 4, \text{ then } 4 = 2a + 4b \quad \ldots \ldots \quad \text{(ii)}$$

Solving these equations, multiplying (i) by 2 and subtracting from (ii),

$$4 = 2a + 4b$$
$$2 = 2a + 2b$$

we obtain $2 = 2b$, i.e. $b = 1$ and $a = 0$
(b) Thus $\qquad\qquad f(x) = 0 + 1.x$, i.e. $f(x) = x$
and $\qquad\qquad F(x) = 0 + 1.x^2$, i.e. $F(x) = x^2$
hence $\qquad\qquad [f(x)]^2 = [x]^2 = x^2 = F(x)$
(c) Since $\qquad\qquad f(x) = x$, $f(2) = 2$.
Since $\qquad\qquad F(x) = x^2$, $F(-1) = (-1)^2 = 1$.

13.4 If $\phi(z) = z^2$, solve the equation $\phi(z) = \phi(2)$.

Since $\phi(z) = z^2$, $\phi(2) = 2^2 = 4$. If $\phi(z) = \phi(2)$, then $z^2 = 4$ and $z = \pm 2$.

13.5 If $F(x) = \dfrac{1}{x^2} - 1$, and $\phi(x) = x^2 - 1$, where $x \neq 0$, show that

(a) $F\left(\dfrac{1}{x}\right) = \phi(x)$, (b) $F(x) = \phi\left(\dfrac{1}{x}\right)$. (c) Solve the equation $F(x) = -\phi(x)$.

(a) Since $F(x) = \dfrac{1}{x^2} - 1$, $F\left(\dfrac{1}{x}\right) = \dfrac{1}{(1/x)^2} - 1 = x^2 - 1 = \phi(x)$.

(b) Since $\phi(x) = x^2 - 1$, $\phi\left(\dfrac{1}{x}\right) = \left(\dfrac{1}{x}\right)^2 - 1 = \dfrac{1}{x^2} - 1 = F(x)$.

(c) If $F(x) = -\phi(x)$ then $\dfrac{1}{x^2} - 1 = -(x^2 - 1)$,

i.e. $\qquad \dfrac{1 - x^2}{x^2} = 1 - x^2$ or $(1 - x^2) = x^2(1 - x^2)$.

i.e. $\quad (1 - x^2) - x^2(1 - x^2) = 0$, i.e. $(1 - x^2)(1 - x^2) = 0$
i.e. $\quad 1 - x^2 = 0$ or $x^2 = 1$, i.e. $x = \pm 1$

EXERCISE 13.1

1 In the following functional relationships, state the symbol which is (a) the independent variable, (b) the dependent variable: (i) $A = l^2$, (ii) $s = 16t^2$, (iii) $V = \frac{88}{21}r^3$.

2 If $f(x) = 3x^2 + 2x + 5$, write down, or calculate and simplify where possible, (a) $f(a)$, (b) $f(b + 3)$, (c) $f(2)$, (d) $f(0)$.

3 If $\phi(z) = 5z + 7$, find (a) $\phi(c)$, (b) $\phi(0)$, (c) $\phi(-1)$.

4 If $F(\theta) = 2 \sin \theta$, find (a) $F(90°)$, (b) $F(0°)$, (c) $F(\pi)$.

5 If $F(x) = x + \dfrac{1}{x}$, and $\phi(x) = x^2 - \dfrac{1}{x^2}$ given that $F(4) = K\phi(2)$, show that $K = 1\frac{2}{15}$.

6 If $A(x) = x^2 + 2x$, solve the equation $A(x) = A(-3)$.

7 If $f(x) = x^3 + 2x^2 - 3x + 4$, prove that (a) $f(0) = f(1)$, (b) $2f(2) = 7f(1)$, (c) $7f(3) = 20f(2)$.

8 If $\phi(x) = a + bx$, and $\phi(1) = 2$, $\phi(-1) = 3$, find a, b and $\phi(2)$.

9 If $f(x) = 1 + x$, $F(x) = 1 + \dfrac{1}{x}$, $\phi(x) = 1 - x$, and $\psi(x) = 1 - \dfrac{1}{x}$, where $x \neq 0$, show that (a) $xF(x) = f(x)$, (b) $x\psi(x) = -\phi(x)$, (c) $f\left(\dfrac{1}{x}\right) = F(x)$, (d) $\phi\left(\dfrac{1}{x}\right) = \psi(x)$, (e) $F(x) + \psi(x) = f(x) + \phi(x)$ (f) $F(x)\phi(x) + f(x)\psi(x) = 0$.
    If $F(a) = \psi(b)$, find the value of a/b. If $F(2) = \psi(c)$, find the value of c.

10 If $\theta(x) = a + bx + cx^2$, and $\theta(1) = 4$, $\theta(2) = 8$, $\theta(-1) = 2$, determine the values of a, b and c. Show that $\theta(7) - \theta(5) = 2[\theta(3) - 1]$.

11 If $F(x) = x^2$, (a) show that $F(x + 2h) - F(x - 2h) = 2[F(x + h) - F(x - h)]$. (b) If also $f(x) = F(x) + x[4 - F(3)] + F(2) + 2$, find $f(x)$, and solve the equation $f(x) = 0$.

12 If $f(x) = 2x^2 - 1$, (a) find $f(x + h)$, (b) find $\dfrac{f(x + h) - f(x)}{h}$, (c) by putting $x = 0$ in (a), find $f(h)$, (d) by putting $x = \frac{3}{4}$ in (b), find $\dfrac{f(\frac{3}{4} + h) - f(\frac{3}{4})}{h}$, (e) If $\dfrac{f(\frac{3}{4} + h) - f(\frac{3}{4})}{h} = f(h)$, find the value of h.

The notion of a limit

Consider the relation $y = f(x) = \dfrac{x^2 - 1}{x - 1}$. If we wish to graph this function and proceed to give x particular values,

$$f(-2) = \frac{(-2)^2 - 1}{-2 - 1} = \frac{3}{-3} = -1, \quad \text{similarly } f(-1) = 0, f(0) = 1.$$

$$f(1) = \frac{(1)^2 - 1}{1 - 1} = \frac{0}{0} \qquad \text{which is indeterminate. } y \text{ is undefined if } x = 1.$$

Then $f(2) = 3, f(3) = 4$, etc.

The calculated values are found to lie on a straight line with a gradient of 1 and an intercept of length 1 on the y-axis, i.e. on the line $y = x + 1$, with the exception that, when $x = 1$, y is indeterminate. It seems natural to investigate the value of y in the neighbourhood of $x = 1$.

Let $x = 1 + h$ . . . . (i)

$$f(1 + h) = \frac{(1 + h)^2 - 1}{(1 + h) - 1} = \frac{1 + 2h + h^2 - 1}{1 + h - 1} = \frac{2h + h^2}{h}$$

$$= 2 + h$$ . . . . (ii)

providing that $h \neq 0$, since division is then not possible.

Thus, when $h = 1$, $x = 2$, $y = 3$, using (i) and (ii)
$\qquad h = 0{\cdot}5$, $x = 1{\cdot}5$, $y = 2{\cdot}5$
$\qquad h = 0{\cdot}01$, $x = 0{\cdot}01$, $y = 2{\cdot}01$
$\qquad h = 0{\cdot}000\,01$, $x = 1{\cdot}000\,01$, $y = 2{\cdot}000\,01$

Thus, although it is known that, when $x = 1$, y is indeterminate, we can say from the above results that, as h tends to 0, x tends to 1 and y tends to 2. We say that

$$\lim_{x \to 1} \frac{x^2 - 1}{x - 1} = 2$$

meaning that, if the value of x is taken sufficiently close to 1, the value of $\frac{x^2 - 1}{x - 1}$ will be determined to be as close to 2 as we please, although, when $x = 1, \frac{x^2 - 1}{x - 1}$ has not got any value, i.e. is undefined. 'Lim' is an abbreviation of the word 'limit'.

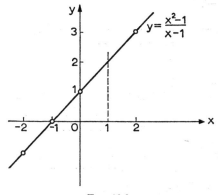

Fig. 13.2

This is illustrated in Fig. 13.2. The graph of the function $\frac{x^2 - 1}{x - 1}$, or $\frac{(x - 1)(x + 1)}{(x - 1)}$, is seen to be identical to the graph of the linear (straight line) function $(x + 1)$, except at $x = 1$. (It is not possible to illustrate graphically the exact behaviour of the function in the neighbourhood of $x = 1$, since any gap in the linear graph must be too large.)

Examples

13.6 (*a*) What is the value of $f(-2)$ if $f(x) = \frac{x^3 + 8}{x + 2}$?

(*b*) Calculate $f(-2 + h)$, where $h \neq 0$.

(*c*) Calculate $f(-2 - 0{\cdot}1)$ and $f(-2 - 0{\cdot}000\,01)$.

(*d*) Evaluate $\lim\limits_{x \to -2} \dfrac{x^3 + 8}{x + 2}$.

(*a*) $f(-2) = \dfrac{(-2)^3 + 8}{(-2) + 2} = \dfrac{-8 + 8}{-2 + 2} = \dfrac{0}{0}$: $f(-2)$ is indeterminate.

(*b*) $f(-2 + h) = \dfrac{(-2 + h)^3 + 8}{(-2 + h) + 2}$

$$= \frac{(-2)^3 + 3(-2)^2 h^1 + 3(-2)\,h^2 + h^3 + 8}{h}$$

$$= \frac{-8 + 12h - 6h^2 + h^3 + 8}{h} = \frac{12h - 6h^2 + h^3}{h}$$

$$= 12 - 6h + h^2 \qquad \text{(since division by } h \text{ is possible if } h \neq 0).$$

(*c*) Putting $h = -0{\cdot}1$,

$$\begin{aligned} f(-2 - 0{\cdot}1) &= 12 - 6(-0{\cdot}1) + (-0{\cdot}1)^2 \\ &= 12 + 0{\cdot}6 + 0{\cdot}01 \\ &= 12{\cdot}61 \end{aligned}$$

Putting $h = -0{\cdot}000\,01$

$$\begin{aligned} f(-2 - 0{\cdot}000\,01) &= 12 - 6(-0{\cdot}000\,01) + (-0{\cdot}000\,01)^2 \\ &= 12 + 0{\cdot}000\,06 + 0{\cdot}000\,000\,000\,1 \\ &= 12{\cdot}000\,060\,000\,1 \end{aligned}$$

(*d*) Since $f(-2 + h) = 12 - 6h + h^2$,

$$\text{as } h \to 0, \ 12 - 6h + h^2 \to 12,$$

ı.e. $\lim\limits_{x \to -2} \dfrac{x^3 + 8}{x + 2} = 12$

13.7 If $f(x) = x^3 + x^2 + x + 1$,   (a) calculate $\dfrac{f(x+h) - f(x)}{h}$,

(b) evaluate $\lim\limits_{h \to 0} \dfrac{f(x+h) - f(x)}{h}$.

(a) $f(x+h) = (x+h)^3 + (x+h)^2 + (x+h) + 1$
$$= x^3 + 3x^2h + 3xh^2 + h^3 + x^2 + 2xh + h^2 + x + h + 1$$
$$f(x) = x^3 \qquad\qquad\qquad\quad + x^2 \qquad\qquad\quad + x \qquad + 1$$

Subtracting,
$$f(x+h) - f(x) = 3x^2h + 3xh^2 + h^3 + 2xh + h^2 + h$$

$$\frac{f(x+h) - f(x)}{h} = 3x^2 + 3xh + h^2 + 2x + h + 1 \quad \text{(since } h \neq 0\text{)}.$$

(b)   As $h \to 0$, $3x^2 + 3xh + h^2 + 2x + h + 1 \to 3x^2 + 2x + 1$

$$\therefore \qquad \lim_{h \to 0} \frac{f(x+h) - f(x)}{h} = 3x^2 + 2x + 1$$

In subsequent work we shall assume without proof the following theorems on limits:
(a) The limit of a sum is the sum of the limits,

i.e. $\lim\limits_{x \to a} [f(x) + g(x)] = \lim\limits_{x \to a} [f(x)] + \lim\limits_{x \to a} [g(x)]$

(b) The limit of a product is the product of the limits,

i.e. $\lim\limits_{x \to a} [f(x) \cdot g(x)] = \lim\limits_{x \to a} [f(x)] \cdot \lim\limits_{x \to a} [g(x)]$

(c) The limit of a quotient is the quotient of the limits,

i.e. $\lim\limits_{x \to a} \dfrac{f(x)}{g(x)} = \dfrac{\lim\limits_{x \to a} f(x)}{\lim\limits_{x \to a} g(x)}$, if $\lim\limits_{x \to a} g(x)$ is not zero.

EXERCISE 13.2

1 (a) Has the function $\dfrac{x-2}{x^2-4}$ a value when $x = 2$?   (b) By putting

$x = 2 + h$, and letting $h \to 0$, evaluate $\lim\limits_{x \to 2} \dfrac{x-2}{x^2-4}$.

2 Evaluate $\dfrac{f(x+h) - f(x)}{h}$ and $\lim\limits_{h \to 0} \dfrac{f(x+h) - f(x)}{h}$ for $y = f(x)$

when   (a) $y = ax + b$,   (b) $y = ax^2 + bx + c$,   (c) $y = ax^3$.

3 If $y = \dfrac{1}{x^2}$, show that $\dfrac{f(x+h) - f(x)}{h} = \dfrac{-2x - h}{x^2(x+h)^2}$

and $\lim\limits_{h \to 0} \dfrac{f(x+h) - f(x)}{h} = \dfrac{-2}{x^3}$.

Meaning of δy, δx and $\dfrac{\delta y}{\delta x}$

If the value of a variable changes from a to b, then the change in the variable is defined as the value at the end of the change minus the value at the beginning of the change, i.e. $b - a$. We use δx to denote the change (or increment) of the variable x. δ, the small Greek letter delta, is a prefix used to denote change: thus δx does not mean the product of δ and x, but the change in x.

13.8 Find δx if x changes (a) from 3 to 7, (b) from -1 to 2, (c) from 3 to 1, (d) from -1 to -3.

Since δx is equal to (the value of x at the end of the change) — (the value of x at the beginning of the change):

(a) $\delta x = 7 - 3 = 4$
(b) $\delta x = 2 - (-1) = 3$
(c) $\delta x = 1 - 3 = -2$
(d) $\delta x = -3 - (-1) = -2$

If y is a function of x, i.e. $y = f(x)$, then a change δx in x will produce a corresponding change δy in y.

13.9 If $y = x^2$ and x changes from $+2$ to -1, calculate δx and δy.

$\delta x =$ (value of x at the end of the change)
   $-$ (value of x at the beginning of the change)
   $= (-1) - (2) = -3$
$\delta y =$ (value of y at the end of the change)
   $-$ (value of y at the beginning of the change)
   $= (-1)^2 - (2)^2 = 1 - 4 = -3$

The ratio $\dfrac{\delta y}{\delta x} = \dfrac{\text{corresponding change in } y}{\text{change in } x}$
   $=$ average rate of change of y with respect to x over the interval δx.

Note that rate does not necessarily involve time: it involves the ratio of two related changes. Thus for a spring

$\dfrac{\delta P}{\delta x} = \dfrac{\text{corresponding change in spring force}}{\text{change of deflection}}$
   $=$ average rate of change of spring force with respect to deflection over the deflection interval δx.

13.10 If $y = x^2$ and x changes from x to $x + \delta x$ while y changes from y to $y + \delta y$, (a) derive a formula for the change produced in y, i.e. δy in terms of x and δx, (b) express $\dfrac{\delta y}{\delta x}$ in terms of x and δx.

(a)
$$y + \delta y = (x + \delta x)^2 = x^2 + 2x\delta x + \delta x^2$$
also
$$y \qquad\qquad = x^2$$

Subtracting, $\qquad \delta y \qquad = \qquad 2x\delta x + \delta x^2 \qquad$ (i)

(b)
$$\frac{\delta y}{\delta x} = \frac{\text{change in } y}{\text{change in } x} = \frac{2x\delta x + \delta x^2}{\delta x}$$
$$= 2x + \delta x \quad \text{(if } \delta x \neq 0\text{)}.$$

(i) is a perfectly general formula. Note that, if $\delta x = 0$, i.e. there is no change in x, then $\delta y = 2x(0) + (0)^2 = 0$, i.e. there is no change in y, and then $\dfrac{\delta y}{\delta x} = \dfrac{0}{0}$ which is indeterminate.

We could use formula (i) to obtain the graph of $y = x^2$ starting from any value of x. Suppose $x = 1$, then $y = 1^2 = 1$, and the point $(1, 1)$ is a point on the graph.

$\therefore \quad \delta y = 2(1)\delta x + \delta x^2 = 2\delta x + \delta x^2$

When $\delta x = 1, 2, 3, 4, \delta y = 2(1) + (1)^2 = 3$, and similarly 8, 15, 24 respectively.

When $\delta x = -1, -2, -3, -4, \delta y = 2(-1) + (-1)^2 = -1$, and similarly 0, 3, 8 respectively.

x	1	1	1	1	1	1	1	1	1
δx	-4	-3	-2	-1	0	1	2	3	4
$x + \delta x$	-3	-2	-1	0	1	2	3	4	5
y	1	1	1	1	1	1	1	1	1
δy	8	3	0	-1	0	3	8	15	24
$y + \delta y$	9	4	1	0	1	4	9	16	25

Figure 13.3 illustrates that for the function $y = x^2$, when $x = 1$, $y = 1$ and $\delta x = 1$, then $\delta y = 3$, so that $x + \delta x = 2$, and $y + \delta y = 4$. Figure 13.4 illustrates that for the function $y = x^2$, when $x = 1$, $y = 1$, $\delta x = 2$,

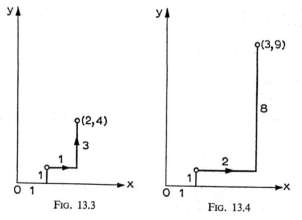

FIG. 13.3   FIG. 13.4

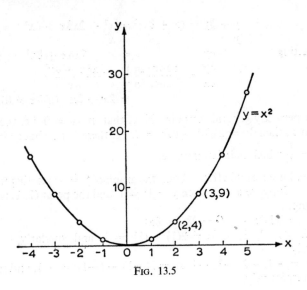

Fig. 13.5

then $\delta y = 8$, so that $x + \delta x = 3$ and $y + \delta y = 9$. Figure 13.5 is the locus of the points $(x + \delta x, y + \delta y)$.

Meaning of $\dfrac{dy}{dx}$ (the differential coefficient of y with respect to x)

We use $\dfrac{dy}{dx}$ as an abbreviation for $\displaystyle\lim_{\delta x \to 0} \dfrac{\delta y}{\delta x}$, since it is obviously more convenient. (At this stage we do not intend $\dfrac{dy}{dx}$ to mean $dy \div dx$. At a later stage we can suitably define the differentials dy and dx such that $dy \div dx = \dfrac{dy}{dx}$, or $dy = \left(\dfrac{dy}{dx}\right) dx$. This is the explanation of the term *differential coefficient*, i.e. the coefficient by which the differential of x is multiplied to give the differential of y.)

Thus $\dfrac{\delta y}{\delta x} = \dfrac{\text{corresponding change in } y}{\text{change in } x}$

       $=$ average rate of change of y with respect to x over the interval δx.

and $\dfrac{dy}{dx} = \displaystyle\lim_{\delta x \to 0} \dfrac{\delta y}{\delta x} =$ rate of change of y with respect to x at x.

Alternative form of $\displaystyle\lim_{\delta x \to 0} \dfrac{\delta y}{\delta x} = \dfrac{dy}{dx}$

If $y = f(x)$,   then $y + \delta y = f(x + \delta x)$

and $\delta y = f(x + \delta x) - f(x)$,   by subtraction

Thus $\quad \lim\limits_{\delta x \to 0} \dfrac{\delta y}{\delta x} = \lim\limits_{\delta x \to 0} \dfrac{f(x + \delta x) - f(x)}{\delta x} = \dfrac{dy}{dx}.$

Example

13.11 If $y = 3x^2$, find (*a*) δy, (*b*) $\dfrac{\delta y}{\delta x}$, (*c*) $\dfrac{dy}{dx}$.

(*a*) Let x change to $x + \delta x$ and, correspondingly, y change to $y + \delta y$, then, since $y = 3x^2$, replacing y by $y + \delta y$ and x by $x + \delta x$,

$$y + \delta y = 3(x + \delta x)^2 = 3(x^2 + 2x\delta x + \delta x^2)$$
$$y \qquad\qquad\qquad\qquad = 3x^2$$

Subtracting $\qquad \delta y \qquad\quad = \qquad 6x\delta x + 3\delta x^2$
(*b*) Dividing by δx (if $\delta x \neq 0$), then

$$\dfrac{\delta y}{\delta x} = 6x + 3\delta x$$

As $\delta x \to 0 \qquad \dfrac{\delta y}{\delta x} \to 6x,$

i.e. $\quad \lim\limits_{\delta x \to 0} \dfrac{\delta y}{\delta x} = 6x$

i.e. $\qquad \dfrac{dy}{dx} = 6x$

Note that, since y is $3x^2$, this could be written as $\dfrac{d(3x^2)}{dx} = 6x.$

This process is known as differentiation from first principles.

Rules for the differentiation from first principles of $y = f(x)$
(*a*) Find $y + \delta y$ by replacing y by $y + \delta y$ and x by $x + \delta x$ in $y = f(x)$.
(*b*) Find δy by subtracting y from $y + \delta y$.
(*c*) Divide δy by δx, giving $\dfrac{\delta y}{\delta x}$ (where $\delta x \neq 0$).
(*d*) Find the limit of $\dfrac{\delta y}{\delta x}$ as $\delta x \to 0$.

Example

13.12 If $y = ax^5$, where a is a constant, find $\dfrac{dy}{dx}.$

(*a*) $\quad y + \delta y = a(x + \delta x)^5 = a\left(x^5 + \dfrac{5}{1}x^4\,\delta x + \dfrac{5.4}{1.2}x^3\delta x^2 \right.$

$$+ \dfrac{5.4.3}{1.2.3}x^2\delta x^3 + \dfrac{5.4.3.2}{1.2.3.4}x\delta x^4$$

$$\left. + \dfrac{5.4.3.2.1}{1.2.3.4.5}\delta x^5 \right)$$

i.e. $y + \delta y = a(x^5 + 5x^4\delta x + 10x^3\delta x^2 + 10x^2\delta x^3 + 5x\delta x^4 + \delta x^5)$

and $y \qquad = ax^5$

(b) Subtracting,

$$\delta y = a(5x^4\delta x + 10x^3\delta x^2 + 10x^2\delta x^3 + 5x\delta x^4 + \delta x^5)$$

(c) Dividing by δx (if $\delta x \neq 0$),

$$\frac{\delta y}{\delta x} = a(5x^4 + 10x^3\delta x + 10x^2\delta x^2 + 5x\delta x^3 + \delta x^4)$$

(d) When $\delta x \to 0$, $10x^3\delta x + 10x^2\delta x^2 + 5x\delta x^3 + \delta x^4 \to 0$, but $5x^4$ is not affected,

i.e.
$$\lim_{\delta x \to 0} \frac{\delta y}{\delta x} = a \cdot 5x^4$$

i.e.
$$\frac{dy}{dx} = 5ax^4$$

Note that, since y is ax^5, this could be written $\dfrac{d(ax^5)}{dx} = 5ax^4$.

EXERCISE 13.3

1 If $y = bx^3$, where b is any constant, show that   (a) $\delta y = 3bx^2\delta x + 3bx\delta x^2 + b\delta x^3$,   (b) $\dfrac{\delta y}{\delta x} = 3bx^2 + 3bx\delta x + b\delta x^2$,   (c) $\dfrac{dy}{dx} = 3bx^2$.

2 If $s = t^4$, show that   (a) $\delta s = 4t^3\delta t + 6t^2\delta t^2 + 4t\delta t^3 + \delta t^4$,   (b) $\dfrac{\delta s}{\delta t} = 4t^3 + 6t^2\delta t^2 + \delta t^3$,   (c) $\dfrac{ds}{dt} = 4t^3$.

3 Differentiate from first principles   (a) $y = ax^2$,   (b) $s = -t^5$ and   (c) $v = u^2 - u$.

To prove that, if $y = ax^n$, where a is a constant and n is a positive integer, then $\dfrac{dy}{dx} = nax^{n-1}$

Since $y = ax^n$ we have $y + \delta y = a(x + \delta x)^n$

and $y \qquad = ax^n$

Subtracting

$$\delta y = a(x + \delta x)^n - ax^n$$

$$= a\left(x^n + \frac{n}{1}x^{n-1}\delta x + \frac{n(n-1)}{1 \cdot 2}x^{n-2}\delta x^2 + \ldots \delta x^n\right) - ax^n$$

$$= a\left(nx^{n-1}\delta x + \frac{n(n-1)}{1 \cdot 2}x^{n-2}\delta x^2 + \ldots \delta x^n\right)$$

$$\therefore \quad \frac{\delta y}{\delta x} = a\left(nx^{n-1} + \frac{n(n-1)}{1 \cdot 2}x^{n-2}\delta x + \ldots \delta x^{n-1}\right)$$

$$\therefore \quad \lim_{\delta x \to 0} \frac{\delta y}{\delta x} = anx^{n-1}$$

i.e. $\dfrac{dy}{dx} = anx^{n-1}$, or $\dfrac{d(ax^n)}{dx} = anx^{n-1}$ . . . . (i)

Note that, when $a = 1$, we have $\dfrac{d(x^n)}{dx} = nx^{n-1}$ . . . . . (ii)

and that, if (ii) is substituted in (i),

$$\frac{d(ax^n)}{dx} = a\frac{d(x^n)}{dx} \qquad \cdots \cdots \cdot \text{(iii)}$$

This is a particular case of the general result that a constant multiplier may be taken through the differentiation sign.

Use of formula $\dfrac{d(x^n)}{dx} = nx^{n-1}$

To find $\dfrac{d(x^7)}{dx}$, put $n = 7$: $\quad \dfrac{d(x^7)}{dx} = 7x^{7-1} = 7x^6$.

To find $\dfrac{d(x^6)}{dx}$, put $n = 6$: $\quad \dfrac{d(x^6)}{dx} = 6x^{6-1} = 6x^5$.

To find $\dfrac{d(x^3)}{dx}$, put $n = 3$: $\quad \dfrac{d(x^3)}{dx} = 3x^{3-1} = 3x^2$.

To find $\dfrac{d(x^2)}{dx}$, put $n = 2$: $\quad \dfrac{d(x^2)}{dx} = 2x^{2-1} = 2x$.

To find $\dfrac{d(x^1)}{dx}$, put $n = 1$: $\quad \dfrac{d(x^1)}{dx} = 1x^{1-1} = 1x^0 = 1$.

Use of formula $\dfrac{d(ax^n)}{dx} = nax^{n-1}$

To find $\dfrac{d(5x^4)}{dx}$, put $a = 5$, $n = 4$: $\quad \dfrac{d(5x^4)}{dx} = 4 \times 5x^{4-1} = 20x^3$.

To find $\dfrac{d(-3x^{10})}{dx}$, put $a = -3$, $n = 10$:

$$\frac{d(-3x^{10})}{dx} = 10 \times -3x^{10-1} = -30x^9.$$

Alternatively, using (iii):

$$\frac{d(5x^4)}{dx} = \frac{5d(x^4)}{dx} = 5 \times 4x^{4-1} = 20x^3$$

$$\frac{d(-3x^{10})}{dx} = -3\frac{d(x^{10})}{dx} = -3 \times 10x^{10-1} = -30x^9$$

This process is sometimes called differentiation by rule.
$\dfrac{d(ax^n)}{dx} = nax^{n-1}$ is known as a standard differential coefficient.

Differentiation of a constant term

If $y = c$, since a constant, by definition, cannot change, it cannot have a rate of change, i.e. $\dfrac{dy}{dx} = 0$.

Alternatively, if $y = c$, $y + \delta y = c$

$$\therefore \quad \delta y = 0 \quad \text{and} \quad \frac{\delta y}{\delta x} = \frac{0}{\delta x} = 0.$$

As
$$\delta x \to 0, \quad \frac{dy}{dx} = 0$$

Thus
$$\frac{d(8)}{dx} = 0, \qquad \frac{d(-5)}{dx} = 0, \qquad \frac{d(0)}{dx} = 0$$

Note that this only applies when the *whole* term is constant, thus $\dfrac{d(8x^2)}{dx} = 16x$, as previously.

Differentiation of a sum

If $y = u + v$, to show that $\dfrac{dy}{dx} = \dfrac{du}{dx} + \dfrac{dv}{dx}$, where u and v are functions of x, let x change to $x + \delta x$, so that u changes to $u + \delta u$, v to $v + \delta v$ and y to $y + \delta y$,

then
$$y + \delta y = u + \delta u + v + \delta v$$
and
$$y \quad = u \quad + v$$

Subtracting,
$$\delta y = \quad \delta u + \quad \delta v$$

hence
$$\frac{\delta y}{\delta x} = \frac{\delta u}{\delta x} + \frac{\delta v}{\delta x}$$

As $\delta x \to 0$,
$$\frac{dy}{dx} = \frac{du}{dx} + \frac{dv}{dx} \qquad \text{(since the limit of a sum is the sum of the limits),}$$

i.e.
$$\frac{d(u + v)}{dx} = \frac{du}{dx} + \frac{dv}{dx}$$

This may be extended to any finite number of terms,

thus
$$\frac{d(u + v - w)}{dx} = \frac{d\{(u + v) + (-w)\}}{dx} = \frac{d(u + v)}{dx} + \frac{d(-w)}{dx}$$

$$= \frac{du}{dx} + \frac{dv}{dx} - \frac{dw}{dx}$$

For example,

$$\frac{d(x^3 + x^2 + x)}{dx} = \frac{d(x^3)}{dx} + \frac{d(x^2)}{dx} + \frac{d(x)}{dx}$$

$$= 3x^2 + 2x + 1$$

$$\frac{d(t^4 - 3t^3)}{dt} = \frac{d(t^4)}{dt} + \frac{d(-3t^3)}{dt}$$

$$= 4t^3 - 9t^2 = t^2(4t - 9)$$

Examples

13.13 If $u = (2 - v)^3$, find $\dfrac{du}{dv}$ by first expanding $(2 - v)^3$ by the binomial theorem.

$$u = [2 + (-v)]^3 = 2^3 + \frac{3}{1} \cdot 2^2(-v) + \frac{3 \cdot 2}{1 \cdot 2} 2^1(-v)^2 + \frac{3 \cdot 2 \cdot 1}{1 \cdot 2 \cdot 3}(-v)^3$$

$$= 8 - 12v + 6v^2 - v^3$$

$$\therefore \quad \frac{du}{dv} = 0 - 12 + 12v - 3v^2 = -3(v^2 - 4v + 4)$$

$$= -3(2 - v)^2$$

13.14 Differentiate with respect to t: $11t^3 - 3t^{11}$.

$$\frac{d(11t^3 - 3t^{11})}{dt} = 33t^2 - 33t^{10} = 33t^2(1 - t^8)$$

EXERCISE 13.4

1 Find $\dfrac{dy}{dx}$ if $y = 1 + \dfrac{x}{1!} + \dfrac{x^2}{2!} + \dfrac{x^3}{3!} + \dfrac{x^4}{4!} + \dfrac{x^5}{5!}$.

2 Find $\dfrac{ds}{dt}$ if $s = 1 - t + \dfrac{t^2}{2} - \dfrac{t^3}{3} + \dfrac{t^4}{4} - \dfrac{t^5}{5}$.

3 By using the binomial theorem where necessary, show that if
(a) $\phi = (1 - \theta)^2$ then $\dfrac{d\phi}{d\theta} = -2(1 - \theta)$, (b) $\phi = (1 - \theta)^3$ then $\dfrac{d\phi}{d\theta} = -3(1 - \theta)^2$, (c) $\phi = (2 - 3\theta)^4$ then $\dfrac{d\phi}{d\theta} = -12(2 - 3\theta)^3$.

4 If $p = Kv^5$, show that $\dfrac{dp}{dv} = \dfrac{5p}{v}$.

5 (a) Find $\dfrac{d(3u^3 + 2u^2 - 3)}{du}$. (b) If $y = \dfrac{(x^2 + x)^2}{x^2}$, simplify y and find $\dfrac{dy}{dx}$. (c) If $f(x) = \dfrac{x^6}{30} - \dfrac{x^5}{20} + \dfrac{x^4}{12}$, find $f'(x)$ where $f'(x)$ is a notation for $\dfrac{df(x)}{dx}$.

Successive differentiation

$\dfrac{d^2y}{dx^2}$ means $\dfrac{d(dy/dx)}{dx}$, $\dfrac{d^3y}{dx^3}$ means $\dfrac{d(d^2y/dx^2)}{dx}$ and $\dfrac{d^ny}{dx^n}$ means $\dfrac{d(d^{n-1}y/dx^{n-1})}{dx}$,

where n is a positive integer greater than 1.

For example, if $\qquad y = x^5$,   then

$$\frac{dy}{dx} = 5x^4$$

$$\frac{d(dy/dx)}{dx} = \frac{d(5x^4)}{dx} = 20x^3$$

i.e. $\qquad\qquad\qquad\qquad \dfrac{d^2y}{dx^2} = 20x^3$

Similarly $\qquad\qquad\qquad \dfrac{d^3y}{dx^3} = 60x^2, \quad \dfrac{d^4y}{dx^4} = 120x, \quad \dfrac{d^5y}{dx^5} = 120$

$$\frac{d^6y}{dx^6} = 0, \qquad \frac{d^7y}{dx^7} = 0$$

In this case we say that the sixth and higher differential coefficients of y with respect to x vanish.

Example

13.15 If $y = \dfrac{4}{3}x^3 - x^2$, find (a) $\dfrac{dy}{dx}$, (b) $\dfrac{d^2y}{dx^2}$, (c) the value of x for which $\dfrac{dy}{dx} = 0$, (d) the value of x for which $\dfrac{d^2y}{dx^2} = 0$.

If $y = \dfrac{4}{3}x^3 - x^2$,

(a) $\dfrac{dy}{dx} = \dfrac{3 \cdot 4}{3}x^2 - 2x = 4x^2 - 2x$

(b) $\dfrac{d^2y}{dx^2} = 8x - 2$

(c) when $\dfrac{dy}{dx} = 0$, then $4x^2 - 2x = 0$,

i.e. $2x(2x - 1) = 0$, i.e. $x = 0$ or $2x - 1 = 0$, i.e. $x = 0$ or $x = \frac{1}{2}$.

(d) When $\dfrac{d^2y}{dx^2} = 0$, then $8x - 2 = 0$, i.e. $8x = 2$ and $x = \dfrac{2}{8} = \dfrac{1}{4}$.

Differentiation of $y = ax^n$, where n is not a positive integer

We may first consider particular examples in which we may proceed from first principles.

Examples

13.16 If $y = \dfrac{1}{x^2}$, find $\dfrac{dy}{dx}$ from first principles.

If $y = \dfrac{1}{x^2}$ then $y + \delta y = \dfrac{1}{(x + \delta x)^2}$

Subtracting, $\delta y = \dfrac{1}{(x + \delta x)^2} - \dfrac{1}{x^2} = \dfrac{x^2 - (x + \delta x)^2}{x^2(x + \delta x)^2}$

i.e. $\delta y = \dfrac{x^2 - (x^2 + 2x\delta x + \delta x^2)}{x^2(x + \delta x)^2} = \dfrac{-2x\delta x - \delta x^2}{x^2(x + \delta x)^2}$

$\dfrac{\delta y}{\delta x} = \dfrac{-2x - \delta x}{x^2(x + \delta x)^2}$

As $\delta x \to 0$, $-2x - \delta x \to -2x$ and $x^2(x + \delta x)^2 \to x^2(x)^2 = x^4$. Since the limit of a quotient is the quotient of the limits, we have

$$\frac{dy}{dx} = \frac{-2x}{x^4} = \frac{-2}{x^3}$$

Writing the result as $\dfrac{d(x^{-2})}{dx} = -2x^{-3}$, we see that, in this case, $\dfrac{d(ax^n)}{dx} = nax^{n-1}$ applies, where n is a negative integer.

13.17 If $y = x^{\frac{1}{2}}$, find $\dfrac{dy}{dx}$ from first principles.

If $y = x^{\frac{1}{2}}$, then $y + \delta y = (x + \delta x)^{\frac{1}{2}}$

Subtracting, $\delta y = (x + \delta x)^{\frac{1}{2}} - x^{\frac{1}{2}}$

hence $\dfrac{\delta y}{\delta x} = \dfrac{(x + \delta x)^{\frac{1}{2}} - x^{\frac{1}{2}}}{\delta x}$

If we let $\delta x \to 0$ at this stage, the indeterminate expression $\dfrac{0}{0}$ is obtained.

The device of multiplying the numerator and denominator by the conjugate expression $(x + \delta x)^{\frac{1}{2}} + x^{\frac{1}{2}}$ may be used, whence

$$\frac{\delta y}{\delta x} = \frac{[(x + \delta x)^{\frac{1}{2}} - x^{\frac{1}{2}}]}{\delta x} \frac{[(x + \delta x)^{\frac{1}{2}} + x^{\frac{1}{2}}]}{(x + \delta x)^{\frac{1}{2}} + x^{\frac{1}{2}}}$$

$$= \frac{(x + \delta x) - x}{\delta x[(x + \delta x)^{\frac{1}{2}} + x^{\frac{1}{2}}]}$$

$$= \frac{\delta x}{\delta x[(x + \delta x)^{\frac{1}{2}} + x^{\frac{1}{2}}]}$$

$$= \frac{1}{(x + \delta x)^{\frac{1}{2}} + x^{\frac{1}{2}}}$$

When $\delta x \to 0$, $(x + \delta x)^{\frac{1}{2}} + x^{\frac{1}{2}} \to 2x^{\frac{1}{2}}$ and $\dfrac{dy}{dx} = \dfrac{1}{2x^{\frac{1}{2}}}$.

Writing the result as $\dfrac{d(x^{\frac{1}{2}})}{dx} = \frac{1}{2}x^{-\frac{1}{2}}$, we see that in this case $\dfrac{d(x^n)}{dx} = nx^{n-1}$ applies, where n is fractional. In fact the standard differential coefficients $\dfrac{d(ax^n)}{dx} = nax^{n-1}$ and $\dfrac{d(x^n)}{dx} = nx^{n-1}$ apply, where n is positive or negative, integral or fractional. The results may be more concisely proved, however, when further methods of differentiation are available.

Examples

13.18 If $\qquad y = \dfrac{1}{x} + \dfrac{2}{x^2} + \dfrac{1}{x^3} - \dfrac{2}{\sqrt{x}} + \dfrac{1}{\sqrt[3]{x^2}}$, find $\dfrac{dy}{dx}$.

Since $\qquad y = x^{-1} + 2x^{-2} + x^{-3} - \dfrac{2}{x^{\frac{1}{2}}} + \dfrac{1}{x^{\frac{2}{3}}}$

i.e. $\qquad y = x^{-1} + 2x^{-2} + x^{-3} - 2x^{-\frac{1}{2}} + x^{-\frac{2}{3}}$,

$$\frac{dy}{dx} = -x^{-1} - 4x^{-3} - 3x^{-4} + x^{-\frac{3}{2}} - \frac{2}{3}x^{-\frac{5}{3}}$$

$$= -\frac{1}{x} - \frac{4}{x^3} - \frac{3}{x^4} + \frac{1}{\sqrt{x^3}} - \frac{2}{3\sqrt[3]{x^5}}$$

13.19 If $v = \left(\sqrt{u^3} - \dfrac{1}{\sqrt{u^3}} \right)^3$, find $\dfrac{dv}{du}$.

$$v = \left(u^{\frac{3}{2}} - \frac{1}{u^{\frac{3}{2}}} \right)^3 = \left(\frac{u^3 - 1}{u^{\frac{3}{2}}} \right)^3 = \frac{(u^3 - 1)^3}{u^{\frac{9}{2}}} = \frac{u^9 - 3u^6 + 3u^3 - 1}{u^{\frac{9}{2}}}$$

i.e. $\qquad v = u^{\frac{9}{2}} - 3u^{\frac{3}{2}} + 3u^{-\frac{3}{2}} - u^{-\frac{9}{2}}$

$$\therefore \quad \frac{dv}{du} = \frac{9}{2}u^{\frac{7}{2}} - \frac{9}{2}u^{\frac{1}{2}} - \frac{9}{2}u^{-\frac{5}{2}} + \frac{9}{2}u^{-11/2}$$

$$= \frac{9}{2}u^{\frac{1}{2}}(u^3 - 1 - u^{-3} + u^{-6}) = \frac{9}{2}u^{\frac{1}{2}}\left[u^3 - 1 - \frac{1}{u^3} + \frac{1}{u^6} \right]$$

$$= \frac{9}{2}u^{\frac{1}{2}}\left[u^3 - 1 - \frac{(u^3 - 1)}{u^6} \right] = \frac{9}{2}u^{\frac{1}{2}}(u^3 - 1)\left(1 - \frac{1}{u^6} \right)$$

$$= \frac{9}{2}u^{\frac{1}{2}}(u^3 - 1)\left(\frac{u^6 - 1}{u^6} \right) = \frac{9}{2u^{11/2}}(u^3 - 1)(u^3 - 1)(u^3 + 1)$$

$$= \frac{9}{2}\frac{(u^3 - 1)^2(u^3 + 1)}{u^{11/2}}$$

EXERCISE 13.5

1 (a) Differentiate the following with respect to x: (i) $2x^5$, (ii) $3\sqrt{x}$, (iii) $\dfrac{5}{\sqrt{x}}$. (b) If $y = 2x^3 - 3x + 6$, find the values of $\dfrac{dy}{dx}$ and $\dfrac{d^2y}{dx^2}$ when $x = 3$. For what values of x is $\dfrac{dy}{dx}$ equal to zero, and what are the corresponding values of y?   NCTEC

2 (a) Determine from first principles the expression for $\dfrac{d\theta}{dt}$, if $\theta = at^2$.

(b) A wheel rotates through θ radians in t seconds, and $\theta = 15 + 16t - t^2$. Find (i) $\dfrac{d\theta}{dt}$ when $t = 4$, (ii) the number of radians through which it rotates in the first 7 seconds.   ULCI

3 (a) Differentiate $y = \dfrac{1}{x}$ from first principles, and check the result using differentiation by rule.

(b) If $y = x^3 - 3x^2$, find $\dfrac{dy}{dx}, \dfrac{d^2y}{dx^2}, \dfrac{d^3y}{dx^3}$ and $\dfrac{d^4y}{dx^4}$.   Show that $\dfrac{d^3y}{dx^2} + \dfrac{d^2y}{dx^2} + \dfrac{dy}{dx} + y = x^3$.

4 (a) If $y = \dfrac{1}{x + 2}$, show that $\dfrac{\delta y}{\delta x} = \dfrac{-1}{(x + \delta x + 2)(x + 2)}$, and that $\dfrac{dy}{dx} = -\dfrac{1}{(x + 2)^2}$.

(b) If $y = \dfrac{x^4}{2} + 6$, find $\dfrac{d^4y}{dx^4}$, and show that $2y - \dfrac{d^4y}{dx^4} = x^4$.

(c) If $p = v^{\frac{1}{2}}$, show that $\dfrac{dp}{dv} = \dfrac{p}{2v}$.

5 Differentiate with respect to the appropriate variable (a) $y = 2x^3 - 3x + 6$, (b) $s = t^2 + 3t - 4$, (c) $y = x^2 + \dfrac{2}{\sqrt{x}}$, (d) $v = u^{1 \cdot 2} - \dfrac{1}{2u^{2 \cdot 5}}$, (e) $y = \dfrac{3}{x^2} + \sqrt{x} + 0 \cdot 2x^{1 \cdot 26}$, (f) $y = 4\sqrt[3]{x^2} + \dfrac{6}{x}$.

6 If $z = \dfrac{t^5}{5!} + \dfrac{t^4}{4!} + \dfrac{t^3}{3!} + \dfrac{t^2}{2!} + \dfrac{t}{1!} + 1$, find, with simplified factorial coefficients, (a) $\dfrac{dz}{dt}$, (b) $\dfrac{d^2z}{dt^2}$, (c) $\dfrac{d^3z}{dt^3}$, (d) $\dfrac{d^4z}{dt^4}$, (e) $\dfrac{d^5z}{dt^5}$, (f) $\dfrac{d^6z}{dt^6}$.

Gradient of the line joining the points (x_1, y_1) and (x_2, y_2)
In Fig. 13.6, P is the point (x_1, y_1), where ON represents the abscissa x_1, and NP the ordinate y_1, and Q is the point (x_2, y_2). Let PQ make an angle θ with PR, which is parallel to OX. Then RQ $= y_2 - y_1$ and PR $= x_2 - x_1$.

Then the gradient of the line $PQ = \tan\theta = \dfrac{RQ}{PR} = \dfrac{\text{increase of } y}{\text{increase of } x} =$

$\dfrac{y_2 - y_1}{x_2 - x_1} = \dfrac{y_1 - y_2}{x_1 - x_2} = \dfrac{\text{difference of } y \text{ values}}{\textit{corresponding} \text{ difference of } x \text{ values}}$

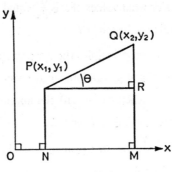

FIG. 13.6

Examples
13.20 Find the gradient of the line joining the points (a) (1, 1) and (2, 3), (b) (−2, −1) and (−3, −4), (c) (−1, 1) and (−3, 4), (d) (1, −1) and (2, −3).

In each case, calculate the angle between the line and OX measured anticlockwise from OX.

(a) The gradient of the line $= \dfrac{3 - 1}{2 - 1} = 2.$

If $\tan\theta = 2$, $\theta = 63°\ 27'$.

(b) The gradient of the line $= \dfrac{(-4) - (-1)}{(-3) - (-2)} = \dfrac{-4 + 1}{-3 + 2} = \dfrac{-3}{-1} = 3.$

If $\tan\theta = 3$, $\theta = 71°\ 30'$.

(c) The gradient of the line $= \dfrac{4 - 1}{(-3) - (-1)} = \dfrac{3}{-3 + 1} = \dfrac{3}{-2} = -1{\cdot}5.$

If $\tan\theta = -1{\cdot}5$, $\theta = 180° - 56°\ 19' = 123°\ 41'$.

(d) The gradient of the line $= \dfrac{(-3) - (-1)}{2 - 1} = \dfrac{-3 + 1}{1} = -2.$

If $\tan\theta = -2$, $\theta = 180° - 63°\ 26' = 116°\ 34'$.

Further note on gradients
(a) A line which slopes *upwards* from *left* to *right* has a positive gradient and makes an acute angle with the positive direction of the x-axis.
(b) A line which slopes *downwards* from *left* to *right* has a negative gradient and makes an obtuse angle with the positive direction of the x-axis.

(c) A line parallel to the y-axis has gradient $= \dfrac{\text{increase of } y}{\text{increase of } x} = \dfrac{\delta y}{0}$: i.e. the gradient is infinite. The line makes an angle of 90° with the positive direction of the x-axis. The equation of such a line is $x = c$, where c is a constant.

(d) A line parallel to the x-axis has gradient $= \dfrac{\text{increase of } y}{\text{increase of } x} = \dfrac{0}{\delta x} = 0,$ and makes an angle of zero with the positive direction of the x-axis. The equation of such a line is $y = c$, where c is a constant (see Fig. 13.7).

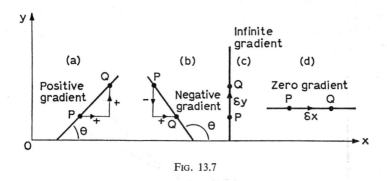

FIG. 13.7

Gradient of the secant joining the points $P(x, y)$ and $Q(x + \delta x\ y + \delta y)$ on the continuous curve $y = f(x)$ and gradient of the curve at $P(x, y)$
Note that the line PQ is called a chord of the curve $y = f(x)$. If it is extended beyond P and Q, it is called a secant (Fig. 13.8).

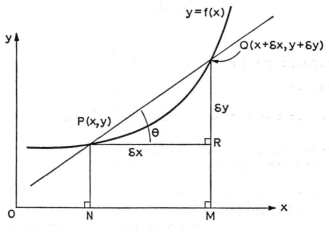

FIG. 13.8

The gradient of the secant PQ $= \dfrac{\text{difference of the } y \text{ values}}{\text{corresponding difference of the } x \text{ values}}$

$$= \frac{(y + \delta y) - (y)}{(x + \delta x) - (x)} = \frac{\delta y}{\delta x} = \frac{RQ}{PR} = \tan \theta$$

Note that if we put $\delta x = 0$, then $\delta y = 0$ and the above gradient becomes indeterminate. If, however, we let $\delta x \to 0$ so that $\delta y \to 0$, remembering that Q always lies on the curve $y = f(x)$, then Q will move towards P as close to P as we choose, and the secant PQ will approach a definite limiting position which is known as the tangent line at P (in the case of the elementary functions with which we shall deal).

Let the tangent line at P make an angle ψ with the positive direction of the x-axis, OX (Fig. 13.9).

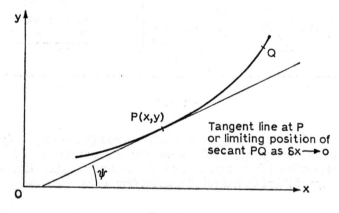

FIG. 13.9

Then

gradient of the secant PQ $= \dfrac{\delta y}{\delta x} = \tan \theta,$

and, as $\delta x \to 0$,

gradient of the tangent line at P $= \displaystyle\lim_{\delta x \to 0} \frac{\delta y}{\delta x} = \frac{dy}{dx} = \tan \psi$

We define the gradient of the curve $y = f(x)$ at the point $P(x, y)$ as being equal to the gradient of the tangent line at the point $P(x, y)$.
Thus

gradient of the curve $y = f(x)$
at the point $P(x, y)$ $= \dfrac{dy}{dx} = \tan \psi$

$\qquad\qquad\qquad = $ rate of increase of y with respect to x
$\qquad\qquad\qquad$ (from previous work).

Examples
13.21 State the values of x for which the curve $y = x^2 - 5x + 6$ has
(a) a positive gradient, (b) a negative gradient and (c) zero gradient.
(d) Calculate the gradient of the curve where $x = -2$, $x = 1$ and $x = 4$.

Since $y = x^2 - 5x + 6$, the gradient function $\dfrac{dy}{dx} = 2x - 5$.

(a) $2x - 5 > 0$ if $2x > 5$ if $x > \dfrac{5}{2}$ (note that $>$ means 'greater than'),

  i.e. the gradient is positive if x is greater than $2\frac{1}{2}$.
(b) $2x - 5 < 0$ if $2x < 5$ if $x < \frac{5}{2}$ (note that $<$ means 'less than'),
  i.e. the gradient is negative if x is less than $2\frac{1}{2}$.
(c) $2x - 5 = 0$ if $2x = 5$, i.e. $x = 2\frac{1}{2}$. The gradient is zero if $x = 2\frac{1}{2}$.
(d) When $x = -2$, gradient $= 2(-2) - 5 = -4 - 5 = -9$.
  When $x = +1$, gradient $= 2(+1) - 5 = +2 - 5 = -3$.
  When $x = +4$, gradient $= 2(+4) - 5 = +8 - 5 = +3$.

13.22 What is the value of $7 - 2x + 3x^2$ when its rate of increase with
respect to x is 4?

Let $$y = 7 - 2x + 3x^2$$

then $$\frac{dy}{dx} = -2 + 6x$$

This is the rate of increase of y with respect to x.

$$\therefore \quad -2 + 6x = 4$$
$$6x = 6$$
and $$x = 1$$

Then the value of $7 - 2x + 3x^2$ when $x = 1$ is $7 - 2 + 3 = 8$.

13.23 A spherical balloon is being blown up, and its radius is increasing at
the rate of 6 mm/s. At what rate is its surface increasing when the radius
is 60 mm?

The surface area $A = 4\pi r^2$ (where $r =$ radius),

$$\therefore \quad \frac{dA}{dr} = 8\pi r$$

$\therefore$ The rate of increase of the area A with respect to r is $8\pi r$—that is the
area is increasing $8\pi r$ times as fast as the radius. But the radius is increasing
at the rate of 6 mm/s.
  $\therefore$ The area A is increasing at the rate of $8\pi r \times 6$ mm²/s.
   But $r = 60$ mm
  $\therefore$ rate of increase of $A = (8\pi \times 60 \times 6)$ mm²/s
        $= 9048$ mm²/s

EXERCISE 13.6

1 Differentiate from first principles

(a) $y = 2x + 3$

(b) $s = 5t^2 - 2t + 3$

(c) $y = 4t^3 + t^2$

(d) $y = \dfrac{1}{x^3}$.

2 Differentiate by inspection:

(a) $y = 5x^4 + 3x - 2$

(b) $s = 6t^3 - 3t^2 + t + 5$

(c) $f(x) = 3\sqrt{x}$

(d) $s = r\theta$ (where r is a constant)

(e) $y = \dfrac{2x^2}{3} - \dfrac{5x}{6} + 7 - \dfrac{2}{x} + \dfrac{4}{x^2}$

(f) $y = \sqrt{x} + \dfrac{1}{\sqrt{x}}$

(g) $y = 2 \cdot 3x^{1 \cdot 2} - 2\sqrt{x} + \dfrac{7}{x}$

(h) $\theta = 18 + 10t - 3t^2$

(i) $y = 0 \cdot 3x^{1 \cdot 35}$.

3 Find the value of:

(a) $\dfrac{d}{du}(3u^2 + u - 5)$,

(b) $\dfrac{d}{dx}\left(x^{1 \cdot 2} - \dfrac{1}{x^{0 \cdot 2}} + 2x^{1 \cdot 5}\right)$

(c) $\dfrac{dp}{dv}$ if $pv^{1 \cdot 4} = 400$

(d) $\dfrac{dh}{dv}$ if $h = 0 \cdot 075v^{1 \cdot 5}$

(e) $\dfrac{d}{dh}\left(0 \cdot 4 + \dfrac{0 \cdot 001}{h}\right)$.

4 Find the value of $\dfrac{d^2s}{dt^2} + \dfrac{3ds}{dt}$ when $s = 6t^3 - 5t^2 + 8t$.

5 Evaluate $\dfrac{2d^2y}{dx^2} + \dfrac{dy}{dx}$ when $y = 5 + \dfrac{3}{x}$.

6 Find $\dfrac{d^3s}{dt^3}$ when $s = t^4 - 3t^3 + 2t - 5$.

7 If $y = x^2$, where x is the side of a square, what is the meaning of $\dfrac{dy}{dx}$?

At what rate is the area increasing when $x = 90$ mm and is growing at the rate of $0 \cdot 25$ mm/s?

8 A spherical toy-balloon is being blown up and its radius is increasing at the rate of 5 mm/s. At what rate is its volume increasing when the radius is 75 mm?

9 Find the side of a square whose area is increasing at the rate of 100 mm² per mm increase in its side.

10 Draw the graph of $6 - 2x^2$ for values of x between -3 and $+3$. Draw, as exactly as possible, the tangents at the points where x is $-1 \cdot 5$ and $+2$. Find the gradient of each tangent by using the scales on the axes, and compare the results with those obtained by differentiation.

11 Draw the graph of $\dfrac{x^2}{2} + 5$ between $x = -3$ and $x = +3$. Find, by drawing the tangents, the gradient of the curve at points whose abscissae (i.e. x values) are -2 and $+1$. Check your answers by differentiation.

12 Find the gradient of the curve $y = 3x^3 - 2x + 7$ at the points where $x = 1$ and $-0 \cdot 5$.

13 For what values of x is the gradient of the curve $x^3 + \dfrac{x^2}{2} - 2x + 6$ equal to (a) 0, (b) −2?

14 Find the gradient of $y = 6x^2 - \dfrac{5}{x}$ when $x = 0\!\cdot\!9$.

15 For what values of x is the gradient of the graph of $6 + 5x - x^2$ equal to (a) −2, (b) 0, (c) 1·5?

16 What is the value of the function $8 - 4x + \dfrac{x^2}{3}$ when its rate of increase with respect to x is 6?

17 (a) Differentiate from first principles $y = 2x^2 - 3x + 4$.

(b) Differentiate by inspection $y = \dfrac{5x^2}{3} - \dfrac{3x}{4} - 2 + \dfrac{3}{x} - \dfrac{2}{3x^2}$.

(c) For what values of x is the gradient of the graph of $2x^2 - 3x + 5$ equal to (i) −7, (ii) 0, (iii) 5? UEI

Motion in a straight line

If the magnitude of the displacement of a point moving in a straight line is given by $s = f(t)$, then the magnitude of the velocity of the point is given by the rate of change of s with respect to time, i.e. $\dfrac{ds}{dt} = f'(t) = v$, where v is the magnitude of the velocity, which may be called the speed.

The magnitude of the acceleration of the point is given by the rate of change of v with respect to time, i.e. $\dfrac{d^2s}{d^2s} = f''(t) = \dfrac{dv}{dt} = a$, where a is the magnitude of the acceleration. It must be noted that we are not differentiating displacement, velocity or acceleration, which are vector quantities, but their magnitudes, which are scalar quantities. Where the term magnitude is left out in the following examples in relation to displacement, velocity or acceleration it is to be understood.

Examples

13.24 A body moves s metres in t seconds in accordance with the equation $s = t^3 - 2t^2 + t + 3$.

Find when its velocity is 10 m/s and when its acceleration is 20 m/s².

$$s = t^3 - 2t^2 + t + 3$$

$$\therefore \quad \text{velocity} = \frac{ds}{dt} = 3t^2 - 4t + 1$$

$\therefore$ When the velocity is 10 m/s,

$$3t^2 = 4t + 1 = 10$$

which gives

$$t = 2\!\cdot\!52 \quad \text{or} \quad -1\!\cdot\!19 \text{ seconds}$$

(the latter answer being inadmissible).

$$\therefore \quad v = 10 \text{ m/s when } t = 2{\cdot}52 \text{ seconds.}$$

$$\text{Acceleration} = \frac{d^2s}{dt^2} = 6t - 4$$

$\therefore$ When the acceleration is 20 m/s²,

$$6t - 4 = 20$$
$$\therefore \quad t = 4 \text{ seconds}$$

13.25 A wheel rotates through θ radians in t seconds, where $\theta = 20 + 15t - t^2$.

Find
(a) the number of radians turned through in the first 6 seconds,
(b) the angular velocity in radians per second at the end of 4 seconds,
(c) the initial angular velocity,
(d) the acceleration in rad/s²,
(e) at what time the wheel will come to rest.

(a) $\theta = 20 + 15t - t^2$.
   When $t = 0$ seconds, $\theta = 20$ radians.
   When $t = 6$ seconds, $\theta = 20 + 90 - 36 = 74$ radians.
   $\therefore$ Number of radians turned through in 6 seconds $= 74 - 20 = 54$ radians.

(b) Angular velocity $(\omega) = \dfrac{d\theta}{dt}$

$$= 15 - 2t$$
$\therefore$ When $t = 4$, $\omega = 15 - 8 = 7$ rad/s.
(c) Initial angular velocity = velocity when $t = 0$

$$\omega = 15 - 2t, \text{ (see above).}$$

$\therefore$ When $t = 0$, initial angular velocity $= 15$ rad/s.
(d) Acceleration = change in velocity with respect to time

$$= \frac{d\omega}{dt}\left(= \frac{d^2\theta}{dt^2}\right)$$

$$\omega = 15 - 2t$$

$$\therefore \quad \frac{d\omega}{dt} = -2$$

$\therefore$ Acceleration $= -2$ rad/s².
(e) Time when wheel comes to rest = time when $\omega = 0$.

$$\text{But} \quad \omega = 15 - 2t,$$
$$\therefore \quad \omega = 0 \text{ when } t = 7\tfrac{1}{2} \text{ seconds.}$$

EXERCISE 13.7

1 If $s = 20 + 100t - 4t^2$, where s metres is the space covered by a moving body in t seconds, what is the speed acquired in (*a*) t seconds, (*b*) 4 seconds?
Show that the acceleration is constant.

2 A body moves s metres in t seconds, where $s = 10 + 5t + 12t^2 - t^3$. Find (*a*) its speed at the end of 2 seconds, (*b*) its acceleration at the end of 3 seconds, (*c*) when its acceleration is zero, (*d*) when its speed is zero.

3 Find the speed and acceleration of a body at the end of 5 seconds if its equation of motion is $s = 3\cdot1 - 5t + 6t^2$.

4 A body moves s metres in t seconds in accordance with the equation $s = 10 + 6t + 13t^2 - t^3$.
Find (*a*) its speed at the end of 3 seconds, (*b*) when its speed is zero, (*c*) its acceleration at the end of 2 seconds, (*d*) when its acceleration is zero.

5 A body moves through a distance s metres in t seconds, where $s = 15 + 8t + 9t^2 - t^3$. Find, by means of the calculus, (*a*) its velocity at the end of 3 seconds, (*b*) its acceleration at the end of 2·5 seconds, (*c*) the value of t when the velocity is 23 m/s, (*d*) the value of t when the velocity is zero, (*e*) the value of t when the acceleration is zero.   UEI

6 The relationship between the displacement, x metres, and the time, t seconds, of a moving body is $x = 2t^3 - 9t^2 + 12t + 4$. Find (*a*) when it comes to rest, (*b*) when its acceleration is zero, (*c*) the velocity when the acceleration is zero, (*d*) the acceleration when the velocity is zero.   NCTEC

7 (*a*) Differentiate with respect to x: $3x^{\frac{4}{3}} + x^{-\frac{1}{2}} + 5x^{1\cdot2} + 4$.   (*b*) A flywheel rotates according to the law $\theta = 28t - 3t^2$, θ giving the angular rotation in radians taking place in t seconds. Find expressions for the angular velocity $\left(\dfrac{d\theta}{dt}\right)$ and the angular acceleration $\left(\dfrac{d^2\theta}{dt^2}\right)$. What is the value of t when $\dfrac{d\theta}{dt} = 0$. Find the value of θ for this value of t.

                                                                    ULCI

8 (*a*) Differentiate the following from first principles: (i) $y = x^2 + 3x$, (ii) $y = \dfrac{1}{x}$.
   (*b*) If $P = 4s^3 - 2s + 4$, find the values of s at which the rate of change of P with respect to s is (i) 0, (ii) 34.
   (*c*) The velocity of a body after time t seconds is given by the expression $v = 3t^2 - 20t$. Calculate the time at which the acceleration of the body is 10 m/s^2.   CRAWLEY

9 (*a*) Differentiate with respect to x (i) $y = (3x^2 + 2x)(x - 4)$,
   (ii) $y = 2x^{10} - \dfrac{12}{5x^4} - 7\sqrt{x^5}$, (iii) $y = \dfrac{3x^3 - 4x + 5}{x^2}$.

  (*b*) If s is measured in metres and t in seconds, and $s = t^4 - 5t^3 + 6t^2 + 7t + 4$, find (i) the times when the acceleration is zero, (ii) the velocity when the acceleration is zero for the first time, (iii) the distance travelled when the acceleration is zero for the first time.

GRAVESEND

10 (*a*) The displacement of a point is given in terms of time t by the equation $s = 4 + 3t - t^2 + t^3$. By differentiating, find the velocity at any time t. By differentiating again, find the acceleration at any time t, and hence find the value of t for which the velocity is neither increasing nor decreasing.

  (*b*) Differentiate by rule (i) $y = 4x^3 - 7x + 2$, (ii) $y = x^2 + \dfrac{1}{x^2}$.

EMEU

Differentiation of sin *x* and cos *x* from first principles

To verify that $\displaystyle\lim_{\theta \to 0} \frac{\sin \theta}{\theta} = 1$, where θ is measured in radians.

  O is the centre of a circle, radius r. Let $\angle AOB = \theta$ (an acute angle in

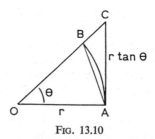

Fig. 13.10

radians). AC is the tangent to the circle at A, cutting OB produced at C so that $\angle OAC = 90°$. $\text{Tan } \theta = \dfrac{AC}{OA} = \dfrac{AC}{r}$ and $AC = r \tan \theta$.

$$\triangle AOB < \text{sector AOB} < \triangle AOC$$

i.e. $\tfrac{1}{2}r^2 \sin \theta < \tfrac{1}{2}r^2\theta < \tfrac{1}{2}r \,.\, r \tan \theta$.

Dividing by $\dfrac{r^2}{2}$,

$$\sin \theta < \theta < \frac{\sin \theta}{\cos \theta}$$

Dividing by $\sin \theta$, which is positive,

$$1 < \frac{\theta}{\sin \theta} < \frac{1}{\cos \theta}$$

Inverting the terms,

$$1 > \frac{\sin \theta}{\theta} > \cos \theta$$

Note that if we invert the terms of $1 < \frac{6}{5} < \frac{5}{4}$ then $1 > \frac{5}{6} > \frac{4}{5}$

i.e. the inequality signs are reversed.

Now as $\theta \to 0$, $\cos \theta \to 1$ and $\frac{\sin \theta}{\theta}$ is between 1 and a value which approaches 1 as closely as we please,

i.e. $$\lim_{\theta \to 0} \frac{\sin \theta}{\theta} = 1$$

Note that if θ is replaced by δx we have

$$\lim_{\delta x \to 0} \frac{\sin \delta x}{\delta x} = 1$$

In all subsequent work in the calculus of trigonometrical functions *radian* measure is to be understood.

To differentiate $y = \sin x$ from first principles

If $y = \sin x$ then $y + \delta y = \sin(x + \delta x) = \sin x \cos \delta x + \cos x \sin \delta x$

and $y - \delta y = \sin(x - \delta x) = \sin x \cos \delta x - \cos x \sin \delta x$

Subtracting $\qquad 2\delta y = 2 \cos x \sin \delta x$

Dividing by $2\delta x \qquad \dfrac{\delta y}{\delta x} = \cos x \dfrac{\sin \delta x}{\delta x}$

As $\delta x \to 0$, $\dfrac{\sin \delta x}{\delta x} \to 1$

$$\therefore \quad \frac{dy}{dx} = \cos x \quad \text{or} \quad \frac{d(\sin x)}{dx} = \cos x$$

To differentiate $y = \cos x$ from first principles

If $y = \cos x$ then $y + \delta y = \cos(x + \delta x) = \cos x \cos \delta x - \sin x \sin \delta x$

and $y - \delta y = \cos(x - \delta x) = \cos x \cos \delta x + \sin x \sin \delta x$

Subtracting $\qquad 2\delta y = -2 \sin x \sin \delta x$

Dividing by $2\delta x \qquad \dfrac{\delta y}{\delta x} = -\sin x \dfrac{\sin \delta x}{\delta x}$

As $\delta x \to 0$, $\dfrac{\sin \delta x}{\delta x} = 1$

$$\therefore \quad \frac{dy}{dx} = \sin x \quad \text{or} \quad \frac{d(\cos x)}{dx} = -\sin x$$

Examples

13.26 (*a*) If $y = \left(\sin \frac{x}{2} + \cos \frac{x}{2} \right)^2$, find $\frac{dy}{dx}$.

(*b*) If $f(\theta) = \theta^3 - 3 + \sin \theta + 2 \cos \theta$, find $f'(\theta)$, i.e. $\frac{df(\theta)}{d\theta}$, and $f''(\theta)$, i.e. $\frac{d^2f(\theta)}{d\theta^2}$.

(*a*) $y = \left(\sin \frac{x}{2} + \cos \frac{x}{2} \right)^2 = \sin^2 \frac{x}{2} + 2 \sin \frac{x}{2} \cos \frac{x}{2} + \cos^2 \frac{x}{2}$

Now $\sin^2 \frac{x}{2} + \cos^2 \frac{x}{2} = 1$ and $2 \sin \frac{x}{2} \cos \frac{x}{2} = \sin x$

$\therefore \quad y = 1 + \sin x$ and $\frac{dy}{dx} = \cos x$

(*b*) $f(\theta) = \theta^3 - 3 + \sin \theta + 2 \cos \theta$

Differentiating with respect to θ, then
$$f'(\theta) = 3\theta^2 + \cos \theta - 2 \sin \theta$$
and $f''(\theta) = 6\theta - \sin - 2 \cos \theta$

13.27 If $y = a \sin x + b \cos x$, where a and b are constants, prove that

$$\frac{d^2y}{dx^2} + y = 0.$$

If $y = a \sin x + b \cos x$ . . . . (i)

then $\frac{dy}{dx} = a \cos x - b \sin x$

and $\frac{d^2y}{dx^2} = -a \sin x - b \cos x$ . . . . (ii)

Adding (i) and (ii),

$$\frac{d^2y}{dx^2} + y = 0$$

The product rule of differentiation, $\frac{d(uv)}{dx} = u \frac{dv}{dx} + v \frac{du}{dx}$, where u and v are functions of x

If $y = uv$, let x increase to $x + \delta x$, then u increases to $u + \delta u$, v to $v + \delta v$ and y to $y + \delta y$,

i.e. $y + \delta y = (u + \delta u)(v + \delta v) = uv + u\delta v + v\delta u + \delta u \delta v$
and $y \qquad\qquad\qquad\qquad\qquad = uv$
Subtracting $\delta y = u\delta v + v\delta u + \delta u \delta v$

Dividing by δx, $\qquad \dfrac{\delta y}{\delta x} = u\dfrac{\delta v}{\delta x} + v\dfrac{\delta u}{\delta x} + \delta u\dfrac{\delta v}{\delta x}$

As $\delta x \to 0$: $\dfrac{\delta y}{\delta x} \to \dfrac{dy}{dx}$, $\dfrac{\delta v}{\delta x} \to \dfrac{dv}{dx}$, $\dfrac{\delta u}{\delta x} \to \dfrac{du}{dx}$, and $\delta u \to 0$

i.e. $\qquad\qquad \dfrac{dy}{dx} = u\dfrac{dv}{dx} + v\dfrac{du}{dx} + 0\dfrac{dv}{dx}$

Replacing y by uv,

i.e. $\qquad\qquad \dfrac{d(uv)}{dx} = u\dfrac{dv}{dx} + v\dfrac{du}{dx}$

Example

13.28 If $y = x^3 \sin x$, find $\dfrac{dy}{dx}$.

Let $u = x^3$ and $v = \sin x$, so that $uv = x^3 \sin x$, $\dfrac{du}{dx} = 3x^2$, and

$\dfrac{dv}{dx} = \cos x$.

Since $\qquad\qquad \dfrac{d(uv)}{dx} = u\dfrac{dv}{dx} + v\dfrac{du}{dx}$, using the above results,

$\qquad\qquad \dfrac{d(x^3 \sin x)}{dx} = x^3 \cos x + \sin x\, 3x^2$

i.e. $\qquad\qquad \dfrac{dy}{dx} = x^2(x \cos x + 3 \sin x)$

EXERCISE 13.8

1 Find (a) $\dfrac{d(x^2 \cos x)}{dx}$, (b) $\dfrac{d(3x \sin x)}{dx}$, (c) $\dfrac{d(\theta^{1.2} \cos \theta)}{d\theta}$.

2 If $y = x^2(1 - x^3)$, find $\dfrac{dy}{dx}$ (a) by first multiplying out the bracket,
(b) by using the product rule. Show that the results are consistent.

3 If $y = x \sin x$, find $\dfrac{dy}{dx}$ and prove that $x\dfrac{dy}{dx} - y = x^2 \cos x$.

4 If $y = x^n(\cos x - \sin x)$, show that $\dfrac{dy}{dx} = x^{n-1}[(n - x)\cos x - (x + n)$
$\sin x]$.

5 If $y = \dfrac{\sin x}{x^5}$, find $\dfrac{dy}{dx}$ by writing y as $x^{-5} \sin x$.

The function of a function rule of differentiation

If $y = (1 - x^2)^3$, then y is a function of the function $1 - x^2$, namely y is the cube of the function $1 - x^2$.

Suppose that y is a function of z, where z is a function of x. If x increases to $x + \delta x$, then z will increase to $z + \delta z$ and y will increase to $y + \delta y$.

Then
$$\frac{\delta y}{\delta x} = \frac{\delta y}{\delta x} \cdot 1 = \frac{\delta y}{\delta x} \cdot \frac{\delta z}{\delta z} = \frac{\delta y}{\delta z} \cdot \frac{\delta z}{\delta x}$$

As $\delta x \to 0$, $\delta z \to 0$

$$\frac{\delta y}{\delta x} \to \frac{dy}{dx}, \frac{\delta y}{\delta z} \to \frac{dy}{dz} \quad \text{and} \quad \frac{\delta z}{\delta x} \to \frac{dz}{dx}$$

$$\therefore \quad \frac{dy}{dx} = \frac{dy}{dz} \cdot \frac{dz}{dx}, \quad \text{which is the function of a function rule of differentiation.}$$

Examples

13.29 If $y = (1 - x^2)^3$, find $\dfrac{dy}{dx}$.

Let $z = 1 - x^2$, then $y = z^3$ since $y = (1 - x^2)^3$.

Differentiating,

$$\frac{dz}{dx} = -2x \quad \text{and} \quad \frac{dy}{dz} = 3z^2$$

Since
$$\frac{dy}{dx} = \frac{dy}{dz} \cdot \frac{dz}{dx}$$

Substituting, $\quad \dfrac{dy}{dx} = 3z^2(-2x) = 3(1 - x^2)^2(-2x) = -6x(1 - x^2)^2$

13.30 If a and b are constants, find $\dfrac{dy}{dx}$ if (a) $y = (ax + b)^n$, (b) $y = \sin(ax + b)$, (c) $y = \cos(ax + b)$.

(a) If $y = (ax + b)^n$, let $z = ax + b$, then $y = z^n$,

$$\frac{dz}{dx} = a \quad \text{and} \quad \frac{dy}{dz} = nz^{n-1}$$

$$\therefore \quad \frac{dy}{dx} = \frac{dy}{dz} \cdot \frac{dz}{dx} = nz^{n-1} a$$

$$= an(ax + b)^{n-1}$$

(b) If $y = \sin(ax + b)$, let $z = ax + b$, then $y = \sin z$.

$$\frac{dz}{dx} = a \quad \text{and} \quad \frac{dy}{dz} = \cos z$$

$$\therefore \quad \frac{dy}{dx} = \frac{dy}{dz} \cdot \frac{dz}{dx} = (\cos z)\, a = a \cos(ax + b)$$

(c) If $y = \cos(ax + b)$, let $z = ax + b$, then $y = \cos z$.

$$\frac{dz}{dx} = a \quad \text{and} \quad \frac{dy}{dz} = -\sin z$$

$$\therefore \quad \frac{dy}{dx} = \frac{dy}{dz} \cdot \frac{dz}{dx} = (-\sin z)\, a = -a \sin(ax + b)$$

If $y = ax^n$, where n is a negative integer, to prove that $\dfrac{dy}{dx} = nax^{n-1}$

If $y = ax^n$, where n is a negative integer, replace n by $-m$, where m is a positive integer: then $y = ax^{-m} = a(x^m)^{-1}$. Let $z = x^m$ and $\dfrac{dz}{dx} = mx^{m-1}$:

then $y = az^{-1}$ and $\dfrac{dy}{dz} = -az^{-2}$ (previously justified from first principles).

Thus
$$\frac{dy}{dx} = \frac{dy}{dz} \cdot \frac{dz}{dx} = -az^{-2} mx^{m-1}$$

$$= -a(x^m)^{-2} mx^{m-1} = -amx^{-2m} x^{m-1}$$
$$= -amx^{-m-1}$$

Replacing $-m$ by n,

$$\frac{dy}{dx} = nax^{n-1}$$

If $y = ax^n$, where n is a positive fraction $\dfrac{p}{q}$, to prove that $\dfrac{dy}{dx} = nax^{n-1}$

Since $y = ax^n$, replacing n by $\dfrac{p}{q}$ then $y = ax^{p/q}$ and $y^q = a^q x^p$ (raising each side to the qth power).

$$\therefore \quad \frac{d(y^q)}{dx} = a^q \frac{d(x^p)}{dx}$$

i.e.
$$\frac{d(y^q)}{dy} \frac{dy}{dx} = a^q p x^{p-1}$$

i.e.
$$q y^{q-1} \frac{dy}{dx} = a^q p x^{p-1}$$

and
$$\frac{dy}{dx} = a^q \cdot \frac{p}{q} \cdot \frac{x^{p-1}}{y^{q-1}} = \frac{a^q p x^{p-1}}{q(ax^{p/q})^{q-1}} = \frac{p}{q} \cdot \frac{a^q \cdot x^{p-1}}{a^{q-1} \cdot x^{p-p/q}} = \frac{p}{q} ax^{p/q-1}$$

Replacing $\dfrac{p}{q}$ by n,

$$\frac{\mathrm{d}y}{\mathrm{d}x} = nax^{n-1}$$

Note that the above argument applies when p/q is a negative fraction, regarding p as a negative integer.

EXERCISE 13.9

1 Find (a) $\dfrac{\mathrm{d}(2 - 3x)^4}{\mathrm{d}x}$, (b) $\dfrac{\mathrm{d}\sin(2 - 3x)}{\mathrm{d}x}$, (c) $\dfrac{\mathrm{d}\cos(2 - 3x)}{\mathrm{d}x}$.

2 If $y = \sin(5x + 3)$, find $\dfrac{\mathrm{d}y}{\mathrm{d}x}$ and $\dfrac{\mathrm{d}^2y}{\mathrm{d}x^2}$, and show that $\dfrac{\mathrm{d}^2y}{\mathrm{d}x^2} + 25y = 0$.

3 If $y = \sin 2x$, find $\dfrac{\mathrm{d}y}{\mathrm{d}x}$ (a) using the function of a function rule, (b) by writing y as $2 \sin x \cos x$ and using the product rule. Show that the results are consistent.

4 Find (a) $\dfrac{\mathrm{d}(1 + \theta^2)^3}{\mathrm{d}\theta}$, (b) $\dfrac{\mathrm{d}\sin(1 + \theta^2)}{\mathrm{d}\theta}$, (c) $\dfrac{\mathrm{d}\cos(1 + \theta^2)}{\mathrm{d}\theta}$.

5 Prove that $\dfrac{\mathrm{d}\sin ax}{\mathrm{d}x} = a \cos ax$ and $\dfrac{\mathrm{d}\cos ax}{\mathrm{d}x} = -a \sin ax$. Find $\dfrac{\mathrm{d}(\cos 3x + \cos 5x + \sin 2x + \sin 4x)}{\mathrm{d}x}$.

6 (a) Show from first principles that $\dfrac{\mathrm{d}(2x^2 + 3)}{\mathrm{d}x} = 4x$.

(b) Differentiate the following with respect to x: (i) $\dfrac{1}{(1 + 2x)^3}$,

(ii) $\sin x \cos ax$.   ULCI

7 Differentiate the following: (a) $3x^2 - 6 + \dfrac{4}{x^2}$, (b) $(3x + 2)^5$, (c) $3x^2 \sin x$.
                                                                                             ULCI

Application of differentiation to small changes

To show that $\delta y \simeq \left(\dfrac{\mathrm{d}y}{\mathrm{d}x}\right)\delta x$, if δx is sufficiently small.

$$\frac{\delta y}{(\mathrm{d}y/\mathrm{d}x)\delta x} = \frac{\delta y/\delta x}{\mathrm{d}y/\mathrm{d}x} \cdot \quad \downarrow \quad \cdot \quad \cdot \quad \cdot \quad \cdot \quad \cdot \quad \cdot \quad \text{(i)}$$

Now, as $\delta x \to 0$, $\dfrac{\delta y}{\delta x} \to \dfrac{\mathrm{d}y}{\mathrm{d}x}$ and the R.H.S. of (i) $\to 1$.

Hence, if δx is sufficiently small, $\dfrac{\delta y}{(\mathrm{d}y/\mathrm{d}x)\delta x} \simeq 1$, or $\delta y \simeq \dfrac{\mathrm{d}y}{\mathrm{d}x}\delta x$.

i.e.   change in $y \simeq \dfrac{\mathrm{d}y}{\mathrm{d}x} \times$ change in x.

Examples
13.31 Find the approximate change in y if (*a*) $y = 2x^3$ and x changes
from 1 to 1·001, (*b*) $y = \dfrac{1}{x^2}$ and x changes from 2 to 1·998.

(*a*) $\delta y \simeq \dfrac{dy}{dx}\delta x$, i.e. $\delta y \simeq 6x^2$

Since $x = 1$ and $\delta x = 0·001$,

$\qquad \delta y \simeq 6 \cdot 1^2 \cdot 0·001 = 0·006$

Check Exact change in $y = 2[(1·001)^3 - 1^3] = 0·006\ 006\ 002$.

(*b*) $\delta y \simeq \dfrac{dy}{dx}\delta x$, i.e. $\delta y \simeq \dfrac{-2}{x^3}\delta x$

Since $x = 2$ and $\delta x = -0·002$,

$\qquad \delta y \simeq \dfrac{-2}{2^3}(-0·002) = +0·0005$

Check Exact change in $y = \dfrac{1}{1·998^2} - \dfrac{1}{2^2} = 0·000\ 500\ 8$ (correct to seven
decimal places).
   When the approximate formula is used, it is necessary to have some estimate of the relative error involved. This will be illustrated in the following
example.

13.32 If $y = x^2$, find (*a*) δy, (*b*) $\dfrac{dy}{dx}\delta x$, (*c*) the error $\delta y - \dfrac{dy}{dx}\delta x$,

(*d*) the relative error, i.e. $\dfrac{\text{error}}{\delta y}$. Show that the relative error tends to zero

with $\dfrac{\delta x}{x}$. Find the percentage relative error if (i) x changes from 2 to
2·001, (ii) x changes from 0·001 to 0·2.

(*a*) $(y + \delta y) - y = (x + \delta x)^2 - x^2$
$\therefore \qquad\qquad \delta y = 2x\delta x + \delta x^2$

(*b*) $\dfrac{dy}{dx}\delta x = 2x\delta x$

(*c*) The error $= (2x\delta x + \delta x^2) - (2x\delta x) = +\delta x^2$

(*d*) The relative error $= \dfrac{\text{error}}{\delta y} = \dfrac{+\delta x^2}{2x\delta x + \delta x^2} = \dfrac{+\delta x}{2x + \delta x} = \dfrac{+(\delta x/x)}{2 + (\delta x/x)}$

   As $\dfrac{\delta x}{x} \to 0$, the relative error $\to 0$, i.e. it is necessary that δx be small
compared with x.

(i) If $x = 2$, $\delta x = 0.001$, relative error $= \dfrac{+(0.001)/2}{2 + (0.001)/2} \times 100 \simeq$ $+0.025\%$.

(ii) If $x = 0.001$, $\delta x = 0.199$, $\dfrac{\delta x}{x} = 199$.

Relative error $= \dfrac{+199}{2 + 199} \, 100 \simeq +99.0\%$.

EXERCISE 13.10

1 If $y = x^4 - x$, find the approximate change in y if x changes from (a) 2 to 2·001, (b) -1 to -1.001. Calculate also the exact changes by arithmetic.

2 If $y = \dfrac{1}{x^2}$ calculate the approximate change in y if x changes from 1 to 0·999. Calculate the change in y correct to six decimal places by arithmetic.

3 (a) If $y = \cos x$, find an approximate formula for the change produced in y due to a small change δx in x.

  (b) Find the change produced in y if x changes from 1·25 to 1·26 radians, given that the sin (1·25 radian) $= 0.9490$.

  (c) Given that $\cos(1.25\ \text{radian}) = 0.3153$, estimate $\cos(1.26\ \text{radian})$.

4 If $y = x^3$, find (a) δy, (b) $\dfrac{dy}{dx}\delta x$, (c) $\delta y - \dfrac{dy}{dx}\delta x$, (d) find the relative error $\dfrac{\delta y - (dy/dx)\delta x}{\delta y}$, (e) calculate, correct to three significant figures, the percentage relative error when x changes from (i) 1 to 1·001, (ii) 0·001 to 0·1.

Chapter Fourteen
Integration

Indefinite integration as the inverse process to differentiation

Consider $\dfrac{d(x^3)}{dx} = 3x^2$, $\dfrac{d(x^3 + 5)}{dx} = 3x^2$, $\dfrac{d(x^3 - 7)}{dx} = 3x^2$, or, more

generally $\dfrac{d(x^3 + C)}{dx} = 3x^2$, where C is any constant. The differential

coefficient of $x^3 + C$ with respect to x is $3x^2$. How is the reversed process described? We say that the indefinite integral of $3x^2$ with respect to x is $x^3 + C$. This is written in the following way:

$$\int 3x^2 dx = x^3 + C$$

and reads

'integral $3x^2 dx$ equals $x^3 + C$'.

$3x^2$ is called the *integrand*, and $x^3 + C$ the *indefinite integral* or *integral function*, where C is the *constant of integration* or the *arbitrary constant*.

The standard integral $\int x^n \cdot dx = \dfrac{x^{n+1}}{n+1} + C$, where $n \neq -1$

Since $\dfrac{dx^N}{dx} = Nx^{N-1}$, multiplying by $\dfrac{1}{N}$ where $N \neq 0$,

$$\frac{d(x^N/N)}{dx} = x^{N-1}$$

Replacing $N - 1$ by n, for convenience, so that $N = n + 1$, and introducing an arbitrary constant,

$$\frac{d[(x^{n+1})/(n+1) + C]}{dx} = x^n$$

hence
$$\int x^n dx = \frac{x^{n+1}}{n+1} + C \quad . \quad . \quad . \quad . \quad . \quad \text{(i)}$$

Since $n = N - 1$ and $N \neq 0$,

$$n \neq -1$$

Note that when $n = -1$, the right hand side of (i) becomes indeterminate, having a zero denominator.

Example

14.1 Evaluate and check (a) $\int x^4 dx$, (b) $\int \frac{1}{x^2}dx$, (c) $\int \frac{1}{z^{2\cdot3}}dz$.

(a) Since $\int x^n dx = \frac{x^{n+1}}{n+1} + C$, replacing n by 4,

$$\int x^4 dx = \frac{x^{4+1}}{4+1} + C = \frac{x^5}{5} + C$$

A convenient way of checking is to denote the integral by I and show that $\frac{dI}{dx}$ is the integrand.

Thus $I = \frac{x^5}{5} + C$, $\frac{dI}{dx} = \frac{5x^4}{5} + 0 = x^4$ (the integrand).

(b) Similarly, $\int \frac{1}{x^2}dx = \int x^{-2}dx = \frac{x^{-2+1}}{-2+1} + C$

$$= \frac{x^{-1}}{-1} + C = \frac{-1}{x} + C$$

Thus $I = -x^{-1} + C$, $\frac{dI}{dx} = +x^{-2} + 0 = \frac{1}{x^2}$ (the integrand).

(c) $\int \frac{1}{z^{2\cdot3}}dz = \int z^{-2\cdot3}dz = \frac{z^{-2\cdot3+1}}{-2\cdot3+1} + C$

$$= \frac{z^{-1\cdot3}}{-1\cdot3} + C$$

Thus $I = - \frac{z^{-1\cdot3}}{1\cdot3} + C$ and $\frac{dI}{dz} = +\frac{1\cdot3z^{-2\cdot3}}{1\cdot3} = z^{-2\cdot3}$

A constant factor of the integrand may be written before the integral sign

We may use the fact that $\frac{d(\int f(x)dx)}{dx} = f(x)$, since integration is the inverse process to differentiation.

Since $$\frac{d(a\int f(x)dx)}{dx} = af(x)$$

integrating both sides with respect to x,

$$a\int f(x)dx = \int af(x)dx$$

i.e. a constant factor of the integrand may be written before the integral sign.

To show that $\int ax^n \cdot dx = \dfrac{ax^{n+1}}{n+1} + C$, **where a is a constant and $n \neq -1$**

$$\int ax^n dx = a\int x^n dx = a\left(\frac{x^{n+1}}{n+1} + A\right)$$

$$= \frac{ax^{n+1}}{n+1} + aA = \frac{ax^{n+1}}{n+1} + C \text{ (replacing } aA \text{ by } C)$$

Thus $\int 5x^5 dx = \dfrac{5x^{5+1}}{5+1} + C = \dfrac{5x^6}{6} + C$

Check $\dfrac{dI}{dx} = \dfrac{6 \cdot 5x^5}{6} + 0 = 5x^5$

$$\int \frac{-3}{\sqrt[5]{\theta^2}}\, d\theta = \int \frac{-3 d\theta}{\theta^{\frac{2}{5}}} = \int -3\theta^{-\frac{2}{5}} d\theta$$

$$= \frac{-3\theta^{-\frac{2}{5}+1}}{-\frac{2}{5}+1} + C = \frac{-3\theta^{\frac{3}{5}}}{\frac{3}{5}} + C$$

$$= -5\theta^{\frac{3}{5}} + C$$

Check $\dfrac{dI}{d\theta} = -3\theta^{-\frac{2}{5}} = \dfrac{-3}{\sqrt[5]{\theta^2}}$

To prove that $\int(u + v - w)dx = \int u dx + \int v dx - \int w dx$, where u, v and w are functions of x

Since $\dfrac{d(\int u dx + \int v dx - \int w dx)}{dx} = u + v - w$, integrating both sides with respect to x,

$$\int u dx + \int v dx - \int w dx = \int(u + v - w)dx$$

This result may be extended to any finite number of functions.

Example

14.2 Evaluate $\displaystyle\int\left(x^3 + x^2 - \frac{1}{x^4} - \frac{4}{\sqrt[4]{x^3}}\right)dx$

$\int(x^3 + x^2 - x^{-4} - 4x^{-\frac{3}{4}})dx = \int x^3 dx + \int x^2 dx - \int x^{-4} dx - \int 4x^{-\frac{3}{4}}dx$

Note that it is usual to include only one arbitrary constant, since several arbitrary constants can always be combined into one arbitrary constant, as follows.

$$\int x^3 dx + \int x^2 dx - \int x^{-4} dx - \int 4x^{-\frac{3}{4}}dx$$

$$= \left(\frac{x^4}{4} + A\right) + \left(\frac{x^3}{3} + B\right) - \left(\frac{x^{-3}}{-3} + C\right) - \left(\frac{4x^{\frac{1}{4}}}{\frac{1}{4}} + D\right)$$

$$= \frac{x^4}{4} + \frac{x^3}{3} + \frac{1}{3x^3} - 16x^{\frac{1}{4}} + A + B - C - D$$

$$= \frac{x^4}{4} + \frac{x^3}{3} + \frac{1}{3x^3} - 16x^{\frac{1}{4}} + E$$

Check $\dfrac{\mathrm{d}I}{\mathrm{d}x} = \dfrac{4x^3}{4} + \dfrac{3x^2}{3} - \dfrac{3x^{-4}}{3} - \dfrac{16x^{-\frac{7}{4}}}{4}$

$= x^3 + x^2 - \dfrac{1}{x^4} - \dfrac{4}{\sqrt[4]{x^3}}$

The integral $\int a\,\mathrm{d}x$, where a is a constant

$$\int a\,\mathrm{d}x = a\int 1\,\mathrm{d}x = a\int x^0\,\mathrm{d}x$$

$$= a\left(\dfrac{x^1}{1} + A\right) = ax + aA$$

$$= ax + C$$

i.e. $\int a\,\mathrm{d}x = ax + C$

Check $\dfrac{\mathrm{d}I}{\mathrm{d}x} = a$

Thus $\int 7\,\mathrm{d}x = 7x + C$
$\int -5\,\mathrm{d}x = -5x + C$
$\int \mathrm{d}x$ or $\int 1\,\mathrm{d}x = x + C$

Conversion of the integrand into indicial form

Where the integrand contains terms in surd form, it is necessary to convert the terms into indicial form.

Example

14.3 Evaluate $\displaystyle\int\left(\sqrt{x} + \dfrac{1}{\sqrt{x}} + \sqrt[3]{x^2} - 5\right)\mathrm{d}x.$

$$I = \int(x^{\frac{1}{2}} + x^{-\frac{1}{2}} + x^{\frac{2}{3}} - 5)\mathrm{d}x$$

$$= \dfrac{x^{\frac{3}{2}}}{\frac{3}{2}} + \dfrac{x^{\frac{1}{2}}}{\frac{1}{2}} + \dfrac{x^{\frac{5}{3}}}{\frac{5}{3}} - 5x + C$$

$$= \dfrac{2x^{\frac{3}{2}}}{3} + 2x^{\frac{1}{2}} + \dfrac{3x^{\frac{5}{3}}}{5} - 5x + C$$

Check $\dfrac{\mathrm{d}I}{\mathrm{d}x} = x^{\frac{1}{2}} + x^{-\frac{1}{2}} + x^{\frac{2}{3}} - 5 = \sqrt{x} + \dfrac{1}{\sqrt{x}} + \sqrt[3]{x^2} - 5$

Expansion of the integrand and term by term integration

If the integrand is in factorised form, it may be necessary to multiply out the factors and integrate term by term.

Example
14.4 Evaluate $\int(2x - 1)(x^2 + 1)dx$.

$$\int(2x - 1)(x^2 + 1)dx = \int(2x^3 - x^2 + 2x - 1)dx$$

$$= \frac{2x^4}{4} - \frac{x^3}{3} + \frac{2x^2}{2} - x + C$$

$$= \frac{x^4}{2} - \frac{x^3}{3} + x^2 - x + C$$

Check

$$\frac{dI}{dx} = \frac{4x^3}{2} - \frac{3x^2}{3} + 2x - 1$$

$$= 2x^3 - x^2 + 2x - 1$$

$$= (2x - 1)(x^2 + 1)$$

To show that $\int(ax + b)^n \cdot dx = \frac{1}{a}\frac{(ax + b)^{n+1}}{(n + 1)} + C$, where $n \neq 1$, a and b are constants, and $a \neq 0$

$$\frac{d}{dx}\left(\frac{1}{a}\frac{(ax + b)^{n+1}}{(n + 1)} + C\right) = \frac{1}{a(n + 1)}\frac{d(ax + b)^{n+1}}{dx}$$

$$= \frac{(n + 1)}{a(n + 1)}(ax + b)^n a = (ax + b)^n$$

Hence

$$\int(ax + b)^n dx = \frac{1}{a}\frac{(ax + b)^{n+1}}{(n + 1)} + C$$

Note that $\dfrac{d(ax + b)^{n+1}}{dx}$ is evaluated by the function of a function rule

since, if $y = (ax + b)^{n+1}$ and $z = ax + b$, then

$$y = z^{n+1}$$

$$\frac{dz}{dx} = a \quad \text{and} \quad \frac{dy}{dz} = (n + 1)z^n$$

and $\quad \dfrac{dy}{dx} = \dfrac{dy}{dz} \cdot \dfrac{dz}{dx} = (n + 1)z^n a = (n + 1)(ax + b)^n a$

Examples
14.5 Evaluate $\int(3x - 2)^6 dx$.

Since $\quad \int(ax + b)^n dx = \dfrac{(ax + b)^{n+1}}{a(n + 1)} + C$, put $a = 3$, $b = -2$,

$$n = 6,$$

then $\quad \int(3x - 2)^6 dx = \dfrac{(3x - 2)^7}{3 \cdot 7} + C = \dfrac{(3x - 2)^7}{21} + C$

14.6 Evaluate $\int(5 - 6x)^{4\cdot4}dx$

Since $\int(ax + b)^n dx = \dfrac{(ax + b)^{n+1}}{a(n + 1)} + C$,   Put $a = -6,\ \ b = 5,$

$$n = 4\cdot4,$$

then $\int(-6x + 5)^{4\cdot4}dx = \dfrac{(-6x + 5)^{5\cdot4}}{-6\cdot5\cdot4} + C = \dfrac{-(5 - 6x)^{5\cdot4}}{32\cdot4} + C$

Note carefully that this standard integral does not apply to the integration of functions of the type $(3x^3 - 6)^7$ or $\left(5 - \dfrac{6}{x^2}\right)^4$. The function contained in the bracket must be a linear function of the type $ax^1 + b$.

EXERCISE 14.1

Integrate the following functions with respect to the appropriate variable and check by differentiation.

1 (a) x^7, (b) x^5, (c) x^3, (d) $\dfrac{1}{x^3}$, (e) $\dfrac{1}{x^5}$, (f) $\dfrac{1}{x^7}$.

2 (a) $z^{3\cdot5}$, (b) $u^{\frac{1}{2}}$, (c) $\dfrac{1}{u^{\frac{1}{3}}}$, (d) $\dfrac{1}{y^{3\cdot5}}$.

3 (a) $4x^8$, (b) $-3x^6$, (c) $\dfrac{-5}{x^4}$, (d) $\dfrac{4}{x^8}$.

4 (a) $0\cdot22z^{5\cdot5}$, (b) $4u^{\frac{2}{3}}$, (c) $\dfrac{-3}{u^{\frac{1}{3}}}$, (d) $\dfrac{-11}{y^{3\cdot2}}$.

5 (a) $\dfrac{2}{\sqrt[5]{p^4}}$, (b) $\dfrac{p^{\frac{1}{3}}}{p^{\frac{1}{4}}}$, (c) $\dfrac{v^5 - v^3}{v^2}$, (d) $\dfrac{\sqrt[3]{\theta^4} - \sqrt{\theta}}{\sqrt[7]{\theta^8}}$.

6 (a) $(2 + 3x)(1 + x^2)$, (b) $\left(x - \dfrac{1}{x}\right)^2$.

7 Show that $\left(\sqrt{t} + \dfrac{1}{\sqrt{t}}\right)^3 = t^{\frac{3}{2}} + 3t^{\frac{1}{2}} + 3t^{-\frac{1}{2}} + t^{-\frac{3}{2}}$ (i), and hence that

$\int\left(\sqrt{t} + \dfrac{1}{\sqrt{t}}\right)^3 dt = \dfrac{2t^{\frac{5}{2}}}{5} + 2t^{\frac{3}{2}} + 6t^{\frac{1}{2}} - 2t^{-\frac{1}{2}} + C$ (ii). Check (ii) by differentiation. Show that each side of (i) has the value $15\frac{5}{8}$ when $t = 4$.

8 Evaluate (a) $\int(5x + 1)^3dx$, (b) $\int(3 - 2\theta)^4 d\theta$, (c) $\int\left(\dfrac{z}{5} + 2\right)^5 dz$.

9 (a) Show that $(2x + 3)^3 = 8x^3 + 36x^2 + 54x + 27$, and hence that
$\int(2x + 3)^3dx = 2x^4 + 12x^3 + 27x^2 + 27x + C$.

(b) By using the appropriate standard integral, show that $\int(2x + 3)^3dx$

$$= \frac{(2x + 3)^4}{8} + A,$$ and expand this result by the binomial theorem.

Show that the results of (a) and (b) are consistent where $C = 10\frac{1}{8} + A$.

The standard integrals $\int \sin (ax + b)dx$ and $\int \cos (ax + b)dx$, where a and b are constants $(a \neq 0)$

We have previously obtained the results

$$\frac{d \cos(ax + b)}{dx} = -a \sin(ax + b)$$

and

$$\frac{d \sin(ax + b)}{dx} = a \cos(ax + b)$$

hence

$$\frac{d\left(-\dfrac{1}{a}\cos(ax + b)\right)}{dx} = \sin(ax + b)$$

and

$$\frac{d\left(\dfrac{1}{a} \sin(ax + b)\right)}{dx} = \cos(ax + b)$$

Hence $$\int\sin(ax + b)dx = -\frac{1}{a} \cos(ax + b) + C$$

and $$\int\cos(ax + b)dx = \frac{1}{a} \sin(ax + b) + C$$

When $b = 0$,

$$\int\sin ax dx = -\frac{1}{a} \cos ax + C \quad \text{and} \quad \int\cos ax dx = \frac{1}{a} \sin ax + C$$

When $a = 1$,

$$\int\sin x dx = -\cos x + C \quad \text{and} \quad \int\cos x dx = \sin x + C$$

Example
14.7 Evaluate (a) $\int(3 \sin x - 2 \cos x)dx$,
(b) $\int(a \sin \omega t - b \cos \omega t)dt$, where a, b and ω are constants,
(c) $\int[\cos(3\theta - 4) + \sin(4\theta + 1)]d\theta$,
(d) $\int(\sin \phi - \cos \phi)^2 d\phi$.

(a) $\int(3 \sin x - 2 \cos x)dx = -3 \cos x - 2 \sin x + C$

Check $\dfrac{dI}{dx} = 3 \sin x - 2 \cos x$

(b) $\int(a \sin \omega t - b \cos \omega t)dt = -\dfrac{a}{\omega} \cos \omega t - \dfrac{b}{\omega} \sin \omega t + C$

Check $\dfrac{dI}{dt} = -\dfrac{a}{\omega}(-\omega) \sin \omega t - \dfrac{b}{\omega}\omega \cos \omega t = a \sin \omega t - b \cos \omega t$

(c) $\int[\cos(3\theta - 4) + \sin(4\theta + 1)]d\theta = \tfrac{1}{3}\sin(3\theta - 4) - \tfrac{1}{4}\cos(4\theta + 1) + C$

Check $\dfrac{dI}{d\theta} = \dfrac{3}{3} \cos(3\theta - 4) - \dfrac{1}{4}(-4)\sin(4\theta + 1) = \cos(3\theta - 4) + \sin(4\theta + 1)$

(d) $\int(\sin \phi - \cos \phi)^2 d\phi = (\sin^2\phi - 2 \sin \phi \cos \phi + \cos^2\phi)d\phi$

   $= \int(1 - \sin 2\phi)d\phi = \phi + \tfrac{1}{2} \cos 2\phi + C$

(Putting $\sin^2\phi + \cos^2\phi = 1$,   and $2 \sin \phi \cos \phi = \sin 2\phi$)

Check $\dfrac{dI}{d\phi} = 1 - \dfrac{2}{2} \sin 2\phi = 1 - \sin 2\phi = \cos^2\phi + \sin^2\phi - 2 \sin \phi \cos \phi$

   $= (\cos \phi - \sin \phi)^2$

EXERCISE 14.2

Integrate the following with respect to the appropriate variable.

1 (a) $\sin x$,   (b) $\sin 2x$,   (c) $\sin 3x$,   (d) $\sin kx$.
2 (a) $\cos \theta$,   (b) $\cos 2\theta$,   (c) $\cos 3\theta$,   (d) $\cos k\theta$.

3 (a) $\sin(3\theta - \pi)$,   (b) $\sin\left(k\phi + \dfrac{\pi}{2}\right)$,   (c) $\sin(\pi - A\theta)$.

4 (a) $\cos(3\theta - \pi)$,   (b) $\cos\left(k\phi + \dfrac{\pi}{2}\right)$,   (c) $\cos(\pi - A\theta)$.

The indefinite integral considered as the solution of a differential equation

If $y = f(x)$, then $\dfrac{dy}{dx} = \dfrac{df(x)}{dx} = F(x)$, say, where $F(x)$ may be found by the rules of differentiation. Conversely, if it is given that $\dfrac{dy}{dx} = F(x)$, this is known as a differential equation which may be solved for y by integrating both sides with respect to x. Thus,

$$y = \int F(x)dx, \quad \text{i.e.} \quad y = f(x) + C$$

which is called the general solution of the differential equation, since it contains an arbitrary constant, C. If corresponding values of x and y are given, C can be determined, and a particular solution of the differential equation can be found.

Examples

14.8 Find the general solution of the differential equation $\frac{dy}{dx} = x^2$. State the particular solution, given that when $x = 1$, $y = 0$.

$\frac{dy}{dx} = x^2$. Integrating both sides with respect to x gives the general solution

$$y = \frac{x^3}{3} + C$$

This represents a family of curves all of which have the gradient function x^2.

If $y = 0$ when $x = 1$, then $0 = \frac{1}{3} + C$

$$\therefore \quad C = -\frac{1}{3}.$$

i.e. $y = \frac{x^3}{3} - \frac{1}{3}$ is the particular solution which represents a curve having the gradient function x^2 and passing through the point $(1, 0)$.

14.9 A particle starts from a point 0 with an initial velocity of 3 m/s and moves in a straight line with an acceleration of a m/s², given after t seconds by $a = 16 + 72t - 12t^2$. Calculate the maximum acceleration of the particle, and its displacement after 4 seconds.   ULCI

Since $\qquad a = 16 + 72t - 12t^2,$

$$\frac{da}{dt} = 72 - 24t \quad \text{and} \quad \frac{d^2a}{dt^2} = -24$$

When $\frac{da}{dt} = 0$, $72 - 24t = 0$, and $t = \frac{72}{24} = 3$ seconds, which must give

a maximum value of a, since $\frac{d^2a}{dt^2}$ is negative. The maximum magnitude of

a is $16 + 72 \times 3 - 12 \times 3^2 = 124$ m/s².

$$a = \frac{dv}{dt} = 16 + 72t - 12t^2$$

$$\therefore \quad v = 16t + \frac{72t^2}{2} - \frac{12t^3}{3} + C$$

$$= 16t + 36t^2 - 4t^3 + C$$

When $t = 0$, $v = 3$, $\therefore 3 = C$

i.e. $$v = 16t + 36t^2 - 4t^3 + 3$$

i.e. $$\frac{ds}{dt} = 16t + 36t^2 - 4t^3 + 3$$

$$\therefore \quad s = \frac{16t^2}{2} + \frac{36t^3}{3} - \frac{4t^4}{4} + 3t + K$$

$$= 8t^2 + 12t^3 - t^4 + 3t + K$$

When $t = 0$, $s = 0$, $\therefore \quad 0 = 0 + 0 - 0 + 0 + K$,   i.e. $K = 0$

$$\therefore \quad s = 8t^2 + 12t^3 - t^4 + 3t$$

When $t = 4$, $s = 8 \times 4^2 + 12 \times 4^4 - 4^3 + 3 \times 4 = 652$ m.

EXERCISE 14.3

1 (a) Integrate with respect to x (i) $4x^3$,   (ii) $5\sqrt{x}$,   (iii) $\dfrac{7}{x^2}$,
(iv) $(2x + 1)^2$.

(b) If $\dfrac{dy}{dx} = 4x^3 + 2x + 7$, find y in terms of x, given that when
$x = 0$, $y = 8$.   NCTEC

2 (a) Determine   (i) $\int(x^3 + 2x^2 + 5)dx$,   (ii) $\int\left(1 + \dfrac{x}{2}\right)^2 dx$.

(b) The acceleration of a body is given by $(2t + 3)$, where t is the time
in seconds. The velocity at the end of 2 seconds is 3 m/s, and the
distance s moved during this time is 15 m. Find the equation connecting s and t, and determine the distance moved during the third
second.   NCTEC

3 (a) Differentiate $3x^{\frac{4}{3}} - \dfrac{2}{x^3} + \sqrt[3]{x}$ with respect to x.

(b) Integrate $\dfrac{1}{3}x^2 - \dfrac{2}{x^{\frac{1}{2}}} + \dfrac{\sqrt[5]{x}}{5}$ with respect to x. (Render with positive indices.)

(c) $\dfrac{dy}{dx} = 15x^2 - 2x + a$, and is numerically equal to 59 when $x = 2$.
If when $x = 2$, $y = 38$, find the expression for y in terms of x.

(d) A body with initial velocity of 4 m/s moves with an acceleration of
$6t^2 - 4t + 2$ m/s^2, where t is the time in seconds after starting.
Find the distance travelled in 3 seconds.   GRAVESEND

4 (a) Differentiate from first principles $y = \sin x$.

(b) Differentiate by inspection $y = \dfrac{5x^3}{6} - \dfrac{2x}{3} + 3 - \dfrac{4}{3x^2} + \dfrac{4}{9x^3}$.

(c) Find expressions for y given (i) $\dfrac{dy}{dx} = 6x^2 - 3$ and $y = 3$ when

$x = 0$, (ii) $\dfrac{dy}{dx} = \dfrac{3}{x^2} + 1$ and $y = 8\frac{1}{2}$ when $x = 6$.

UEI

5 (a) If $v = k \sin kt$, where v m/s represents the speed of a body and k is a constant, calculate the distance moved from rest in the first $\dfrac{\pi}{3k}$ seconds.

(b) Integrate with respect to x (i) $(2 - 3x)^5$,   (ii) $\dfrac{1}{(x + 2)^3}$,

(iii) $\cos(3x - 2)$, (iv) $\sin(3 - 4x)$.

6 (a) Find the value of the following: (i) $\int 5x^3 dx$,   (ii) $\displaystyle\int_1^4 \sqrt{x}\, dx$,

(iii) $\displaystyle\int_{\pi/8}^{\pi/2} 3 \sin 4\theta\, d\theta$.

(b) Determine the equation of the curve which passes through the point $(x = 2, y = 2)$, given that the slope at every point on the curve is equal to $3x^2 - 5x - 1$.

NCTEC

The definite integral $\displaystyle\int_a^b f(x) dx$

If $\dfrac{dF(x)}{dx} = f(x)$, then $\int f(x) dx = F(x) + C$, an indefinite integral.

We define the definite integral $\displaystyle\int_a^b f(x) dx$ to be

$$[F(b) + C] - [F(a) + C], \quad \text{i.e.} \quad F(b) - F(a).$$

Since the arbitrary constant vanishes, it is omitted.

i.e.
$$\int_a^b f(x) dx = F(b) - F(a)$$

b is called the upper limit, and a the lower limit, the word limit being used in the sense of a boundary value.

Example

14.10 Find (a) $\int_1^2 \left(x + \dfrac{1}{x}\right)^2 dx$, (b) $\int_0^{\pi/2p} \sin px\,dx$, (c) $\int_0^{\pi/4K} \cos^2 K\theta\,d\theta$.

(a) $\int_1^2 \left(x + \dfrac{1}{x}\right)^2 dx = \int_1^2 (x^2 + 2 + x^{-2})dx = \left[\dfrac{x^3}{3} + 2x + \dfrac{x^{-1}}{-1}\right]_1^2$

$= \left[\dfrac{2^3}{3} + 2 \times 2 - \dfrac{1}{2}\right] - \left[\dfrac{1^3}{3} + 2 \times 1 - \dfrac{1}{1}\right]$

$= \dfrac{8}{3} + 4 - \dfrac{1}{2} - \dfrac{1}{3} - 2 + 1 = 2\tfrac{1}{3} + 2\tfrac{1}{2} = 4\tfrac{5}{6}$

N.B. that the limits are written on the right after integration and before substitution.

(b) $\int_0^{\pi/2p} \sin px\,dx = \left[\dfrac{-\cos px}{p}\right]_0^{\pi/2p} = -\dfrac{\cos(p\,\pi/2p)}{p} + \dfrac{\cos 0}{p}$

$= \dfrac{-\cos \pi/2}{p} + \dfrac{\cos 0}{p} = 0 + \dfrac{1}{p} = \dfrac{1}{p}$

(c) We first use the identity $\cos 2x = 2\cos^2 x - 1$, i.e. $1 + \cos 2x = 2\cos^2 x$, i.e. $\cos^2 x = \tfrac{1}{2}(1 + \cos 2x)$, thus $\cos^2 K\theta = \tfrac{1}{2}(1 + \cos 2K\theta)$.

$\therefore \int_0^{\pi/4K} \cos^2 K\theta\,d\theta = \tfrac{1}{2}\int_0^{\pi/4K}(1 + \cos 2K\theta)d\theta = \tfrac{1}{2}\left[\theta + \dfrac{\sin 2K\theta}{2K}\right]_0^{\pi/4K}$

$= \tfrac{1}{2}\left\{\left[\dfrac{\pi}{4K} + \dfrac{\sin 2K\pi/4K}{2K}\right] - \left[0 + \dfrac{\sin 0}{2K}\right]\right\}$

$= \tfrac{1}{2}\left\{\dfrac{\pi}{4K} + \dfrac{\sin \pi/2}{2K} - 0\right\} = \tfrac{1}{2}\left\{\dfrac{\pi}{4K} + \dfrac{1}{2K}\right\}$

$= \dfrac{1}{4K}\left\{\dfrac{\pi}{2} + 1\right\}$

It should be noted that a definite integral may not exist in certain cases. Suppose that we attempt to evaluate

$\int_0^1 1/x^2 \ dx$ i.e. $\int_0^1 x^{-2}dx$ or $\left[\dfrac{x^{-1}}{-1}\right]_0^1$ or $\left[-\dfrac{1}{x}\right]_0^1$

Substitution of the limits would yield the indeterminate expression $-\dfrac{1}{1} + \dfrac{1}{0}$. The definite integral does not exist in this case. In some other cases the definite integral can be found as a limit.

EXERCISE 14.4

1 Integrate the following: (a) $\int_0^1 (x^2 - 2\sqrt{x})dx$, (b) $\int_0^r \pi(r^2 - x^2)dx$,

(c) $\int_1^2 \left(x^3 - \frac{1}{x^3}\right)dx$.

ULCI

2 Evaluate the following: (a) $\int_0^1 \sqrt{(\theta)}d\theta$, (b) $\int_0^1 \frac{dz}{\sqrt{z}}$, (c) $\int_0^1 \frac{dz}{\sqrt{3 - 2z}}$.

3 Evaluate the following definite integrals where possible:

(a) $\int_1^\infty \frac{d\theta}{\sqrt{\theta}}$, (b) $\int_{-1}^1 \frac{dz}{\sqrt[3]{z^2}}$, (c) $\int_0^1 \frac{dp}{p^m}$

4 (a) Find $\int_0^1 (1 + x)^4 dx$, (i) using the appropriate standard form, and
(ii) by expanding the integrand using the binomial theorem and integrating term by term.

(b) Similarly find $\int_0^2 (5 - 2x)^3 dx$ by two methods.

5 Find (a) $\int_0^H \frac{B}{H}z^3 dz$, (b) $\int_0^H \frac{B}{H}(H - z)z^2 dz$, (c) $\int_{-H/3}^{2H/3} \frac{B}{H}\left(\frac{2H}{3} - z\right)z^2 dz$.

6 Evaluate (a) $\int_{-1}^1 \left(x - \frac{1}{x}\right)^2 dx$, (b) $\int_0^{\pi/6} \cos 3\theta d\theta$, (c) $\int_0^{\pi/12} \cos^2 3z dz$.

7 (a) Use the expansion of $\cos(A + B)$ to show that $\cos 2\theta = \cos^2 3\theta - \sin^2\theta$.

(b) Show that $\cos 2\theta = 1 - 2\sin^2\theta$ (using $\cos^2\theta + \sin^2\theta = 1$ with the result of the previous section).

(c) Transpose to obtain $\sin^2\theta = \frac{1}{2}(1 - \cos 2\theta)$.

(d) Show that $\int_0^{\pi/4} \sin^2\theta d\theta = \frac{1}{2}\left(\frac{\pi}{4} - \frac{1}{2}\right)$.

(e) Use $\sin^2 p\theta = \frac{1}{2}(1 - \cos 2 p\theta)$ to evaluate $\int_0^{\pi/p} \sin^2 p\theta d\theta$.

8 (a) Using the expansion of $\cos(2\theta + \theta)$, show that $\cos 3\theta = 4\cos^3\theta - 3\cos\theta$, and hence that $\cos^3\theta = \frac{1}{4}(\cos 3\theta + 3\cos\theta)$.

(b) Evaluate $\int_0^{\pi/6} \cos^3\theta d\theta$.

9 (a) Using the expansion of $\sin(2\theta + \theta)$, show that $\sin 3\theta = 3\sin\theta - 4\sin^3\theta$, and hence that $\sin^3\theta = \frac{1}{4}(3\sin\theta - \sin 3\theta)$.

(b) Evaluate $\int_0^{\pi/6} \sin^3\theta d\theta$.

Meaning of $\displaystyle\sum_{x=a}^{x=b} y\delta x$ **and** $\displaystyle\lim_{\delta x \to 0} \sum_{x=a}^{x=b} y\delta x$

Suppose that $y = f(x)$ is a continuous positive increasing function over the range $x = a$ to $x = b$, and let this interval be divided into n strips of equal width δx, with ordinates $y_0, y_1 \ldots y_{n-1}, y_n$, so that $\delta x = \dfrac{b-a}{n}$.

Then the curve, the x-axis, and the ordinates at $x = a$ and $x = b$ enclose n rectangles each of width δx (Fig. 14.1).

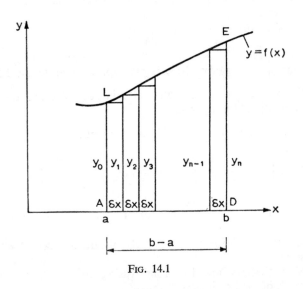

Fig. 14.1

The sum of the areas of the rectangles enclosed from $x = a$ to $x = b$

$$= y_0\delta x + y_1\delta x + \ldots y_{n-1}\delta x$$
$$= (y_0 + y_1 + \ldots y_{n-1})\delta x$$

$$= \sum_{x=a}^{x=b} y\delta x$$

where Σ is used to denote 'the sum of', and the expression reads 'the sum of the $y\delta x$ values from $x=a$ to $x=b$

$\simeq$ the area ADEL, if δx is sufficiently small.

Suppose now that, as $\delta x \to 0$, $\displaystyle\sum_{x=a}^{x=b} y\delta x \to A$ (a certain number); then

we say that the area ADEL $= A = \displaystyle\lim_{\delta x \to 0} \sum_{x=a}^{x=b} y\delta x$

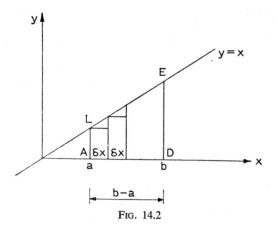

FIG. 14.2

Example

14.11 Find the area enclosed by graph of $y = x$, the x-axis and the ordinates at $x = a$ and $x = b$.

Since the abscissae are $a, a + \delta x, a + 2\delta x, \ldots a + (n - 1)\delta x$, in this case the ordinates are $a, a + \delta x, a + 2\delta x, \ldots a + (n - 1)\delta x$ (Fig. 14.2), and again $\delta x = \dfrac{b - a}{n}$.

$$\therefore \sum_{x=a}^{x=b} y\delta x = [a + (a + \delta x) + (a + 2\delta x) \ldots (a + (n - 1)\delta x]$$

Using $s = \dfrac{n}{2}(2a + [n - 1]d)$, since the terms form an arithmetic progression of n terms, where $d = \delta x$, we have

$$\sum_{x=a}^{x=b} y\delta x = \left(\frac{n}{2}(2a + [n - 1]\delta x)\right)\delta x,$$

$$= \frac{n}{2}(2a + [n - 1]\delta x)\frac{b - a}{n} \quad \left(\text{putting } \delta x = \frac{b - a}{n}\right)$$

$$= \frac{n}{2}\left(2a + [n - 1]\frac{b - a}{n}\right)\frac{b - a}{n}$$

$$= \frac{b - a}{2}[2a + \left(1 - \frac{1}{n}\right)(b - a)]$$

As $\delta x \to 0$, $n \to \infty$, $\dfrac{1}{n} \to 0$

$$\therefore \text{ area ADEL} = \lim_{\delta x \to 0} \sum_{x=a}^{x=b} y\delta x = \frac{b - a}{2}(2a + b - a)$$

$$= \frac{b - a}{2}(b + a) = \frac{b^2 - a^2}{2}$$

We may check this result, for trapezium ADEL is of width $b - a$ and heights a and b, i.e. area ADEL $= \left(\dfrac{b+a}{2}\right)(b-a) = \left(\dfrac{b^2-a^2}{2}\right)$

The student will probably appreciate that this method of 'integration from first principles' is tedious to apply. Consider, however, the definite integral.

$$\int_a^b y\,\mathrm{d}x, \text{ where } y = x$$

i.e. $$\int_a^b x\,\mathrm{d}x = \left[\dfrac{x^2}{2}\right]_a^b = \dfrac{b^2}{2} - \dfrac{a^2}{2} = \dfrac{b^2-a^2}{2}$$

so that for the case considered it appears that

$$\text{area ADEL} = \lim_{\delta x \to 0} \sum_{x=a}^{x=b} y\delta x = \int_a^b x\,\mathrm{d}x,$$

where y is a continuous positive increasing function of x over the interval from $x = a$ to $x = b$. This suggests a much shorter approach to finding the area under a curve.

To find the area enclosed between $y = f(x)$, where $f(x)$ is a positive increasing continuous function, the lines $x = a$, $x = b$ and the x-axis
Let K, B, C, D correspond to the abscissae a, x, $x + \delta x$, and b respectively, then BH or CG represents y, and GF represents δy. If A is a function of x which represents the area KBHL such that $A = 0$ when $x = a$, we require to find the area KDEL, i.e. the value of A when $x = b$ (Fig. 14.3).
The area BCFH represents δA, the change in A as x changes to $x + \delta x$.

Now area BCGH < area BCFH < area BCFK

$$\therefore \quad y\delta x < \delta A < (y + \delta y)\delta x$$

$$\therefore \quad y < \dfrac{\delta A}{\delta x} < (y + \delta y).$$

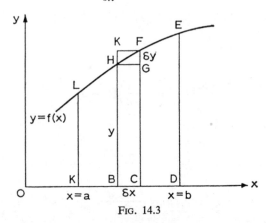

Fig. 14.3

As $\delta x \to 0$, $\delta y \to 0$, $\dfrac{\delta A}{\delta x} \to \dfrac{\mathrm{d}A}{\mathrm{d}x}$, and since $\dfrac{\delta A}{\delta x}$ is between y and an expression which tends to y,

$$\therefore \qquad \frac{\mathrm{d}A}{\mathrm{d}x} = y.$$

Integrating with respect to x,

$$A = \int y\,\mathrm{d}x = F(x) + C, \quad \text{where } \frac{\mathrm{d}F(x)}{\mathrm{d}x} = y$$

When $x = a$, $A = 0$, $\therefore$ $0 = F(a) + C$, or $C = -F(a)$

$$\therefore \quad A = F(x) - F(a)$$

Putting $x = b$, we have

$$\text{area KDEL} = F(b) - F(a)$$

$$= [F(x)]_a^b = \int_a^b y\,\mathrm{d}x$$

It can be shown that this result can be extended to apply to elementary functions which are continuous within the range of integration and are

(*a*) non-negative within the range,
(*b*) non-positive within the range, see Example 14.13,
(*c*) change sign within the range, see Example 14.14.

Examples
14.12 Find the area enclosed between the given curve, the x-axis and the given lines   (*a*) $y = x^2$ from $x = 1$ to $x = 3$,   (*b*) $y = \sin 2x$ from $x = 0$ to $x = \dfrac{\pi}{4}$.

(*a*) The function being positive in the range, the area enclosed

$$= \int_1^3 x^2\,\mathrm{d}x = \left[\frac{x^3}{3}\right]_1^3 = \left[\frac{3^3}{3} - \frac{1^3}{3}\right]$$

$$= \frac{27 - 1}{3} = \frac{26}{3} = 8\tfrac{2}{3} \text{ units.}$$

(*b*) The function being positive or zero in the range, the area enclosed

$$= \int_0^{\pi/4} \sin 2x\,\mathrm{d}x = [-\tfrac{1}{2}\cos 2x]_0^{\pi/4}$$

$$= -\tfrac{1}{2}\left[\cos 2\left(\frac{\pi}{4}\right) - \cos 0\right] = -\tfrac{1}{2}\left[\cos\left(\frac{\pi}{2}\right) - \cos 0\right]$$

$$= -\tfrac{1}{2}[0 - 1] = \tfrac{1}{2} \text{ unit.}$$

If y is always negative or zero, i.e. non-positive in the range considered, we evaluate the definite integral and ignore its sign.

14.13 Find the area enclosed between the curve $y = (1 - x)^3$, the x-axis, and the lines $x = 2$, $x = 3$.

$$\int_2^3 (1 - x)^3 dx = \left[\frac{1}{-1} \cdot \frac{(1 - x)^4}{4} \right]_2^3 = \left[-\frac{(1 - 3)^4}{4} + \frac{(1 - 2)^4}{4} \right]$$

$$= \left[-\frac{16}{4} + \frac{1}{4} \right] = -\frac{15}{4} = -3\tfrac{3}{4}$$

i.e. $3\tfrac{3}{4}$ units of area are enclosed.

14.14 Find the area enclosed between the curve $y = x^2 - 3x + 2$, the x-axis, and the ordinates at $x = 0$ and $x = 3$.

We first sketch the graph of $y = x^2 - 3x + 2$. Since $\dfrac{dy}{dx} = 2x - 3$,

and $\dfrac{d^2y}{dx^2} = 2$, we have a minimum point where $2x - 3 = 0$, i.e. $x = 1\tfrac{1}{2}$.

Put $x^2 - 3x + 2 = 0$, then $(x - 2)(x - 1) = 0$, i.e. at $x = 1$ and $x = 2$ the curve crosses the x-axis.

Thus, when $x = 0$, $y = 2$; $x = 1$, $y = 0$; $x = 2$, $y = 0$.

When $x = 1\tfrac{1}{2}$ or $\tfrac{3}{2}$, $y = \tfrac{9}{4} - \tfrac{9}{2} + 2 = -\tfrac{1}{4}$; when $x = 3$, $y = 9 - 9 + 2 = 2$. We divide the integration up into the ranges $x = 0$ to $x = 1$, $x = 1$ to $x = 2$, and $x = 2$ to $x = 3$ (Fig. 14.4).

$$\int_0^1 (x^2 - 3x + 2)dx = \left[\frac{x^3}{3} - \frac{3x^2}{2} + 2x \right]_0^1 = \frac{1}{3} - \frac{3}{2} + 2 = \frac{5}{6}$$

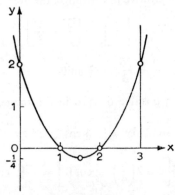

FIG. 14.4

$$\int_1^2 (x^2 - 3x + 2)dx = \left[\frac{x^3}{3} - \frac{3x^2}{2} + 2x\right]_1^2 = \frac{8}{3} - \frac{3 \times 4}{2} + 2 \times 2$$

$$- \frac{5}{6} = -\frac{1}{6}$$

$$\int_2^3 (x^2 - 3x + 2)dx = \left[\frac{x^3}{3} - \frac{3x^2}{2} + 2x\right]_2^3 = \frac{27}{3} - \frac{27}{2} + 6 - \frac{8}{3}$$

$$+ \frac{3 \times 4}{2} - 4$$

$$= \frac{5}{6}$$

$\therefore \ \dfrac{5}{6} + \dfrac{1}{6} + \dfrac{5}{6}$, i.e. $1\frac{5}{6}$ units of area are enclosed.

Suppose that integration had been carried out over the full range

$$\int_0^3 (x^2 - 3x + 2)dx = \left[\frac{x^3}{3} - \frac{3x^2}{2} + 2x\right]_0^3 = \frac{27}{3} - \frac{27}{2} + 6 = 1\frac{1}{2}$$

Now $\dfrac{5}{6} - \dfrac{1}{6} + \dfrac{5}{6} = 1\frac{1}{2}$, i.e. this latter definite integral gives the **magni-**
tude of the *difference* between the areas above and below the x-axis.
Hence it is always advisable to sketch the graph of the integrand before
using definite integration to calculate area.

EXERCISE 14.5

1 (a) Evaluate the following, simplifying answers where possible.

(i) $\int(4t + 3)(t^2 - 2)dt$, (ii) $\int\left(\sqrt{9x} + \dfrac{1}{\sqrt[3]{(8x^6)}}\right)dx.$

(b) Sketch the curve $y = x^2 + 1$, and calculate the area enclosed by
the curve, the x-axis and the ordinates where $y = 2$.

BROOKLANDS

2 (a) Evaluate

$$\int_1^4 \left(3x^2 + \frac{3}{\sqrt{x}} + \frac{3}{x^2} - x^{\frac{3}{2}}\right)dx.$$

(b) If $\int_2^4 (9x^2 + K)dx = 200$, find K, which is a constant.

(c) Calculate the area enclosed between the curve $3x^2 - 7x + 4$, the
x-axis, and the ordinates at $x = 1$ and $x = 3$.

(d) Find the area of the segment of the curve $y = -x^2 + x + 6$ cut
off by the x-axis.

GRAVESEND

3 (*a*)   (i) Integrate with respect to *x*: $x^2 + \sqrt{x} - \dfrac{2}{x}$.

(ii) Obtain the $\displaystyle\int \frac{6 - t^3}{2t^3} dt$.

(iii) Evaluate $\displaystyle\int_{-\frac{1}{2}}^{\frac{1}{2}} (1 + 2y)^2 dy$.

(*b*) Sketch the curve $y = 3x - x^2$. (Accurate plotting is not required.) At what values does the curve cut the *x*-axis? Calculate the area enclosed by the curve and the *x*-axis.   MEDWAY

4 (*a*) By integrating the function x^2 from $x = 0$ to $x = C$, show that the area of the parabolic fillet enclosed between the curve, the *x*-axis, and the ordinate $x = C$ is one third of the circumscribing rectangle.

(*b*) By integrating the function $x(a - x)$ from 0 to *a*, show that the area of the parabolic fillet enclosed between the curve and the *x*-axis is two thirds of the circumscribing rectangle.

Illustrate the results with a sketch in each case.

5 Calculate the coordinates of the points of intersection of the curves $y = x^2 - 6$ and $y = 4 - 3x$. Sketch the curves, show that the area between them can be expressed by

$$\int_{-5}^{2} (10 - 3x - x^2)dx$$

and determine its value.   BROOKLANDS

6 (*a*) Sketch the curves $y = x^2$ and $y = 4x - x^2$ on the same base, between $x = 0$ and $x = 2$.

(*b*) Calculate by integration the area enclosed by the *x*-axis the ordinates at $x = 0$, $x = 2$ and the curve   (i) $y = x^2$, (ii) $y = 4x - x^2$.

(*c*) Find the area enclosed between the curves $y = x^2$ and $y = 4x - x^2$ using the above results. Check all results using Simpson's rule, which is exact for polynomials of the second degree.

MID-CHESHIRE

7 (*a*) Sketch the curves $y = x^3$ and $y = \dfrac{13x^2}{4} - \dfrac{9}{4}$ on the same base, between $x = 0$ and $x = 4$.

(*b*) Show that the area enclosed by these curves in the interval $x = 1$ to $x = 3$ may be expressed as

$$\int_{1}^{3} \left(-\frac{9}{4} + \frac{13x^2}{4} - x^3 \right) dx$$

and evaluate it.

(*c*) Check the result of (*b*) using Simpson's rule, which is exact for polynomials of the third degree.   MID-CHESHIRE

8 The area under the graph of $y = \dfrac{x}{2}$ from $x = 0$ to $x = 2a$ is divided into n strips of equal width $\dfrac{2a}{n}$, and the corresponding rectangles are constructed.

Show (a) that the heights of the rectangles so formed are the n terms of an arithmetic series $0, \dfrac{a}{n}, \dfrac{2a}{n} \ldots, a - \dfrac{a}{n}$, (b) that the sum of these heights is $\dfrac{a(n-1)}{2}$, and the area of the n rectangles is given by $a^2\left(1 - \dfrac{1}{n}\right)$, (c) that the area of 1000 such rectangles is $0\cdot999a^2$, and of 1 000 000 such rectangles is $0\cdot999\,999a^2$, (d) that the required area under the curve is a^2.

Chapter Fifteen
Maxima and Minima

The sign of the differential coefficient—stationary values

We have seen that, if $y = x^2$

$$\frac{dy}{dx} = 2x$$

This is true for all values of x.

Hence, if x is positive, $\frac{dy}{dx}$ is positive,

if x is negative, $\frac{dy}{dx}$ is negative,

if x is 0, $\frac{dy}{dx}$ is 0.

Let us examine these by the help of the graph of $y = x^2$ (Fig. 15.1).

First the student should recall the convention used in the representation of numbers when drawing a graph. Numbers which are shown on the selected scale on the x-axis are regarded as *increasing* from an infinite distance on the left or negative side of the origin, through zero, to an infinite distance on the right, or positive, side. Similarly, on the y-axis the numbers are regarded as *increasing* from $-\infty$ below the x-axis to $+\infty$ above the x-axis.

If from points such as P on the curve of $y = x^2$ (Fig. 15.1) perpendiculars are drawn to the x-axis, the lengths of these perpendiculars represent the values of x^2 or y at these points. It is evident that *as x increases from $-\infty$ to 0, the values of y are decreasing.*

At the origin the value of y is zero. At that point the curve touches the x-axis, which is a tangent to it, and momentarily *the value of y is stationary*, i.e. it is neither increasing nor decreasing.

Similarly, *as x increases from 0 to $+\infty$, the values of y are increasing.*

Comparing these facts with the sign of the differential coefficient, stated above,

(*a*) when x is negative,

$$\frac{dy}{dx} \text{ is negative, and } y \text{ is decreasing.}$$

(*b*) when x is positive,

$$\frac{dy}{dx} \text{ is positive, and } y \text{ is increasing.}$$

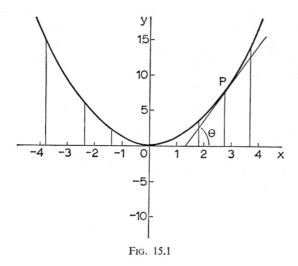

Fig. 15.1

(c) when x is 0,

$$\frac{dy}{dx} \text{ is 0, and } y \text{ is stationary.}$$

It should also be noted that, when x is positive, the angle of slope, θ, is an acute angle, and tan θ, which is $\frac{dy}{dx}$, is positive.

When x is negative, θ is clearly an obtuse angle and tan θ is negative. We conclude therefore that

(a) if $\frac{dy}{dx}$ is positive,

y is increasing as x is increasing,

(b) if $\frac{dy}{dx}$ is negative,

y is decreasing as x is increasing,

(c) if $\frac{dy}{dx}$ is 0,

y is stationary.

As another example we might consider the curve of

$$y = -x^2 \quad \text{(Fig. 15.2)}$$

Differentiating,

$$\frac{dy}{dx} = -2x$$

When x is negative, $\frac{dy}{dx}$ is positive.

Considering the lengths of the perpendiculars, such as PQ (Fig. 15.2), for these negative values of x, it is seen that they are decreasing in length as x increases, but, as they represent negative numbers, their absolute value is increasing: for example, -1 is greater than -5. Hence, as shown, for negative values of x, as x increases

$$\frac{\mathrm{d}y}{\mathrm{d}x}$$ is positive and y is increasing.

Similarly, for positive increasing values of x,

$$\frac{\mathrm{d}y}{\mathrm{d}x}$$ is negative and y is decreasing.

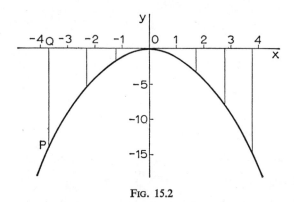

FIG. 15.2

At the origin, as before,

$$\frac{\mathrm{d}y}{\mathrm{d}x}$$ is zero, and y has a stationary value.

It will readily be seen that similar results will be obtained in the cases of other functions, and so we may conclude generally that, when y is a function of x,

(a) If y is increasing as x increases, $\frac{\mathrm{d}y}{\mathrm{d}x}$ is positive,

(b) If y is decreasing as x increases, $\frac{\mathrm{d}y}{\mathrm{d}x}$ is negative,

(c) If $\frac{\mathrm{d}y}{\mathrm{d}x}$ is zero, y is stationary.

Turning-points

Comparing the stationary points in the curves of $y = x^2$ and $y = -x^2$, we see that there are important differences.

In the case of $y = x^2$, we note

(a) The curve is changing direction: the slope θ is changing from an *obtuse* angle, through 0 to an *acute* angle.

(b) The values of x^2 are *decreasing* before the stationary point and *increasing* after.

But in $y = -x^2$, the opposite of these is happening.

(a) The curve is changing direction, but θ is changing from an *acute* angle, through 0, to an *obtuse* angle.

(b) The values of x^2 are *increasing* before the stationary point and *decreasing* after.

Such points on a curve are called turning-points.

It should be noted that not all stationary points are turning-points. Consider the case of $y = x^3$ (Fig. 15.3), in which

$$\frac{dy}{dx} = 3x^2$$

We see that $3x^2$ is always positive, hence the function must be always increasing.

Since, however, $\dfrac{dy}{dx} = 0$ when $x = 0$, there is a *stationary point* at the

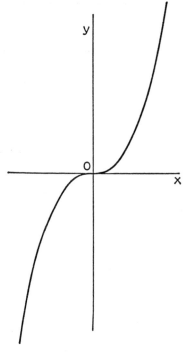

Fig. 15.3

origin, but the function is not increasing before this and decreasing afterwards, nor vice versa. The slope θ remains as an acute angle throughout, except at the origin, where θ is zero: hence the *stationary point is not a turning-point*.

It may also be deduced that, if the differential coefficient of a function is always negative, the function itself is always decreasing, and there can be no turning-points: only stationary points are possible.

Examples

15.1 For what value of x is there a turning-point on the curve of $y = 2x^2 - 6x + 9$, and what is the value of the function at the turning-point?

If
$$y = 2x^2 - 6x + 9$$
then
$$\frac{dy}{dx} = 4x - 6$$

For a turning-point the differential coefficient must vanish.

$$\therefore \quad 4x - 6 = 0$$
and
$$x = 1.5$$

Thus there is one turning-point on the curve, where

$$x = 1.5$$

To find the value of the function at the turning-point, substitute $x = 1.5$ in the function.

Then
$$y = 2(\tfrac{3}{2})^2 - 6(\tfrac{3}{2}) + 9$$
$$= 4.5 - 9 + 9$$
$$= 4.5$$

15.2 For what values of x are there turning-points on the curve of $y = 4x^3 - x^2 - 2x + 1$?

Now
$$\frac{dy}{dx} = 12x^2 - 2x - 2$$

For turning-points,
$$\frac{dy}{dx} = 0$$

$$\therefore \quad 12x^2 - 2x - 2 = 0$$
or
$$6x^2 - x - 1 = 0$$
$$\therefore \quad (3x + 1)(2x - 1) = 0$$
and
$$x = -\tfrac{1}{3} \quad \text{or} \quad +\tfrac{1}{2}$$

$\therefore$ There are turning-points on the curve when

$$x = -\tfrac{1}{3}$$
and
$$x = +\tfrac{1}{2}$$

15.3 For what values of x are there turning or stationary points on the curve of $y = 1 - x^3$?

$$\frac{dy}{dx} = -3x^2$$

Now $-3x^2$ is negative for all real values of x.

∴ There are no turning-points,

but, if $$\frac{dy}{dx} = -3x^2 = 0$$

then $$x = 0$$

∴ there is a stationary point when $x = 0$.

Maximum and minimum points

An examination of the curves of $y = x^2$ and $y = -x^2$ (Figs. 15.1 and 15.2) reveals a very important difference in the respective turning-points.

(*a*) In $y = x^2$, the turning-point is the lowest point on the curve. If other points are taken close to it and on either side of it, the value of the function at each of these is *greater* than at the turning-point. Such a value at a turning-point is called a *minimum* value.

(*b*) In $y = -x^2$, examining the turning-point, we see that values just before and after it are *less* than at the turning-point itself. Such a value is called a *maximum* value.

It is very important to be able to distinguish between maximum and minimum values, even before the curve is drawn, especially when a function has more than one turning-point. If the curve of the function has been drawn, it will generally be easy to see which is which, but an algebraical method must be found which will apply in all cases.

Let us first consider a function which has both maximum and minimum values.

Consider

$$y = (x - 1)(x - 2)(x - 3)$$

or, when multiplied out,

$$y = x^3 - 6x^2 + 11x - 6$$

It will be seen that the function will vanish when $x - 1 = 0$, $x - 2 = 0$, $x - 3 = 0$, or when $x = 1$, 2 and 3.

Clearly, if the curve cuts the axis at $x = 1$ and $x = 2$, and a small increase in x produces a corresponding small increase in y, then there must be a turning-point between $x = 1$ and $x = 2$. Similarly there must be a turning-point between $x = 2$ and $x = 3$.

The shape of the curve is shown in Fig. 15.4. This shows that the turning-point between $x = 1$ and $x = 2$ is a maximum, and that between $x = 2$ and $x = 3$ a minimum.

Differentiating
$$y = x^3 - 6x^2 + 11x - 6$$
$$\frac{dy}{dx} = 3x^2 - 12x + 11$$

For turning-points, $3x^2 - 12x + 11 = 0$.
Solving, we find the roots of these equations are

$$x = 2 \cdot 58 \quad \text{or} \quad 1 \cdot 42$$

$\therefore$ There are turning-points when $x = 2 \cdot 58$ and $x = 1 \cdot 42$.

We must proceed to find an algebraical method of determining which is a maximum and which a minimum.

It should be noted that the value at a maximum point such as P (Fig. 15.4) is not necessarily the greatest value of the function.

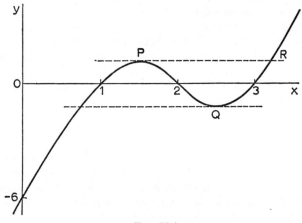

FIG. 15.4

If PR be drawn from P parallel to the x-axis, then all points on the curve beyond R give values which are greater than that at P. Similarly there are values less than the minimum at Q.

To distinguish between maximum and minimum values

First method
We have seen that at a *maximum* point the value of the function is *greater* than at points very close to it, i.e. for values of x a little greater and a little less. Similarly at a *minimum* point such values are *less*. We could therefore test whether the point is a maximum or minimum by substituting in the function values of x which are slightly greater or slightly less than the value at the turning-point, and then comparing the resulting values of the function with its value at the turning-point. This method is, however, usually somewhat tedious.

Second method

At a maximum point, since the function is increasing before it and decreasing afterwards, $\dfrac{dy}{dx}$ must be positive before the point and negative afterwards. Similarly at a minimum point, $\dfrac{dy}{dx}$ must be negative before and positive afterwards. Thus, if we substitute in the expression for $\dfrac{dy}{dx}$ values of x a little greater and a little less than at the turning-point, then

(a) if $\dfrac{dy}{dx}$ is positive before and negative after, the point is a maximum;

(b) if $\dfrac{dy}{dx}$ is negative before and positive after, the point is a minimum.

Example

15.4 Examine $y = 6x - x^2$ for maximum and minimum values,

If
$$y = 6x - x^2$$
$$\frac{dy}{dx} = 6 - 2x$$

∴   For   turning-points, $6 - 2x = 0$
$$\therefore \quad x = 3$$

∴   There is one turning-point only, at $x = 3$.

When $x = 2 \cdot 9$, $\dfrac{dy}{dx} = 6 - 2(2 \cdot 9)$, which is positive.

When $x = 3 \cdot 1$, $dy = 6 - 2(3 \cdot 1)$, which is negative.
∴   When $x = 3$, $6x - x^2$ has a *maximum* value.

Third method

We have seen that at a *maximum* point

(a) $\dfrac{dy}{dx}$ is zero;

(b) $\dfrac{dy}{dx}$ is positive before and negative afterwards.

$$\therefore \quad \frac{dy}{dx} \text{ must be } \textit{decreasing.}$$

But if $\dfrac{dy}{dx}$ is decreasing, its differential coefficient must be *negative*.

∴   At a maximum point, $\dfrac{d^2y}{dx^2}$ is negative.   (i)

At a *minimum* point, similarly

(a) $\dfrac{dy}{dx}$ is zero;

(b) $\dfrac{dy}{dx}$ is negative before and positive after.

$$\therefore \quad \dfrac{dy}{dx} \text{ is } \textit{increasing}.$$

$\therefore$ its differential coefficient must be *positive*.

$\therefore$ **At a minimum point $\dfrac{d^2y}{dx^2}$ is positive**   (ii)

Exceptions to statements (i) and (ii) can arise when $\dfrac{d^2y}{dx^2}$ is zero.

Consider $y = x^6$: $\dfrac{dy}{dx} = 6x^5$, $\dfrac{d^2y}{dx^2} = 30x^4$. When $\dfrac{dy}{dx} = 0$ and $6x^5 = 0$,

then $x = 0$ and $\dfrac{d^2y}{dx^2} = 0$. A rough sketch of the function x^6 shows that there is indeed a minimum point $(0, 0)$, which is verified by the first and second methods.

These results may be summarised as follows:

	Maximum	*Minimum*
$y = f(x)$	Increasing before. Decreasing after.	Decreasing before. Increasing after.
$\dfrac{dy}{dx}$	Positive before, negative after, $\therefore$ decreasing	Negative before, positive after, $\therefore$ increasing.
$\dfrac{d^2y}{dx^2}$	Negative.*	Positive.*

* If both $\dfrac{dy}{dx}$ and $\dfrac{d^2y}{dx^2}$ are zero at the point, use the first or second method.

Examples
15.5 We will now apply these tests to the curve $y = (x - 1)(x - 2)(x - 3)$, where we have previously seen there was a maximum point at $x = 1\cdot42$ and a minimum point at $x = 2\cdot58$.

First method
When $x = 1\cdot4$, $\quad y = (1\cdot4 - 1)(1\cdot4 - 2)(1\cdot4 - 3)$
$\qquad\qquad\qquad = (0\cdot4) \times (-0\cdot6) \times (-1\cdot6)$
$\qquad\qquad\qquad = +0\cdot384$

When $x = 1\cdot42$, $\quad y = (1\cdot42 - 1)(1\cdot42 - 2)(1\cdot42 - 3)$
$\qquad\qquad\qquad = (0\cdot42)(-0\cdot58)(-1\cdot58)$
$\qquad\qquad\qquad = +0\cdot3849$

When $x = 1.5$, $\quad y = (1.5 - 1)(1.5 - 2)(1.5 - 3)$
$$= (0.5)(-0.5)(-1.5)$$
$$= +0.375$$

Thus in front of $x = 1.42$ (from $x = 1.4$ to $x = 1.42$), y is increasing (from 0.384 to 0.3849), while after $x = 1.42$ (from $x = 1.42$ to $x = 1.5$), y is decreasing (from 0.3849 to 0.375).

$$\therefore \quad x = 1.42 \text{ gives a } maximum \text{ point.}$$

Again, when $\quad x = 2.5$, $\quad y = (2.5 - 1)(2.5 - 2)(2.5 - 3)$
$$= (1.5)(0.5)(-0.5)$$
$$= -0.375$$

when $\quad x = 2.58$, $\quad y = (2.58 - 1)(2.58 - 2)(2.58 - 3)$
$$= (1.58)(0.58)(-0.42)$$
$$= -0.3849$$

when $\quad x = 3$, $\quad y = (3 - 1)(3 - 2)(3 - 3)$
$$= 0$$

Thus in front of $x = 2.58$ (from $x = 2.5$ to $x = 2.58$), y is decreasing (from -0.375 to -0.3849), while after $x = 2.58$ (from $x = 2.58$ to $x = 3$), y is increasing (from -0.3849 to 0).

$$\therefore \quad x = 2.58 \text{ gives a } minimum \text{ point.}$$

Second method
Since $\qquad\qquad\qquad y = (x - 1)(x - 2)(x - 3)$
or $\qquad\qquad\qquad\quad y = x^3 - 6x^2 + 11x - 6$
$$\therefore \quad \frac{dy}{dx} = 3x^2 - 12x + 11$$

When $x = 1.4$, $\qquad \dfrac{dy}{dx} = 3(1.4)^2 - 12(1.4) + 11$
$$= 5.88 - 16.8 + 11$$
$$= 16.88 - 16.8$$
$$= 0.08, \text{ which is } positive$$

When $x = 1.5$ $\qquad \dfrac{dy}{dx} = 3(1.5)^2 - 12(1.5) + 11$
$$= 6.75 - 18 + 11$$
$$= 17.75 - 18$$
$$= -0.25, \text{ which is } negative$$

$\therefore \quad \dfrac{dy}{dx}$ is positive before $x = 1.42$ and negative after.

$$\therefore \quad x = 1.42 \text{ gives a } maximum \text{ point.}$$

When $x = 2 \cdot 5$, $\dfrac{dy}{dx} = 3(2 \cdot 5)^2 - 12(2 \cdot 5) + 11$

$$= 18 \cdot 75 - 30 + 11$$
$$= 29 \cdot 75 - 30$$
$$= -0 \cdot 25, \text{ which is } \textit{negative}$$

When $x = 3$, $\dfrac{dy}{dx} = 3(3^2) - 12(3) + 11$

$$= 27 - 36 + 11$$
$$= 37 - 36$$
$$= 2, \text{ which is } \textit{positive}$$

$\therefore \quad \dfrac{dy}{dx}$ is negative before $x = 2 \cdot 58$ and positive after.

$$\therefore \quad x = 2 \cdot 58 \text{ gives a } \textit{minimum} \text{ point.}$$

Third method

Since $\dfrac{dy}{dx} = 3x^2 - 12x + 11$

$$\dfrac{d^2y}{dx^2} = 6x - 12$$
$$= 6(x - 2)$$

When $x = 1 \cdot 42$, $\dfrac{d^2y}{dx^2} = 6(1 \cdot 42 - 2)$

$$= (6)(-0 \cdot 58)$$
$$= \text{a } \textit{negative} \text{ value}$$

When $x = 2 \cdot 58$, $\dfrac{d^2y}{dx^2} = 6(2 \cdot 58 - 2)$

$$= (6)(0 \cdot 58)$$
$$= \text{a } \textit{positive} \text{ value}$$

Thus

$$x = 1 \cdot 42 \text{ gives a } \textit{maximum} \text{ point.}$$
$$x = 2 \cdot 58 \text{ gives a } \textit{minimum} \text{ point.}$$

The student thus has the choice of three methods which have been shown to lead to the same conclusions, but the third method (of using the second differential coefficient) is usually to be preferred owing to its simplicity.

15.6 Examine $y = 3x - x^3$ for maximum and minimum values.

$$y = 3x - x^3$$
$$\therefore \quad \dfrac{dy}{dx} = 3 - 3x^2$$

For turning-points, $\dfrac{dy}{dx} = 3 - 3x^2 = 0$

$$\therefore \quad 3(1 - x^2) = 0$$
$$\therefore \quad x = \pm 1$$

But $$\frac{d^2y}{dx^2} = -6x$$

When $$x = +1, \frac{d^2y}{dx^2} = (-6)(+1)$$

$$= -6 \quad \text{(i.e. } negative)$$

When $$x = -1, \frac{d^2y}{dx^2} = (-6)(-1)$$

$$= +6 \quad \text{(i.e. } positive)$$
$$\therefore \quad 3x - x^3 \text{ has}$$

a *maximum* value when $x = +1$,
a *minimum* value when $x = -1$.

15.7 Examine $x^3 - 6x^2 - 15x + 10$ for maximum and minimum values. Also find the maximum and minimum values of the function.

$$x = x^3 - 6x^2 - 15x + 10$$
$$\therefore \quad \frac{dy}{dx} = 3x^2 - 12x - 15$$
$$= 3(x^2 - 4x - 5)$$

$\therefore$ For turning-points

$$x^2 - 4x - 5 = 0$$
$$\therefore \quad x = 5 \quad \text{or} \quad -1$$

But $$\frac{d^2y}{dx^2} = 6x - 12$$
$$= 6(x - 2)$$

When $x = 5,$ $$\frac{d^2y}{dx^2} = 6(5 - 2) = +18$$

When $x = -1,$ $$\frac{d^2y}{dx^2} = 6(-1 - 2) = -18$$

$\therefore \quad x^3 - 6x^2 - 15x + 10$ has a *maximum* value when $x = -1$, a *minimum* value when $x = 5$.

The actual maximum and minimum values are found by substituting $x = -1$ and $+5$, respectively, in

$$x^3 - 6x^2 - 15x + 10$$
When $x = -1,$ $\quad x^3 - 6x^2 - 15x + 10 = +18$
When $x = 5,$ $\quad x^3 - 6x^2 - 15x + 10 = -90$

Thus the *maximum* value of $x^3 - 6x^2 - 15x + 10$ is 18 and occurs when $x = -1$, the *minimum* value of $x^3 - 6x^2 - 15x + 10$ is -90 and occurs when $x = 5$.

Practical applications

15.8 An open tank is to be made of sheet iron; it must have a square base and vertical sides, with a capacity of 8 m³. Find its width and depth so as to use the least amount of sheet iron.

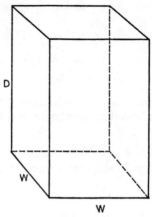

FIG. 15.5

Let its width and length $= w$ m (Fig. 15.5)
Let its depth $= D$ m
∴   Volume V $= Dw^2$ m³
and area of sheet iron A $= 4Dw + w^2$ m²

∴ $A = 4Dw + w^2$ must be a minimum.
∴   The differential coefficient of $4Dw + w^2$ must be zero.

As $4Dw + w^2$ contains two variables, D and w, we cannot differentiate at this stage, but

$$V = Dw^2 = 8$$
$$\therefore\ D = \frac{8}{w^2}$$

Substituting this value in $4Dw + w^2$,

$$A = 4 \cdot \frac{8}{w^2} \cdot w + w^2$$
$$= \frac{32}{w} + w^2$$
$$= 32w^{-1} + w^2$$
$$\therefore\ \frac{\mathrm{d}A}{\mathrm{d}w} = -32w^{-2} + 2w$$
$$= -\frac{32}{w^2} + 2w$$

∴   For a maximum or minimum value of A,

$$-\frac{32}{w^2} + 2w = 0$$

$$\therefore \quad 2w^3 = 32$$
$$\therefore \quad w = \sqrt[3]{16}$$
$$\therefore \quad w = 2\!\cdot\!52$$

But
$$\frac{d^2A}{dw^2} = +\frac{64}{w^3} + 2$$

and when
$$w = 2\!\cdot\!52$$

$$\frac{d^2A}{dw^2} = +\frac{64}{(2\!\cdot\!52)^3} + 2, \text{ which is positive}$$

$$\therefore \quad w = 2\!\cdot\!52 \text{ gives a minimum value to } A.$$

Also, since
$$D = \frac{8}{w^2}$$
$$D = \frac{8}{(2\!\cdot\!52)^2} = 1\!\cdot\!26$$

$\therefore$ For a minimum amount of sheet iron to be used the width and length $= 2\!\cdot\!52$ m, and the depth $= 1\!\cdot\!26$ m.

The above figures are correct mathematically. Since, however, this is a practical application, we shall not leave the answer in that form without comment.

To state the answer as $2\!\cdot\!52 \times 2\!\cdot\!52 \times 1\!\cdot\!26$ m³ implies that the container should be made to these dimensions. But the makers will need a 'tolerance', and, unless the requirement as to capacity is very strict, it will be best to employ simple figures, and so avoid the suggestion that the last hundredth of a millimetre, or even one millimetre, is of vital importance. Since the round figure 8 m³ is used, the requirement is unlikely to be strict. If the dimensions are taken as $2\!\cdot\!5$ m $\times$ $2\!\cdot\!5$ m $\times$ $1\!\cdot\!25$ m, the corresponding capacity will be $7\!\cdot\!8125$ m³, and the divergence of this calculated capacity from the specified 8 m³ might be comparable with that due to unavoidable error.

Note that in practical applications of mathematical methods any substitution of a round number for a calculated one must be made as a final step. As long as the problem remains one of mathematics, the calculation should be as precise as the tables permit: any approximation must then be related to the known facts of the case. Although in the making of a sheet-iron box, measurements to $\frac{1}{100}$ mm have no place, in the origination of an angle by means of a sine bar, for which accurate plane surfaces, accurate thickness blocks and accurate rollers are available, the ordinary four-figure tables are inadequate to match the accuracy of the mechanical equipment.

15.9 A rectangular sheet of tin 750 × 600 mm has four equal squares cut out at the corners, and the sides are then turned up to form a rectangular box. What must be the size of the squares cut away so that the volume of the box may be as great as possible?

Let the side of each square = x mm (Fig. 15.6).

After the box is formed

$$\text{its length} = (750 - 2x) \text{ mm,}$$
$$\text{its breadth} = (600 - 2x) \text{ mm,}$$
$$\text{its depth} = x \text{ mm.}$$

$$\therefore \quad \text{Volume of box, } V = (750 - 2x)(600 - 2x)(x) \text{ mm}^3$$

$$\therefore \quad = 450\,000x - 2700x^2 + 4x^3 \text{ mm}^3$$

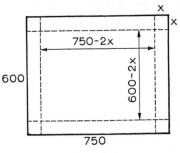

Fig. 15.6

Thus, for V to be a maximum (or a minimum),

$$\frac{dV}{dx} = 450\,000 - 5400x + 12x^2 = 0$$

$$\therefore \quad 37\,500 - 450x + x^2 = 0$$

$$\therefore \quad x = 339 \cdot 6 \quad \text{or} \quad 110 \cdot 4 \text{ mm}$$

But
$$\frac{d^2V}{dx^2} = -5400 + 24x$$

$$= 24(x - 225)$$

When $x = 339 \cdot 6$, $\dfrac{d^2V}{dx^2} = 24(339 \cdot 6 - 225)$, which is positive

When $x = 110 \cdot 4$, $\dfrac{d^2V}{dx^2} = 24(110 \cdot 4 - 225)$, which is negative

$\therefore$ V, the volume of the box, is as great as possible *when the sides of the squares cut out are* 110·4 mm. Clearly, however, the sizes 750 mm and 600 mm given for the original sheet are nominal, so that a tinsmith might well be asked to cut out small squares of 100 mm side.

It should be noted that 339·6 is an impossible value in this problem.

15.10 The cost, £C, per km of an electric cable is given by

$$C = \frac{72\,000}{x} + x$$

where x is its cross-section in mm². Find the cross-section for which the cost is least, and also the least cost per km.

$$C = \frac{72\,000}{x} + x$$

$$\therefore \quad \frac{\mathrm{d}C}{\mathrm{d}x} = \frac{-72\,000}{x^2} + 1$$

$\therefore$ C has a maximum or a minimum value when

$$\frac{72\,000}{x^2} = 1$$

$$\therefore \quad x = \sqrt{(72\,000)}$$
$$= 258 \text{ mm}^2$$

the negative value being inadmissible.

$$\frac{\mathrm{d}^2 C}{\mathrm{d}x^2} = \frac{144\,000}{x^2}, \text{ which is positive when } x = 258.$$

$\therefore$ When $x = 258$ mm², the cost is a minimum.
The actual minimum cost in pounds

$$= \frac{72\,000}{258} + 258$$

$\therefore$ Minimum cost per km = £537

EXERCISE 15.1

Find the turning-point on each of the following, and state whether the function has a maximum or a minimum value in each case:

1 $3x^2 - 2x$.   **3** $10x - 2x^2$.   **5** $24x - x^2$.
2 $x^2 + 3x$.   **4** $2x^2 + x$.   **6** $12x + 4x^2$.

Find the turning-points of each of the following, and also the maximum and minimum values.

7 $3x^3 - 9x^2 - 27x + 10$.
8 $x^3 - 9x^2 + 15x - 7$.
9 $24x + 5x^2 - 2x^3$.
10 $4x^3 - 21x^2 + 36x + 3$.
11 $2 - 9x + 6x^2 - x^3$.
12 Divide 50 into 2 parts such that their product is a maximum.

13 Find a number which exceeds its square by the greatest possible amount.

14 What number added to its reciprocal gives a minimum sum?

15 A closed cylindrical tank is required to contain 200 m³. Find the ratio of the height to the diameter so as to use the minimum amount of metal.

16 If the strength of a rectangular beam of wood varies as its breadth and the square of its depth, find the dimensions of the strongest beam that can be cut out of a round log of diameter d.

17 The relation between C, the cost per hour of running a certain ship, and its speed, v knots, is given by

$$C = 1·25 + 0·0008\ v^3$$

Find the average speed which is most economical for a voyage of 800 international nautical miles.

EXERCISE 15.2 (miscellaneous)

1 (*a*) Differentiate the following three functions with respect to x:

(i) $\dfrac{3x^4 + \sqrt{x}}{2x^2}$, (ii) $-\tfrac{1}{3}x^{-\frac{7}{8}}$, (iii) $x(2x + 1)^3$.

(*b*) Find the coordinates of the points on the graph of the function $x^2 + 2x - 3$ at which the slope is $+6$ and -2. What is the distance between these two points?   HANDSWORTH

2 Find the coordinates of the point on the curve $y = x^2 - 2x - 3$ at which the tangent to the curve is parallel to the x-axis. Find the gradient of the curve at the points where it intersects the x-axis. WEST RIDING

3 Find the coordinates of the maximum and minimum points on the curve $y = x^3 - 6x^2 + 9x + 1$. Sketch the curve.   HANDSWORTH

4 (*a*) Differentiate with respect to x: (i) $7x$, (ii) $5\sqrt{x}$, (iii) $\dfrac{1}{x^3}$,

(*b*) If a body is thrown upwards at 50 m/s under gravity, its height h m after t seconds is given by $h = 50t - 5t^2$. Find its velocity at $t = 2$ and $t = 6$ seconds. When is its velocity zero, and how high is it then?

(*c*) Draw a rough graph of $y = 9x^2 + 6x + 3$ and find the maximum or minimum values of y. Check your answer by a calculus method.

HANDSWORTH

5 (*a*) Find: (i) $\displaystyle\int \frac{a}{x^2}\,dx$, (ii) $\displaystyle\int \sqrt{x} - \frac{1}{2\sqrt{x}}\,dx$, (iii) $\int(x^2 + ax + b)dx$.

(*b*) The slope of a curve is given by $\dfrac{dy}{dx} = x^2 - 2$. Find the equation of this curve, given that it passes through $(0, 2)$. Make a neat freehand sketch of the curve.   HANDSWORTH

6 (*a*) Find from first principles the differential coefficient of x^3 with respect to x.

(b) Find, by rule, the differential coefficients, with respect to x, of the following expressions:

(i) $x^2(2x + 1)$, (ii) $\dfrac{(x + 1)^2}{x}$, (iii) $x^5(2x - 1)(x + 2)$.

(c) Find the maximum and minimum values of the expression $x^2(2x + 1)$ and distinguish between them.

EMEU

7 (a) Given the curve $y = x^2 - x - 6$,

(i) calculate the coordinates of the turning-point, (ii) prove whether this turning-point is a maximum or a minimum. (iii) What are the roots of the equation $x^2 - x - 6 = 0$? (iv) What is the gradient of the curve at the point $(-3, +6)$? (v) Sketch the curve.

(b) Differentiate the following, where x is the only variable: (i) ax^a

(ii) $\dfrac{a}{bx^2}$.

HANDSWORTH

8 (a) If $y = 3x^2$, find the value of $\dfrac{dy}{dx}$ from first principles.

(b) Find $\dfrac{dy}{dx}$ if $y = (3x - 1)(x - 4)$.

(c) The equation of a curve is

$$y = ax^2 + bx + 2,$$

where a and b are constants.

When $x = 1, y = 4$, and $\dfrac{dy}{dx} = -2$, find the values of a and b.

ULCI

9 (a) Find from first principles an expression for $\dfrac{dy}{dx}$, given that

$$y = \frac{x}{x + 1}.$$

(b) Make a graph of $y = \dfrac{x}{x + 1}$ for values of x from 0 to 3, and, by constructing a tangent at the point where $x = 2$, estimate the slope of the curve. Use the result of the part (a) to obtain a precise value for this slope.

COVENTRY

10 A wire 250 mm long is bent so as to form the boundary of a sector of a circle. Express the area of the sector in terms of the radius. Calculate the maximum area of the sector.

HANDSWORTH

11 If $y = 4x^3 - 3x$, find $\dfrac{dy}{dx}$ from first principles. Check by the rule for differentiating x^n. What is the true increase in y as x increases from 2 to 2·01?

COVENTRY

12 (a) Differentiate with respect to x $\dfrac{2x^2 - x + 4}{\sqrt{x}}$.

(b) Sketch the graph of the function $y = (x - 1)(x - 3)(x - 5)$, and find, by differentiation, the abscissae of the points on the curve at which the tangents are parallel to the axis of x.   SUNDERLAND

13 (a) Working from first principles, find $\dfrac{dy}{dx}$ if $y = 4x^2$.

(b) Using the rule, find $\dfrac{dy}{dx}$ when:

(i) $y = ax^{n+1}$,   (ii) $y = \dfrac{1}{\sqrt[3]{x}}$,   (iii) $y = 10x^5 - \dfrac{1}{x^2} + \dfrac{1}{4x^3}$.

(c) If $w = \dfrac{2V^2 + 3}{V}$,   find $\dfrac{dw}{dV}$ and $\dfrac{d^2w}{dV^2}$.   HANDSWORTH

14 The displacement $V\,\text{m}^3$ of water by a vessel immersed to a depth of h m is given by
$$V = 200h^2 - 10h^3$$

Plot the values of V against values of h between the limits $h = 3$ and $h = 8$. Determine graphically the value of $\dfrac{dV}{dh}$ at a depth of 6·5 m.

NCTEC

15 (a) Differentiate $x^3 - 2$.

(b) What is the gradient of the curve $y = x^2 - x - 2$ at the point on the curve where $x = 3$? At what point on this curve is the tangent parallel to the x-axis?   WEST RIDING

16 If the distance S m travelled by a motor car in t seconds after the brakes are applied is given by $S = 22t - 3t^2$, what is (i) the speed just when the brakes are applied, (ii) the distance travelled by the car before it stops, (iii) the retardation?   HANDSWORTH

17 What is the meaning of $\dfrac{dy}{dx}$ when considering graphs?

Find the equation of the straight line whose gradient is 3 and which is a tangent to the parabola $y = 3x^2 - 9x + 4$.   WORCESTER

18 The table below gives the velocity of a car between two points at given time intervals. Draw the velocity-time graph and determine the instantaneous acceleration of the car at times 18 and 48 seconds. What is the average acceleration of the car in the second and fourth periods of 6 seconds given?

Show, by use of Simpson's rule, that the car travels just over half its total distance in the first 36 seconds.

Time s	0	6	12	18	24	30	36	42	48	54	60
Velocity m/s	0	1·0	3·0	6·0	10·7	12·7	13·3	12·3	10·7	8·0	0

19 The distance of a moving body from its starting-point is S m after t seconds, given by $S = t^3 - 2t^2 + t - 10$. Show that the body is twice at rest, and find the accelerations at these instants.   NUNEATON

20 (*a*) If $y = \dfrac{1}{5 - x}$, find the value of $\dfrac{dy}{dx}$ from first principles, and evaluate it when $x = 3$.

(*b*) Draw a graph of the equation $y = \dfrac{1}{5 - x}$ for values of x from 0 to 4, and check the result of the calculations of part (*a*).

COVENTRY

21 (*a*) Differentiate with respect to x (i) $x^5 - 3x^4 + 4x^3 - 7x^2 +$,

(ii) $\dfrac{1}{\sqrt{x}} + \sqrt{x}$.

(*b*) An open rectangular container has a square base and its volume is 4 litres. Show that the surface area is $x^2 + \dfrac{16 \times 10^6}{x}$ if the side of the base is x mm. Hence find the dimensions when a minimum amount of material is used to make the container.   RUGBY

22 (*a*) If $y = 2x^3$ and x increases by a small amount, find the corresponding increase in y. Find also the ratio $\dfrac{\text{increase in } y}{\text{increase in } x}$, and hence find the derivative (or differential coefficient) of $2x^3$ with respect to x.

(*b*) If $s = t^3 - 6t^2 + 9t + 4$, find the values of t for which the gradient of the graph of s against t is zero, and in each case determine whether the corresponding value of s is a maximum or a minimum.   CANNOCK

23 (*a*) If $y = 3x^2$, and y increases to $y + \delta y$ when x increases to $x + \delta x$, find δy and $\dfrac{\delta y}{\delta x}$ in terms of δx.

Hence deduce the differential coefficient of y with respect to x in this case.

(*b*) Find for what values of x the gradient of $y = 2x^3 - 15x^2 + 36x$ is zero, and in each case determine whether the corresponding value of y is a maximum or a minimum.   CANNOCK

24 (*a*) If $y = x\left(1 - \dfrac{1}{x}\right) + \dfrac{1}{x}(1 - x)^2$, find $\dfrac{dy}{dx}$ when $x = \frac{1}{2}$.

(*b*) If $S = 8t^3 - 10t^2 - 4t - 5$, with the usual notation, find the speed and acceleration at the end of (i) 2 seconds, (ii) $\frac{1}{2}$ second. Find the positive value of t for which the velocity is zero.

(*c*) Given that $\dfrac{dy}{dx} = (x - 2)(x + 1)$, find the equation for y, given that when $x = 1$, $y = 5$.   NUNEATON

25 (*a*) Differentiate   (i) $y = 1 \cdot 3x^{0 \cdot 4}$,   (ii) $y = \sqrt{x^5}$.

(*b*) When a stone is thrown vertically upwards with velocity 30 m/s, its height s m after t seconds is given by $s = 30t - 5t^2$.
Find (i) its velocity when $t = 1\frac{1}{2}$ seconds, (ii) its velocity at a height of 25 m, (iii) the greatest height reached.   DUDLEY

26 Differentiate from first principles $y = 8x^2$, and state what you understand by the result.   WORCESTER

27 (*a*) Differentiate with respect to x the function

$$4x^3 + \frac{3}{x^2} + 2\sqrt{x} + 7.$$

(*b*) The displacement s m of a body after time t seconds is given by $s = 16t^2 - t^3$.   Find   (i) the velocity after 3 seconds, and (ii) when the acceleration is zero.   SUNDERLAND

28 (*a*) Differentiate from first principles $(x^2 + 3x)$.

(*b*) If a stone is thrown upwards at 25 m/s, its height h m after t seconds is given by $h = 25t - 5t^2$.   Find its velocity when $t = 2$ and $t = 3$.   At what time is its velocity zero, and how high is it then?   SHREWSBURY

29 (*a*) Differentiate with respect to x: (i) $y = 3x^4 - 2x^3 - 5x^2 + 7$, (ii) $y = 4\sqrt{x} - 3$.

(*b*) Find from first principles the differential coefficient of $\frac{1}{x^2}$ with respect to x.   RUGBY

30 (*a*) The rise of current in an inductive resistance is given by the equation $V = Ri + L\dfrac{di}{dt}$.   The current at any time t seconds is given by $i = 5 + 6t - 10t^2$.   Find $\dfrac{di}{dt}$ and the value of V when $t = 0 \cdot 1$ second, $R = 2$ and $L = 0 \cdot 05$.

(*b*) Find the maximum and minimum values of $2x^3 - 3x^2 - 12x + 10$.   NUNEATON

31 (*a*) Find $\dfrac{dy}{dx}$ in the following cases:

(i) $y = 4ax^2$,   (ii) $y = \dfrac{1}{3x^2}$,   (iii) $y = \dfrac{1}{\sqrt{x}} + 2\sqrt{x}$

(*b*) If s is distance in metres, and t is time in seconds, and $s = 3t^3 - 19t^2 + 8t - 3$, calculate (i) velocity when $t = 4$ and $t = 5$ seconds, (ii) acceleration when $t = 2$ seconds, (iii) times when the velocity is zero.   HANDSWORTH

32 If $f(x) = ax^3 + bx^2 + cx + d$, and $f(-1) = 3$, $f(0) = -2$, $f(1) = 9$, and $f(2) = 60$, find the values of a, b, c, d, and the values of x which will make the function zero.   COVENTRY

33 (*a*) Differentiate $7x^3 + 3x^2 - 2 + \dfrac{5}{x^2}$.

(*b*) A certain wheel, rotating through θ radians in t seconds, has $\theta = 18 + 20t - 2t^2$ as its equation of motion.

Find (i) its angular velocity after $3\frac{1}{2}$ seconds, (ii) its angular acceleration, (iii) when it comes to rest, (iv) its initial angular velocity.

(c) A square sheet of metal of edge 1 m, has a square of side x m cut out of each corner, and the remainder is folded to form an open box. Prove that the volume of the box is $x(1 - 2x)^2$. Find the maximum volume of the box. NUNEATON

34 Draw the graph of $y = \frac{1}{4}x^2$ for values of x from 0 to 4, and use it to illustrate the meanings of $\frac{\delta y}{\delta x}$ and $\frac{dy}{dx}$.

Taking the point of the graph where $x = 2$, find successive values of $\frac{\delta y}{\delta x}$ obtained by taking a range of values of δx, and hence deduce the probable value of $\frac{dy}{dx}$ at the point.

Obtain $\frac{dy}{dx}$ at the point $(3, 2\frac{1}{4})$ for the curve $y = \frac{x^2}{4}$.

Hence insert very carefully on your graph the tangent to $y = \frac{x^2}{4}$ at $(3, 2\frac{1}{4})$. EMEU

35 The angle θ radians turned through by a flywheel in t seconds is given by $\theta = 18 + 15t - 9t^2 + t^2$.

If the angular velocity ω and the angular acceleration α are given by $\frac{d\theta}{dt}$ and $\frac{d\omega}{dt}$ respectively, express ω and α in terms of t. Hence find the values of ω and α when t is 2. For what values of t is ω zero? ULCI

36 (a) Differentiate with respect to x (i) $(\frac{1}{2}x^4 - 3x) \div 10$, (ii) $(1 - 2x)^2$, (iii) $\frac{1}{2}x$, (iv) $10\sqrt{x}$

(b) Distance S and time t are related by the equation $S = 100 + 88t - 16t^2$. Find the expressions for velocity and acceleration. Find the maximum value of S. COVENTRY

37 (a) Find the differential coefficient of $2x^3 - x$ from first principles.

(b) Derive the expression for the slope of the curve $y = 4x^3 + 3x^2 - 5x + 2$ at any point. Calculate the values of the slope at the points $x = 0.6$ and $x = 0.25$. Hence find the acute angle included between the tangents to the curve at these points. HANDSWORTH

Chapter Sixteen
The Laws of Probability

If a ball is selected at random from a bag containing four black balls and five white balls, the probability of obtaining a black ball is defined as the number of ways a black ball may be picked divided by the number of ways any ball may be picked, i.e. $\frac{4}{9}$. Similarly, the probability of picking a white ball is $\frac{5}{9}$. The probability of picking either a black ball or a white ball is the number of ways either a black ball or a white ball may be picked divided by the number of ways any ball may be picked, i.e. $\frac{(4+5)}{9}$, i.e. $\frac{9}{9}$ or 1, which denotes certainty. Similarly, the probability of picking neither a black ball nor a white ball is $\frac{0}{9}$, i.e. 0, which denotes impossibility.

Random selection implies selection in such a way that each and every ball has an equal chance of being selected. The events of selecting a black ball and of selecting a white ball are said to be mutually exclusive, since, if a black ball is selected in a given trial, then a white ball is automatically excluded in that particular trial, and vice versa.

The events of selecting a white ball and of selecting a black ball are said to be independent, since, if after picking say a black ball, it is replaced, the chance of then picking a white ball is $\frac{5}{9}$, which is precisely what it would have been had the black ball not been selected before it. If, however, the black ball had not been replaced after selection, the chance of picking a white ball from five white balls out of eight remaining balls would then be $\frac{5}{8}$. The selection of the black ball would have altered the chance of selecting a white ball from $\frac{5}{9}$ to $\frac{5}{8}$, thus the events would not be independent if selection were made without replacement.

The addition law of probability
Consider the mutually exclusive events A and B. If the probability of the event A is $\frac{r}{n}$, and of the event B is $\frac{s}{n}$, then the probability of either event A or event B happening is given by the number of ways that either A or B may happen divided by n, i.e.

$$\frac{r+s}{n} = \frac{r}{n} + \frac{s}{n} = \text{probability of A} + \text{probability of B}$$

Example

16.1 Find the probability that in cutting a pack of cards (*a*) either a seven or a six or a black five appears, (*b*) either an Ace of Hearts or a King of Diamonds or a red Knave appears.

(*a*) Probability of a seven $= \dfrac{4}{52}$

probability of a six $= \dfrac{4}{52}$

probability of a black five $= \dfrac{2}{52}$

∴ probability of either a seven or a six or a black five $= \dfrac{4}{52} + \dfrac{4}{52} + \dfrac{2}{52} = \dfrac{10}{52} = \dfrac{5}{26}$

(*b*) Probability of an Ace of Hearts $= \dfrac{1}{52}$

probability of a King of Diamonds $= \dfrac{1}{52}$

probability of a red Knave $= \dfrac{2}{52}$

∴ probability of either an Ace of Hearts or a King of Diamonds or a red Knave $= \dfrac{1}{52} + \dfrac{1}{52} + \dfrac{2}{52} = \dfrac{1}{13}$

The multiplication law of probability

If the probability of an event A is $\dfrac{r}{n}$, and of an independent event B is $\dfrac{s}{n}$, then the probability of the event A followed by the event B is given by the number of ways that A then B can occur divided by the total number of pairs of events, i.e. $\dfrac{r \times s}{n \times n}$, which is the probability of the event A multiplied by the probability of the event B.

The following may serve as an illustration of the multiplication law of probability and of independent events.

Suppose that a bag contains two black balls and one white ball, and that a ball is selected at random from the bag, noted and then replaced. Then a second ball is selected. We can show that the probabilities of obtaining (*a*) a black then a white ball, (*b*) a white then a black ball, (*c*) two white balls and (*d*) two black balls obey the multiplication law of probability.

In each case the probability of a black ball is $\dfrac{2}{3}$, and of a white ball is $\dfrac{1}{3}$.

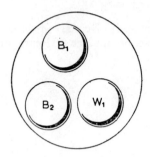 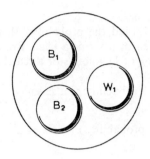

Fig. 16.1

Figure 16.1 illustrates the bag before each selection. It can be seen   that B_1 can be linked with B_1 or B_2 or W_1,
    that B_2 can be linked with B_1 or B_2 or W_1,
and that W_1 can be linked with B_1 or B_2 or W_1.

Thus the pairs of events may be represented by

$$B_1B_1, \ B_1B_2, \ B_1W_1, \ B_2B_1, \ B_2B_2, \ B_2W_1, \ W_1B_1, \ W_1B_2, \ W_1W_1.$$

There are nine of these events.

(a) (B_1W_1 and B_2W_1). There are two events of a black then a white ball,

$$\therefore \quad \text{probability of a black then a white ball} = \frac{2}{9} = \frac{2}{3} \times \frac{1}{3}$$

= probability of a black ball $\times$ probability of a white ball.

(b) (W_1B_1 and W_1B_2). There are two events of a white ball then a black ball,

$$\therefore \quad \text{probability of a white ball then a black ball} = \frac{2}{9} = \frac{1}{3} \times \frac{2}{3}$$

= probability of a white ball $\times$ probability of a black ball.

(c) (W_1W_1). There is one event of a white ball followed by a white ball,

$$\therefore \quad \text{probability of a white ball followed by a white ball} = \frac{1}{9} = \frac{1}{3} \times \frac{1}{3}$$

= probability of a white ball $\times$ probability of a white ball.

(d) ($B_1B_1, B_1B_2, B_2B_1, B_2B_2$). There are four events of a black ball followed by a black ball,

$$\therefore \quad \text{probability of a black ball followed by a black ball} = \frac{4}{9} = \frac{2}{3} \times \frac{2}{3}$$

= probability of a black ball $\times$ probability of a black ball.

N.B. The sum of the probabilities in (a), (b), (c) and (d) is $\frac{2}{9} + \frac{2}{9} + \frac{1}{9} + \frac{4}{9}$

= 1. The probabilities associated with the second choice are independent only because the first ball is replaced after selection. Suppose that the example is repeated, but that the ball is not replaced after the first selection.

Then $\qquad$ B_1 can be linked with B_2 or W_1,
$\qquad\qquad$ B_2 can be linked with B_1 or W_1,
$\qquad\qquad$ W_1 can be linked with B_1 or B_2,

Thus the pairs of events can be represented by

$$B_1B_2,\ B_1W_1,\ B_2B_1,\ B_2W_1,\ W_1B_1,\ W_1B_2.$$

There are six of these pairs of events
(a) $(B_1W_1,\ B_2W_1)$. There are two events of a black ball followed by a white ball,

$\therefore$ $\quad$ probability of a black then a white ball $= \dfrac{2}{6} = \dfrac{2}{3} \times \dfrac{1}{2}$

$=$ $\quad$ probability of a black ball (two out of three) $\times$ probability of a white ball (one out of two).

(b) $(W_1B_1,\ W_1B_2)$. There are two events of a white ball followed by a black ball,

$\therefore$ $\quad$ probability of a white then a black ball $= \dfrac{2}{6} = \dfrac{1}{3} \times \dfrac{2}{2}$

$=$ $\quad$ probability of a white ball (one out of three) $\times$ probability of a black ball (two out of two).

(c) The probability of obtaining a white ball followed by a white ball is zero.
(d) $(B_1B_2,\ B_2B_1)$. The probability of obtaining two black balls

$= \dfrac{2}{6} = \dfrac{2}{3} \times \dfrac{1}{2} =$ probability of obtaining a black ball (two out of three) $\times$ probability of obtaining a black ball (one out of two).

N.B. The sum of the probabilities in (a), (b), (c) and (d) is

$$\frac{2}{6} + \frac{2}{6} + 0 + \frac{2}{6} = 1.$$

Examples
16.2 An electrician has 25 electric light bulbs in a box; 7 of them are coloured and the remaining 18 are plain. If he selects two bulbs at random in the dark, calculate the chance or probability that he selects (a) two plain bulbs, (b) one plain then one coloured bulb, (c) one coloured then one plain bulb, (d) two coloured bulbs, if (i) the bulb is not replaced after selection, (ii) the bulb is replaced after selection. Verify in each case that the total probability is one.

(a) (i) The chance of the selection of a plain bulb from 18 plain bulbs out of 25 bulbs, followed by the selection of a plain bulb from 17 remaining plain bulbs out of 24 remaining bulbs $= \dfrac{18}{25} \times \dfrac{17}{24}$.

(b) (i) The chance of the selection of a plain bulb from 18 plain bulbs out of 25 bulbs, followed by the selection of a coloured bulb from 7 coloured bulbs out of 24 remaining bulbs $= \dfrac{18}{25} \times \dfrac{7}{24}$.

(c) (i) The chance of the selection of a coloured bulb from 7 coloured bulbs out of 25 bulbs, followed by the selection of a plain bulb from 18 plain bulbs out of 24 remaining bulbs $= \dfrac{7}{25} \times \dfrac{18}{24}$.

(d) (i) The chance of the selection of a coloured bulb from 7 coloured bulbs out of 25 bulbs, followed by the selection of a coloured bulb from 6 coloured bulbs out of a remaining 24 bulbs $= \dfrac{7}{25} \times \dfrac{6}{24}$.

The total chance $= \dfrac{18 \times 17 + 18 \times 7 + 7 \times 18 + 7 \times 6}{25 \times 24} = \dfrac{600}{600} = 1.$

(a) (ii) The chance of the selection of a plain bulb from 18 plain bulbs out of 25 bulbs, followed by the selection of a plain bulb from 18 plain bulbs out of 25 bulbs $= \dfrac{18}{25} \times \dfrac{18}{25}$.

(b) (ii) Similarly, the chance of one plain bulb then one coloured bulb $= \dfrac{7}{25} \times \dfrac{18}{25}$.

(c) (ii) The chance of a coloured bulb then a plain bulb $= \dfrac{18}{25} \times \dfrac{7}{25}$.

(d) (ii) The chance of a coloured bulb then a coloured bulb $= \dfrac{7}{25} \times \dfrac{7}{25}$.

The total chance $= \dfrac{18 \times 18 + 7 \times 18 + 18 \times 7 + 7 \times 7}{25 \times 25} = \dfrac{625}{625} = 1.$

16.3 An unbiased six-sided die is thrown twice. Find the probability that (a) a four is obtained once, (b) a four is obtained at least once, (c) a total score of four is obtained.

The chance of a four in a given throw is $\dfrac{1}{6}$, since there are six possible numbers all equally likely. The chance of not obtaining a four in a given throw is $1 - \dfrac{1}{6} = \dfrac{5}{6}$.

The chance of four then not four $= \dfrac{1}{6} \times \dfrac{5}{6} = \dfrac{5}{36}$.

The chance of not four then four $= \dfrac{5}{6} \times \dfrac{1}{6} = \dfrac{5}{36}.$

The chance of four then four $= \dfrac{1}{6} \times \dfrac{1}{6} = \dfrac{1}{36}.$

The chance of not four then not four $= \dfrac{5}{6} \times \dfrac{5}{6} = \dfrac{25}{36}.$

(a) The chance that four is obtained once $= \dfrac{5}{36} + \dfrac{5}{36} = \dfrac{5}{18}.$

(b) The chance that four is obtained at least once $= \dfrac{5}{36} + \dfrac{5}{36} + \dfrac{1}{36} = \dfrac{11}{36}$

(since this includes obtaining four twice).

(c) A total score of four may be obtained in the following ways: 1, 3; 3, 1; 2, 2.

The probability of a total score of four $= \dfrac{1}{6} \times \dfrac{1}{6} + \dfrac{1}{6} \times \dfrac{1}{6} + \dfrac{1}{6} \times \dfrac{1}{6} = \dfrac{3}{36} = \dfrac{1}{12}.$

Relation of the laws of probability to experience

The work of this chapter so far has not been based upon experience. If we state that the probability of obtaining a head when an unbiased coin is tossed is $\dfrac{1}{2}$, we do not mean to imply that in every 10 tosses of a coin 5 heads will be obtained. Our previous experience of coin tossing may possibly guide us to the prediction that, if a coin is tossed a very large number of times, then the relative frequency of heads will be approximately $\dfrac{1}{2}$.

The following data was obtained in an experiment with an undamaged coin using a coin shaker. A graph was plotted of the results, showing the relative frequency of heads plotted against the total number of tosses. The points were joined solely for visual clarity (Fig. 16.2).

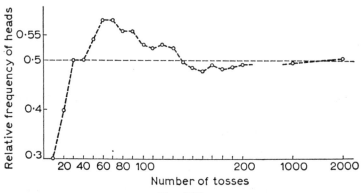

Fig. 16.2

Total number of heads	3	18	15	20	27	34	40	44
Total number of tosses	10	20	30	40	50	60	70	80
Relative frequency $= \dfrac{\text{total no. of heads}}{\text{total no. of tosses}}$	0·3	0·4	0·5	0·5	0·54	0·57	0·57	0·55

Total number of heads	49	53	57	63	67	70	73	77
Total number of tosses	90	100	110	120	130	140	150	160
Relative frequency $= \dfrac{\text{total no. of heads}}{\text{total no. of tosses}}$	0·55	0·53	0·52	0·53	0·52	0·5	0·49	0·48

Total number of heads	84	87	92	98	494	1011
Total number of tosses	170	180	190	200	1000	2000
Relative frequency $= \dfrac{\text{total no. of heads}}{\text{total no. of tosses}}$	0·49	0·48	0·48	0·49	0·494	0·506

In this experiment there appeared to be a tendency for the relative frequency of heads to settle down nearer to 0·5 as the number of tosses was increased. Suppose that this tendency was maintained as the total of tosses mounted into millions: then it might have been concluded that no evidence had been obtained to refute the suggestion that useful probability models for the tossing of unbiased coins may be based on the laws of probability, using $p = \frac{1}{2}$ as the probability of obtaining a head. The number of trials should be large and any predictions compared with experimental results.

Example
16.4 A symmetrical coin is tossed without bias four times. Show that the probability of 0, 1, 2, 3, 4 heads is $\dfrac{1}{16}, \dfrac{4}{16}, \dfrac{6}{16}, \dfrac{4}{16}, \dfrac{1}{16}$ respectively. Compare these values with the experimental probabilities obtained from the results of 550 such trials, in which 0, 1, 2, 3, 4 heads were obtained on 25, 142, 225, 136 and 22 occasions respectively.

Probability of 0 heads, i.e. 4 tails, which may be obtained in one way only,

$$\text{TTTT,} = \frac{1}{2} \times \frac{1}{2} \times \frac{1}{2} \times \frac{1}{2} = \frac{1}{16}.$$

The ways in which 1 head may be obtained are illustrated as follows:

$$\text{HTTT, THTT, TTHT, TTTH.}$$

There are four such ways, and the probability of each way is

$$\frac{1}{2} \times \frac{1}{2} \times \frac{1}{2} \times \frac{1}{2} = \frac{1}{16}.$$

Hence the probability of 1 head is $\frac{4}{16}$.

The ways in which 2 heads may be obtained are illustrated as follows:

HHTT, HTHT, HTTH, THHT, THTH, TTHH.

There are six such ways, and the probability of each way is $\frac{1}{16}$.

Hence the probability of 2 heads is $\frac{6}{16}$.

The ways in which 3 heads may be obtained are illustrated as follows:

HHHT, HHTH, HTHH, THHH.

Hence the probability of 3 heads is $\frac{4}{16}$.

4 heads may be obtained in one way only, as HHHH. Hence the probability of 4 heads is $\frac{1}{16}$.

A table may now be drawn up as follows.

Number of heads	0	1	2	3	4
Theoretical probability	$\frac{1}{16} = 0 \cdot 0625$	$\frac{4}{16} = 0 \cdot 25$	$\frac{6}{16} = 0 \cdot 375$	$\frac{4}{16} = 0 \cdot 25$	$\frac{1}{16} = 0 \cdot 0625$
Experimental probability	$\frac{25}{550} = 0 \cdot 0455$	$\frac{142}{550} = 0 \cdot 2582$	$\frac{225}{550} = 0 \cdot 4091$	$\frac{136}{550} = 0 \cdot 2473$	$\frac{22}{550} = 0 \cdot 0400$

EXERCISE 16.1

1 A bag contains five black balls and five white balls. If three balls are selected at random from the bag (a) with replacement, (b) without replacement, calculate the probability of selecting (i) three black balls, (ii) two black and one white ball, (iii) balls alternating in colour.

2 In cutting a pack of cards, calculate the probability that (a) a red card, (b) a black deuce, (c) an Ace, King or Queen of Hearts, (d) a card between two and ten is obtained.

3 A box of twelve components contains two defective components. If two components are selected at random, calculate the probability that (a) no defective, (b) 1 defective, (c) 2 defective components are selected if selection is (i) with replacement, (ii) without replacement. Verify that in both (i) and (ii) the total probability is 1.

4 If two unbiased six-sided dice are thrown, and the score recorded, calculate the probabilities of the various possible scores. Compare these theoretical probabilities with the experimental probabilities obtained from the following experimental results of 144 trials.

Score	2	3	4	5	6	7	8	9	10	11	12
Frequency	3	9	13	18	19	22	20	15	14	7	4

Carry out the experiment yourself, and compare your results with the above.

5 Repeat the experiment described in the text to obtain the relative frequency of heads for as many tosses as you have time for. Draw your graph in a similar way to Fig. 16.2.

6 A symmetrical coin is tossed without bias three times. Calculate the probability of 0, 1, 2, 3 heads respectively. Compare these values with the experimental probabilities based on 80 such trials. (The time required is approximately 20 minutes to toss and record results.)

Iterative methods
To find the square root of a given number N to a required number of decimal places by an iterative method.

Suppose that x is an approximation to the square root of N, then $\dfrac{N}{x}$ must also be an approximation to the square root of N. Further, one of the approximations, x or $\dfrac{N}{x}$, must be in excess of the square root of N, and the other one must be less than the square root of N. It follows that the mean of these approximations, $\frac{1}{2}\left(x + \dfrac{N}{x}\right)$, must be a closer approximation to the square root than either of these approximations. This may be written as an iterative formula thus:

$$x_{n+1} = \tfrac{1}{2}\left(x_n + \frac{N}{x_n}\right)$$

where iteration may be carried out as in the following example.

Example
16.5 Find the square root of 29 by an iterative method correct to 6D (six decimal places).

A linear approximation to the square root is given by $5 + \lambda$, where $\dfrac{\lambda}{4} = \dfrac{1}{11}$, i.e. $\lambda = \dfrac{4}{11}$, and the approximation is $5\frac{4}{11}$ (Fig. 16.3).

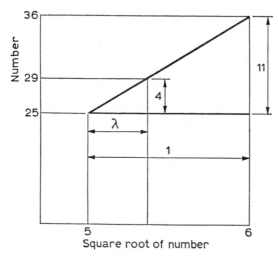

Number

36
29
25

Square root of number

5   6

11

4

λ

1

Fig. 16.3

Thus $x_0 = \dfrac{59}{11}$, and the iterative formula is $x_{n+1} = \frac{1}{2}\left(x_n + \dfrac{29}{x_n}\right)$.

Replacing n by 0, $x_1 = \frac{1}{2}\left(x_0 + \dfrac{29}{x_0}\right)$, where $x_0 = \dfrac{59}{11}$

$$\therefore \quad x_1 = \frac{1}{2}\left(\frac{59}{11} + \frac{29 \times 11}{59}\right) = \frac{1}{2}\left(\frac{59^2 + 29 \times 11^2}{11 \times 59}\right)$$

$$= \frac{6990}{1298} = \frac{3495}{649} \simeq 5 \cdot 385\,21$$

Replacing n by 1, $x_2 = \frac{1}{2}\left(x_1 + \dfrac{29}{x_1}\right)$, where $x_1 = \frac{3495}{649}$

$$\therefore \quad x_2 = \frac{1}{2}\left(\frac{3495}{649} + \frac{29 \times 649}{3495}\right) = \frac{1}{2}\left(\frac{3495^2 + 29 \times 649^2}{649 \times 3495}\right)$$

$$= \frac{24\,429\,854}{45\,365\,10} \simeq 5 \cdot 385\,164\,807$$

x_1 and x_2 have four significant figures in common, thus we expect x_1 to be correct to four significant figures. In this method the number of significant figures is about doubled at each stage. x_2 may be expected to be correct to eight significant figures.

Check $5 \cdot 385\,164\,8^2 = 28 \cdot 999\,999\,923\,159\,04$
                          $5 \cdot 385\,164\,9^2 = 29 \cdot 000\,001\,000\,192\,01$

Thus $\sqrt{29} = 5 \cdot 385\,165$, correct to 6D.

The calculation is carried out here in fractional form, and division avoided in the iterative process: the iterative formula for the fractional approximation to the square root of N could be written as

$$\frac{a_{n+1}}{b_{n+1}} = \frac{1}{2}\left(\frac{a_n}{b_n} + \frac{Nb_n}{a_n}\right) = \frac{1}{2}\left(\frac{a_n^2 + Nb_n^2}{b_n \cdot a_n}\right)$$

Alternatively, if division is used throughout the stages,

$$x_0 = 5\tfrac{4}{11} \simeq 5 \cdot 36$$

$$x_1 = \tfrac{1}{2}\left(5 \cdot 36 + \frac{29}{5 \cdot 36}\right) = \tfrac{1}{2}(5 \cdot 36 + 5 \cdot 4104) = 5 \cdot 3852$$

$$x_2 = \tfrac{1}{2}\left(5 \cdot 3852 + \frac{29}{5 \cdot 3852}\right) = \tfrac{1}{2}(5 \cdot 3852 + 5 \cdot 385\ 129\ 6)$$

$$= 5 \cdot 385\ 164\ 8$$

When x_0 and x_1 have been obtained, x_0 is observed to be correct to two significant figures: thus x_1 is expected to be correct to four significant figures. This is confirmed when x_2 is obtained, and then x_2 is expected to be correct to eight significant figures.

To find the reciprocal of a given number N to a required number of decimal places by an iterative method

Let x be an approximation to $\frac{1}{N}$ such that $x = \frac{1}{N} - \Delta$ (i), where Δ is a small number. Can we find a better approximation to $\frac{1}{N}$ involving Δ^2?

$$x + \Delta = \frac{1}{N}$$

Squaring both sides,

$$x^2 + 2x\Delta + \Delta^2 = \frac{1}{N^2}$$

Multiplying both sides by N,

$$Nx^2 + 2x\Delta N + N\Delta^2 = \frac{1}{N}$$

i.e. $$Nx^2 + 2x\Delta N = \frac{1}{N} - N\Delta^2$$

Eliminating Δ from the left-hand side of this result, since $\Delta = \frac{1}{N} - x$ from (i),

$$Nx^2 + 2xN\left(\frac{1}{N} - x\right) = \frac{1}{N} - N\Delta^2$$

i.e. $$Nx^2 + 2x - 2Nx^2 = \frac{1}{N} - N\Delta^2$$

i.e. $$2x - Nx^2 = \frac{1}{N} - N\Delta^2$$

i.e. $$x(2 - Nx) = \frac{1}{N} - N\Delta^2 \qquad \text{(ii)}$$

The expression (ii) will give a better approximation than (i) provided that the numerical value of $N\Delta^2$ is less than that of Δ. Equation (ii) may be written in the iterative form $x_{n+1} = x_n(2 - Nx_n)$, then $x_1 = x_0(2 - Nx_0)$, $x_2 = x_1(2 - Nx_1)$, etc. is used.

Example
16.6 Find the reciprocal of 129 correct to 8D.

Since $\dfrac{1}{125} = \dfrac{8}{1000} = 0{\cdot}008$, take $x_0 = 0{\cdot}008$, and $N = 129$.

$$x_1 = 0{\cdot}008(2 - 129 \times 0{\cdot}008) = 0{\cdot}007\ 744$$
$$x_2 = 0{\cdot}007\ 744(2 - 129 \times 0{\cdot}007\ 744) = 0{\cdot}007\ 751\ 929\ 856$$

As x_1 is observed to be correct to two significant figures, x_2 may be expected to be correct to about four significant figures.

$$x_3 = 0{\cdot}007\ 751\ 9(2 - 129 \times 0{\cdot}007\ 751\ 9)$$
$$= 0{\cdot}007\ 751\ 937\ 984$$

As x_2 is observed to be correct to five significant figures, x_3 may be expected to be correct to about ten significant figures.

Check $$0{\cdot}007\ 751\ 937 \times 129 = 0{\cdot}999\ 999\ 873$$
$$0{\cdot}007\ 751\ 938 \times 129 = 1{\cdot}000\ 000\ 002$$

Thus $\dfrac{1}{129} = 0{\cdot}007\ 751\ 94$, correct to 8D.

To apply the method of iteration to the solution of linear simultaneous equations in which the leading diagonal coefficients are dominant
The method of iteration may be applied to the solution of linear simultaneous equations in two unknowns, $ax + by = c$
$$dx + ey = f$$
in which the leading diagonal coefficients a and e are large in magnitude compared with other coefficients. The equations may be transposed to

$x = \dfrac{(c - by)}{a}$ and $y = \dfrac{(f - dx)}{e}$, and written in the iterative form

$x_{n+1} = \dfrac{(c - by_n)}{a}$ and $y_{n+1} = \dfrac{(f - dx_{n+1})}{e}$. The method may be extended to the solution of n equations in n unknowns, provided that the leading diagonal coefficients are dominant.

Example
16.7 Solve the equations $100x - y = 169\cdot4$
$3x + 50y = 185\cdot2$

correct to 3D.

Since $$x_{n+1} = \frac{(169\cdot4 + y_n)}{100} \quad . \quad . \quad . \quad . \quad . \quad . \quad \text{(i)}$$

and $$y_{n+1} = \frac{(185\cdot2 - 3x_{n+1})}{50} \quad . \quad . \quad . \quad . \quad . \quad \text{(ii)}$$

an approximation for y_0 may be obtained by ignoring the term in x in (ii),

i.e. $y_0 = \dfrac{185\cdot2}{50} \simeq 3\cdot7$.

Using $x_1 = \dfrac{(169\cdot4 + y_0)}{100}$, $y_1 = \dfrac{(185\cdot2 - 3x_1)}{50}$, etc.,

$x_1 = (169\cdot4 + 3\cdot7)/100 = 1\cdot731$
$y_1 = (185\cdot2 - 3 \times 1\cdot731)/50 = 3\cdot600\ 14$
$x_2 = (169\cdot4 + 3\cdot600\ 14)/100 = 1\cdot730\ 001\ 4$
$y_2 = (185\cdot2 - 3 \times 1\cdot730\ 001\ 4)/50 = 3\cdot600\ 199\ 916$
$x_3 = (169\cdot4 + 3\cdot600\ 199\ 916)/100 = 1\cdot730\ 001\ 999\ 16$

Retaining the figures common to two successive approximations, the solutions are approximately

$x = 1\cdot730\ 001$ and $y = 3\cdot6001$
i.e. $x = 1\cdot730$ and $y = 3\cdot600$, correct to 3D.

EXERCISE 16.2

1 Show that a linear approximation to the square root of 41 is $6\frac{5}{13}$ or $\frac{83}{13}$, and that further approximations are 13 818/2158 and 381 872 648/ 59 638 488 or 6·403 15 and 6·403 124 237 approximately. Verify that

$6\cdot403\ 124\ 2^2 = 40\cdot999\ 999\ 520\ 625\ 64$, and that
$6\cdot403\ 124\ 3^2 = 41\cdot000\ 000\ 801\ 250\ 49$, and hence that
$6\cdot403\ 124\ 2 < \sqrt{41} < 6\cdot403\ 124\ 3$.

2 Show that a linear approximation to the square root of 3 is $1\frac{2}{3}$ or $\frac{5}{3}$, and that further approximations are 26/15 and 1351/780, or $1\cdot733\ 33\ldots$ and $1\cdot732\ 051\ 3$ approximately. How many significant figures do you expect to be correct in this latter approximation?

3 Taking $\frac{1}{20}$ as an approximation to the reciprocal of 19, show that better approximations are 21/400 and 8421/160 000, or 0·0525 and 0·052 631 25.

Verify that $0\cdot052\ 631 \times 19 = 0\cdot999\ 989$, and that
$0\cdot052\ 632 \times 19 = 1\cdot000\ 008$, and hence that
$0\cdot052\ 631 < \frac{1}{19} < 0\cdot052\ 632$.

4 Taking $\dfrac{8}{100} = \dfrac{1}{12\cdot5}$ as an approximation to the reciprocal of 13, obtain better approximations 0·0768, 0·076 922 88 and 0·076 923 076 922 572 8.

5 Given the equations $\quad 80x + 5y = 17\cdot3$
$$2x - 40y = 29\cdot1$$

or $x = (17\cdot3 - 5y)/80$ and $y = (2x - 29\cdot1)/40$, starting with the value of y when $x = 0$, i.e. $y_0 = -0\cdot7275$, show that x_1 may be taken as $0\cdot262$, y_1 as $-0\cdot7144$, x_2 as $0\cdot2609$, y_2 as $-0\cdot714\,455$, x_3 as $0\cdot260\,903\,25$, y_4 as $-0\cdot714\,454\,837\,5$. Solve the equations by elimination, and verify that x_2 is correct to 6D and y_4 is correct to 7D.

6 Given the equations $\quad 50x - y = 101\cdot2$
$$x + 40y = 81\cdot7$$

or $x = (101\cdot2 + y)/50$ and $y = (81\cdot7 - x)/40$, starting with the value of y when $x = 0$, i.e. $y_0 = 2\cdot0425$, show that x_1 may be taken as $2\cdot065$, y_1 as $1\cdot9909$, x_2 as $2\cdot063\,818$, y_2 as $1\cdot990\,904\,55$, x_3 as $2\cdot063\,818\,091\,0$.

Evaluate the determinants

$$\begin{vmatrix} 101\cdot2 & -1 \\ 81\cdot7 & 40 \end{vmatrix} \quad \begin{vmatrix} 50 & 101\cdot2 \\ 1 & 81\cdot7 \end{vmatrix} \quad \begin{vmatrix} 50 & -1 \\ 1 & 40 \end{vmatrix}$$

and hence verify that x_2 is correct to 6D and y_2 to 7D.

7 Find by an iterative method, and check, (*a*) the square root correct to 6D, (*b*) the reciprocal correct to 8D of (i) 47, (ii) 71, (iii) 97.

8 Solve the following equations correct to 3D by an iterative method, and check.

(*a*) $47x - 2y = 139\cdot42$ (*b*) $70x + 5y = 278\cdot80$
$3x + 51y = 72\cdot25$ $x - 80y = 178\cdot0$

Chapter Seventeen
Statistics

In the plural the word statistics has been taken to mean numerical facts systematically collected, in the singular to mean the scientific method of treating quantitative facts to aid decision making in situations of uncertainty. This can involve:

(a) collecting data, which may require the previous design of suitable experiments,

(b) arranging, condensing, tabulating and graphically displaying this data,

(c) mathematical analysis of the data, and the evaluation of certain statistical measures, of which the mean, median, mode and standard deviation will be encountered later in this chapter,

(d) the estimation of the characteristics of a large collection of data from suitably chosen samples,

(e) the study of the mutual relation or interdependence of numerical facts which is called correlation.

In this book we shall be mainly considering sections (a), (b) and (c).

Necessity for the collection of data

Historically, the development of statistics has involved the collection of state facts. Estimates of the quantities of money, food, weapons, transport and manpower required for defence or offence were required, and also of the location and resources of the taxpayers who financed these requirements. The insurance actuary uses large collections of data concerning death and disease and its relation to the consumption of tobacco and alcohol, to occupation and to age. Figures concerning climate; floods; fires; motor, industrial and domestic accidents, etc. are used to give a factual basis from which, with statistical processing, predictions can be made enabling premiums to be fixed which should be profitable. A recent advertisement for the Government Statistical Service states, 'the critical interpretation of information about a great range of events and trends is essential, not only to Government administration and formulation of policy, but also to commerce, industry and the general public. Some examples are: forecasting manpower needs and supplies, forward planning of the educational system, assessing future pattern of demands for energy, research into socio-economic problems, forecasting demands for freight and passenger transport. Techniques employed include sample surveys, input/output analysis, model building, econometric methods, cost/benefit analysis. A wide range of automatic data processing equipment is used.' Statistical methods are applied in the glass, rubber, textile, chemical, metallurgical, agricultural and engineering industries.

In science, where experimental errors are inevitable, and where a great multiplicity of factors operates, statistical techniques may be the only methods of handling experimental data. It is often stated in error that 'statistics can prove anything'. It might be more properly said 'the science of statistics cannot prove anything, but properly applied it can make estimates of probability in more precise measures than "likely" or "not likely", i.e. numerical measures'. An extreme view was held by Lord Kelvin, the great physicist: 'When you can measure what you are speaking about and express it in numbers, you know something about it; but when you cannot measure it, when you cannot express it in numbers, your knowledge is of a meagre and unsatisfactory kind.' Numerical data is the raw material: statistics provides a method of interpreting this data.

Limitations in the collection of data

A major limitation in the collection of data is the expense and time required to obtain this information. This leads to the idea of sampling. Representative sample information is obtained and statistically interpreted so as to provide predictions about the parent body of information. Railway and telephone undertakings, for example, possess a great mass of capital equipment for which a complete assessment of the state of depreciation is required. This equipment varies in complexity and is widely scattered. Complete inspection for the purpose of collecting this data would be too costly and slow. By choosing a relatively small sample of this equipment according to statistical principles, useful predictions about the state of the capital equipment of the whole organisation can be made at a fraction of the expense of time and money. It is interesting to note that statistical treatment of information from intelligence reports and aerial surveys about German V2 bomb production towards the end of the Second World War enabled estimates to be made which were later found to be surprisingly accurate, particularly in view of the fact that the figures were not known to the Germans.

Sampling forms the basis for the statistical control of production processes. A great variety of fluctuating factors, such as temperature, illumination, fatigue and vibrations, may contribute to the production of defective items, producing random errors. A sample from a batch of given size will often contain defective items due to the operation of these chance factors. The problem is to be able to decide when the number of defectives produced cannot be explained by chance, indicating that there is some systematic error in the process which must be investigated and corrected. When the process is judged to be operating satisfactorily after repeated samples have been taken, warning and action limits are calculated, and a control chart drawn. Observed values from further samples are plotted on this chart, and, if the results are within these limits, the process s said to be under statistical control. Otherwise action can be taken. Thus, with a limited amount of data, a record is provided of the process as it is taking place, and control achieved. It should be noted that a small number of measurements taken by more skilled and careful observers may be more reliable than a large number of measurements taken hurriedly by many less skilful observers. Another limitation on the collection of data is that measurement sometimes involves destruction of the item, thus making sampling essential.

Accuracy and approximation

Statistical data consisting of a set of values of a variable, usually called the variate, cannot be exact where measurement rather than counting is employed. Any statistical measure calculated from this data will tend to be less, rather than more, precisely known than the data.

Example

17.1 The lengths of 10 rods were measured to the nearest millimetre using a rule. Between what limits will the mean rod length lie? The rod lengths were 230, 233, 232, 234, 229, 231, 233, 232, 229 and 233 mm.

If we take the mean of these numbers, we obtain $\frac{2316}{10}$, i.e. 231·6, which, if quoted as the mean rod length in mm, involves spurious accuracy.

$$\text{Mean rod length} = \frac{(230 \pm 0·5) + (233 \pm 0·5) + \ldots + (233 \pm 0·5)}{10}$$

$$= \frac{2316 \pm 10 \times 0·5}{10}$$

$$= 231·6 \pm 0·5 \text{ mm}$$

i.e. the mean rod length lies between $231·6 - 0·5$ mm and $231·6 + 0·5$ mm, or 231·1 and 232·1 mm, and is 230 mm correct to two significant figures. In the above example the possible error is $\pm 0·5$ mm. This is greater than the probable error, which is often taken to be the possible error divided by the square root of the number of items, in this case $0·5/\sqrt{10}$ or 0·2 approximately. On this basis the mean rod length is $231·6 \pm 0·2$ mm, and probably lies between 231·4 and 231·8 mm, due to the tendency of the errors in measurement to be compensating if they are assumed to be random rather than systematic.

In many instances, where past records rather than direct collections are the source of data, it is necessary to have details of the methods used to collect the information, and to be satisfied that it is sufficiently accurate for the present purpose.

It may be noted that the word accuracy in connection with counting implies that no error has been made, but with reference to measurement that the measurement is within stated limits.

Relative error

The significance of the error in a quantity is better measured by the relative error, which is the ratio of the error in a quantity to the approximate value of the quantity. Previously we obtained a mean rod length of $231 \cdot 6 \pm 0 \cdot 5$ mm. The greatest possible error here is 1 mm, which may seem small, but the relative error is given by

$$\text{relative error} = \frac{\text{magnitude of error in quantity}}{\text{approximate value of the quantity}}$$

$$= \frac{1}{230}$$

$$= 0 \cdot 0043 \text{ correct to two significant figures}$$
$$\text{or } 0 \cdot 43 \%$$

Note that the numerator is probably less than 1, and the denominator is greater than 230, so that our estimate of the relative error tends to be in excess, and therefore on the safe side.

Suppose that a distance is measured as 351 km, whilst the distance is known by a very accurate method to be 352 km to the nearest 100 mm. Then the error in this case is 1 km, compared with 1 mm at most in the previous case. The relative error, however, is given by

$$\text{relative error} = \frac{\text{magnitude of error in quantity}}{\text{approximate value of the quantity}}$$

$$= \frac{1}{351}$$

$$= 0 \cdot 0028 \text{ correct to two significant figures}$$
$$\text{or } 0 \cdot 28 \%$$

which indicates greater accuracy than in the previous case.

Note that 351 is used as the denominator because it tends to over-estimate the relative error, and because the more exact value 352 is not usually known: the answer is not affected correct to two significant figures.

Discrete and continuous values of the variate

Suppose that x, the number of defective components in 4 batches of 10 components, is 0, 1, 2, 2, then x is said to be a *discrete* or *discontinuous variate*, since values of the variate cannot differ by less than a finite amount (1 in this case).

Suppose next that a particular dimension, l units, of each component is measured as $13 \cdot 0$, $13 \cdot 1$, $13 \cdot 1$, $13 \cdot 2$, ... The measurements again provide a discrete variate which do not differ by less than $0 \cdot 1$. If, however, we express the results as $13 \cdot 0 \pm 0 \cdot 5$, $13 \cdot 1 \pm 0 \cdot 5$, etc., then the variate is *continuous*. $13 \cdot 0 \pm 0 \cdot 5$ can include any value between $12 \cdot 5$ and $13 \cdot 5$ and *there is no limit to the number of possible values* in this range which the variate could take. This is characteristic of a continuous variate.

Introduction of the median and the mode

In addition to the familiar arithmetic mean, the median and the mode are two other types of average used in statistics.

(1) Suppose that the number of heads H obtained when 4 coins are tossed is recorded, the experiment repeated 5 times, and the results arranged in ascending order:

$$0 \ 1 \ 2 \ 3 \ 3$$

The *median* is *that value of the variate which divides this distribution into two equal parts*. Since the middle value is 2, by inspection, the median value is two heads.

The *mode* is *that value of the variate which has the highest frequency density*. Less generally, here it is the value of the variate which occurs most often, i.e. 3. Consider now

$$0 \ 0 \ 1 \ 1 \ 2 \ 2 : 2 : 3 \ 3 \ 3 \ 4 \ 4 \ 4$$

The median is 2, the modes are 2, 3 and 4. In general there is only 1 median but there may be more than 1 mode.

(2) Suppose that the total number of values is even:

$$0 \ 0 \ 1 \ 2 : 2 \ 3 : 3 \ 4 \ 4 \ 4$$

In this case the modal value is 4 heads. There is no middle value, so the mean of the two central values is taken. The median value is $\dfrac{2+3}{2} = 2 \cdot 5$ heads.

More generally, if N numbers are arranged in order of magnitude, *when* N *is odd the median is the* $\dfrac{N+1}{2}$*th value.* In case (i) the $\dfrac{5+1}{2} = 3$rd value. *When N is even the median is the mean of the* $\dfrac{N}{2}$ *and* $\left(\dfrac{N}{2}+1\right)$*th* values.

In case (2) the median is the mean of the $\frac{10}{2}$th and $\left(\dfrac{10}{2}+1\right)$th value, i.e. of the 5th and 6th value.

The arithmetic mean is familiar in application. Whilst a style of footwear or clothing 'à la mode' can expect to have the largest sales and to require the largest stocks, similarly the size required by the 'average' man uses average in the sense of the mode, rather than the arithmetic mean. Average consumption of bread per head in the UK, on the other hand, uses average in the sense of the arithmetic mean. The median is less familiar in application. If an exact problem was worked through six times and reasonable answers A, B, A, A, C, A were obtained, the modal answer A would be preferred to the mean answer $(4A + B + C)/6$.

A simple practical application of the median

Along a road are petrol stations P_1, P_2, P_3, P_4, P_5, P_6, P_7 situated at 0, 3, 12, 20, 30, 32, and 134 miles from P_1. It is required to site a stores depot at one of the stations so that the sum of its distances from the other stations is a minimum. The median distance from P_1 is 20 miles, at station P_4, which is the correct site as the following table shows.

Proposed depot station		P_1	P_2	P_3	P_4	P_5	P_6	P_7
	P_1	0	3	12	20	30	32	132
	P_2	3	0	9	17	27	29	131
Distance from the depot station	P_3	12	9	0	8	18	20	122
	P_4	20	17	8	0	10	12	114
	P_5	30	27	18	10	0	2	104
	P_6	32	29	20	12	2	0	102
	P_7	134	131	122	114	104	102	0
Sum of distances from the depot station		231	216	189	181	191	197	707

Frequency table for a discrete variate – median and mode
Suppose that the variate x takes the value 0, 24 times; 1, 12 times; 2, 9 times; 3, 4 times; 4, once. We can draw up a frequency table:

x	0	1	2	3	4
f	24	12	9	4	1

where f represents the frequency, or the number of times that the corresponding value of the variate occurs. Notice that in this case we have exact information about the values of x; the only information lost is the order in which the values occurred.

The modal value is 0, corresponding to the greatest frequency, namely 24.

The total number of items, i.e. the total frequency, is denoted by Σf where $\Sigma f = 24 + 12 + 9 + 4 + 1 = 50$.

The median is the mean of the $\frac{50}{2}$th and $(\frac{50}{2} + 1)$th, i.e. of the 25th and 26th items. Inspections shows that the 25th and 26th items correspond to $x = 1$; hence the median value is 1.

Grouped frequency table for a continuous variate
Suppose that certain errors may be measured to the nearest 0·5 mm, and that errors of 0 to 1 mm occur 8 times, 1 to 2 mm occur 12 times, 2 to 3 mm occur 5 times. We may draw up the following grouped frequency table.

Error interval	0/1	1/2	2/3	
Mid-value of the group, x	0·5	1·5	2·5	
Frequency f	8	12	5	$\Sigma f = 25$

For the purpose of calculating the arithmetic mean approximately, we can replace each interval by a discrete mid-value x, since we expect random errors to be spread fairly uniformly throughout each interval (there is an analogy in mechanics, where a uniformly distributed load is replaced by a point load), but in the estimation of the median and the mode we have difficulty. The median item of 25 discrete items is the $\left(\dfrac{25+1}{2}\right)$th $= 13$th item, which would suggest the discrete value 1·5; but since the variate is actually continuous, the best we can do is to estimate the median in the group 1/2 by a method of proportion which seeks to divide the total frequency into two equal parts.

Similarly, we can state that the modal group would be the group 1/2 for a discrete variate. Since the variate is continuous, we cannot base the notion of a modal value on maximum frequency, since we have no reason to expect that a particular value need occur more than once. Later we may form an estimate of the mode based upon the notion of frequency density and the frequency curve as the limiting form of a histogram.

Finally a grouped frequency table may be formed for convenience when a large set of values of a discrete variate is grouped in a frequency table. If the original set of values is retained, the median and mode may be assessed exactly; but if only the grouped frequency table is retained, then the best we can do is to estimate the median and mode by treating the variate as a continuous one.

Tabulation of data and frequency tables

In an industrial manufacturing process, a count of the number of defective components per sample of 100 components was taken over 60 samples, as follows.

```
0 2 0 1 0 1 0 1 1 0 2 1 1 2 0 0 1 0
3 0 2 1 3 0 1 2 0 1 0 3 2 0 1 2 0 0
1 1 0 1 2 0 0 1 0 2 0 1 0 1 2 0 1 2
0 1 1 0 0 4
```

The data may be classified in the following table by first entering a tally mark in the appropriate column, and at the same time encircling the data in the above table. Any neglected data will be apparent at the end of the process. x is used to denote the number of defectives per sample, and f the frequency with which it occurs. Σf is used to denote the sum of the frequency values. Note that the variate is *discrete*, and hence the median and mode may be obtained exactly.

The relative frequency is obtained by dividing the corresponding frequency by the total frequency. The cumulative frequency at a given column represents the sum of all the frequencies occurring to the left of that column.

The f-row shows at a glance that the *mode*, i.e., the *number of defectives with the highest frequency* is 0, and that the frequency decreases as the number of defectives increases.

x	0	1	2	3	4	
	卌 卌 卌 卌 卌	卌 卌 卌 卌	卌 卌 I	III	I	
f	25	20	11	3	1	$\Sigma f = 60$
Relative frequency	25/60 = 0·42	20/60 = 0·33	11/60 = 0·18	3/60 = 0·05	1/60 = 0·02	
Cumulative frequency	25	45	56	59	60	
Cumulative relative frequency	0·42	0·75	0·93	0·98	1·00	

The median is seen to be 1, since the $\frac{60}{2}$th and $(\frac{60}{2}+1)$th, i.e. the 30th and 31st, items correspond to 1 defective.

The relative frequency row shows that 0·42 or 42% of samples have 0 defective items, and that 0·02 or 2% of samples have 4 defective items.

The cumulative frequency row shows, for example, that 45 samples have 1 or less defectives, and that 59 samples have 3 or less defectives.

The cumulative relative frequency row shows, for example, that 75% of the samples have 1 or less defectives and that 93% of the samples have 2 or less defectives.

It is also more quickly seen using the table that the range of defectives extends from 0 to 4 defectives.

Graphical representation of data—histograms

A frequency table may be represented by a histogram by marking on the horizontal axis intervals corresponding to the values taken by the variate. The mid-point of each of these intervals is labelled with the corresponding value of the variate. The frequency is represented by the area of a rectangle or cell with the corresponding interval as base. Using the data of the previous section,

x	0	1	2	3	4
f	25	20	11	3	1

In this case (Fig. 17.1), where the horizontal intervals are all equal, the ordinates of the histogram do represent frequency. When the horizontal intervals are not equal, the ordinates of the histogram do not represent frequency. *In all cases, however, frequency is represented by the area of the corresponding cell of the histogram, and the ordinates represent frequency density, or frequency per unit variate.*

Suppose that the table is now constructed as follows, so that the last interval is three times the width of the other two.

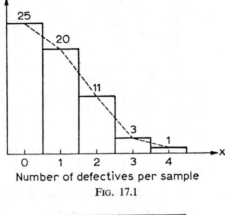

Number of defectives per sample

FIG. 17.1

x	0	1	2, 3 or 4
f	25	20	15

In this case (Fig. 17.2) the ordinate of the last cell does not represent 15 to the same scale that the other ordinates represent 20 and 25. Since the base is three times that of the other two cells, the ordinate of the last cell represents 5 to the same scale that the other ordinates represent 20 and 25.

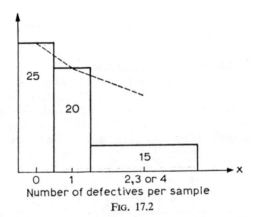

Number of defectives per sample

FIG. 17.2

Frequency polygons
If at the midpoint of each interval an ordinate is erected, representing the frequency to scale, and if the points so formed are joined by straight lines, then a frequency polygon is obtained. In the previous two diagrams the frequency polygon is represented by a dotted line figure. Frequency polygons are a little clearer than histograms when several sets of data are plotted on the same base. They have little to recommend them otherwise, as areas are not proportional to frequencies.

Types of average—the mean
Calculation of the mean of a frequency distribution (discrete or continuous)
If the values $x_1, x_2, x_3 \ldots$ occur $f_1, f_2, f_3 \ldots$ times respectively, then the
mean $\bar{x}$ is given by

$$\bar{x} = \frac{f_1x_1 + f_2x_2 + f_3x_3 \ldots}{f_1 + f_2 + f_3 \ldots} = \frac{\text{total of all values}}{\text{number of values}} = \frac{\Sigma fx}{\Sigma f}$$

Calculation of the mean of a frequency distribution using a false mean
It is often convenient to reduce the variate by a false mean M, which may
be any convenient number. Often the most convenient number may be a
rough estimate of the mean, hence the use of the term false mean. After
M has been deducted, the mean of the resulting reduced variate is found.
If the false mean M is added back to this mean, the true mean results.
Proof N.B. $\Sigma fM = f_1M + f_2M + \ldots = M(f_1 + f_2 + \ldots) = M\Sigma f$
a constant may be taken outside the Σ sign)
The mean of the reduced variate is given by

$$\frac{\Sigma f(x - M)}{\Sigma f} = \frac{\Sigma fx}{\Sigma f} - \frac{\Sigma fM}{\Sigma f} = \bar{x} - \frac{M\Sigma f}{\Sigma f} = \bar{x} - M$$

$$\therefore \quad \bar{x} = \frac{\Sigma f(x - M)}{\Sigma f} + M$$

i.e. true mean = $\begin{array}{c}\text{mean of variate reduced}\\\text{by the false mean } M\end{array}$ + $\begin{array}{c}\text{the false}\\\text{mean } M\end{array}$

Note that these methods for the calculation of the mean apply when x is
a discrete variate *or* the mid-interval value of a group, but in the latter case
the result is approximate.

Example
17.2 Find the mean of the following frequency distribution using two
methods. x is a mid-interval value.

x	12·25	13·25	14·25	15·25	16·25
f	2	16	27	15	3

First method

x	12·25	13·25	14·25	15·25	16·25	
f	2	16	27	15	3	$\Sigma f = 63$
fx	24·50	212·00	384·75	228·75	48·75	$\Sigma fx = 898·75$

$$\bar{x} = \frac{898·75}{63} = 14·265\ 87$$

Second method
Using a false mean of, say, 14·25 (since this variate of 14·25 has the highest frequency and the distribution is fairly symmetrical, 14·25 should be an approximation to the mean),

x	12·25	13·25	14·25	15·25	16·25	
$x - 14·25$	−2	−1	0	1	2	
f	2	16	27	15	3	$\Sigma f = 63$
$f(x - 14·25)$	−4	−16	0	15	6	$\Sigma f(x - 14·25)$ $= 21 - 20$ $= 1$

$$\frac{\Sigma f(x - 14·25)}{\Sigma f} = \frac{1}{63} = 0·015\ 87 \qquad \bar{x} = 14·25 + 0·015\ 87$$
$$= 14·265\ 87$$

The results agree exactly as quoted to five decimal places. The number of valid significant figures in this result will be limited by the approximation involved in using mid-interval values, $\bar{x} \simeq 14·3$.

Grouped distributions—calculation of mean, mode and median
The following data relates to the time interval in hundredths of a second prior to the emission of a radioactive particle.

```
2·7  3·5  3·1  2·4  1·5  2·4  1·6  2·5  3·4  2·5  3·4  2·5  3·1  2·7  3·3
3·2  2·8  1·8  3·2  2·9  3·0  1·2  3·5  2·6  1·0  3·2  2·2  1·0  1·7  2·4
3·5  2·4  3·6  2·1  2·1  3·7  2·9  3·0  3·8  2·3  3·6  2·0  2·6  2·6  2·4
2·1  3·9  2·5  3·4  2·9  3·4  0·3  2·1  1·7  2·2  1·9  2·1  1·4  1·4  1·3
3·5  1·0  2·8  1·2  2·1  1·5  2·2  1·5  2·2  0·7
```

A tally of the data may be drawn up as follows, using the groups or classes 0 to 1, 1 to 2, 2 to 3, 3 to 4. It may be noted that a value such as 2·0 is entered as $\frac{1}{2}$ in the group 1 to 2, and as $\frac{1}{2}$ in the group 2 to 3. From the tally a frequency table is drawn up in which x, the variate, is taken as the value at the centre of each class, as follows. (It may be restated that,

	0/1	1/2	2/3	3/4	
	‖	‖‖‖‖ ‖‖‖‖ ‖‖‖	‖‖‖ ‖‖‖ ‖‖‖ ‖‖‖ ‖‖‖ ‖‖‖	‖‖‖ ‖‖‖‖ ‖‖‖ ‖‖‖	
		$\frac{1}{2}\frac{1}{2}$ $\frac{1}{2}\frac{1}{2}$ $\frac{1}{2}\frac{1}{2}$	$\frac{1}{2}\frac{1}{2}$	$\frac{1}{2}\frac{1}{2}$ $\frac{1}{2}\frac{1}{2}$	
x in 0·01 s units	0·5	1·5	2·5	3·5	
f	$3\frac{1}{2}$	15	$31\frac{1}{2}$	20	$\Sigma f = 70$
fx	1·75	22·5	78·75	70	$\Sigma fx = 173$

although the measurements are discrete, due to limited precision the variate is actually continuous.)

The mean calculated from the grouped frequency table is given by $\bar{x} = \dfrac{173}{70} = 2\cdot471$ in 0·01 second units. The mean calculated from the original data is

$(2\cdot7 + 3\cdot5 + \ldots 2\cdot2 + 0\cdot7)/70 = 170\cdot2/70 = 2\cdot431$ in 0·01 second units.

$$\text{Relative error} = \frac{2\cdot471 - 2\cdot431}{2\cdot431} \times 100 = \frac{4\cdot0}{2\cdot431} = 1\cdot6\%$$

It can be seen that the mean is obtained with sufficient accuracy by the grouped frequency method.

Estimation of the median for the grouped frequency distribution

The median is that value of the variate which divides the histogram into two parts equal in area, representing an equal division of frequency,

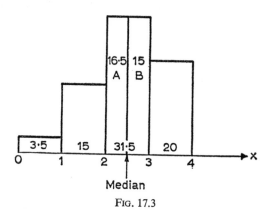

Median

Fig. 17.3

Fig. 17.3. Since the total frequency is 70, the areas on each side of the median must represent a frequency of $\frac{70}{2} = 35$. Thus area A represents $35 - (15 + 3\frac{1}{2}) = 16\cdot5$, and area B represents $35 - 20 = 15\cdot0$, whilst as a check we find A + B represents $16\cdot5 + 15\cdot0 = 31\cdot5$, as required. Thus the interval between 2 and 3 must be divided in the ratio of 16·5 to 31·5.

The median value $= 2 + \dfrac{16\cdot5}{31\cdot5} \simeq 2\cdot5$.

The calculation or graphical estimation of the *mode* for a grouped frequency table of data differs from the discrete case. The mode is a value of the variate in the class with the highest frequency. The frequencies in the cells to the left and to the right of the modal class are used in the estimation of the mode.

Estimation of the mode from the histogram of a grouped frequency distribution with equal class intervals
The intersection point Y of BD and AC serves to locate the mode on the *x* scale (Fig. 17.4). It will be noted that the mode is closer to the class

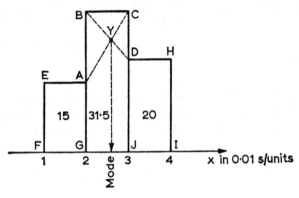

FIG. 17.4

boundary adjacent to the cell DHIJ than to the class boundary adjacent to cell AEFG. It is weighted towards the adjacent cell with the higher frequency, which in this case is DHIJ. The mode is scaled as 2.6.

Comparison of mean, mode and median for a frequency distribution
For the distribution of emission time intervals considered, with averages measured in numbers of hundredths of a second, the values of the mean, mode and median are 2·5, 2·6 and 2·5, correct to one decimal place. The mean is the most useful if the total time interval for the population of seventy particles is required. It has many arithmetic properties not possessed by the mode or median but requires the use of all the data for its calculation, and cannot be found when the frequency distribution contains open-ended classes.

The mode in this case indicates the most common time interval. One of its uses is in selecting the correct value after a series of repetitions of the same calculation. It requires a minimum of data for its calculation.

The median requires the use of more data than that used for the calculation of the mode, and less than that used for the calculation of the mean. It can be calculated for a frequency distribution with open-ended classes. In the case considered, it gives the typical individual time interval in the sense that there are as many such intervals above it as below it.

When the histogram is constructed for a very large population with a large number of groups, it approximates closely to a smooth frequency curve. The mean is the value of the variate corresponding to the centroid of this curve. The median corresponds to the ordinate dividing the area under the curve into two parts equal in area. The mode is the value of the variate corresponding to the peak of this curve. Such a curve may have more than one peak or hump, and is then said to be multi-modal, whilst still possessing only one mean and median.

It must be emphasised that the values of the median and mode obtained by these methods for a grouped frequency distribution are no more than *estimates*. They are based upon the *assumption* that the histogram of the distribution is representative of the smooth frequency curve which would be obtained in the limit, if the number of measurements were increased indefinitely and the width of each group tended to zero. Thus the reliability of these estimates tends to increase with the size of the population considered. The main justification of their use with small populations is to illustrate the method of obtaining these estimates. If the form of the histogram changes markedly as the population increases, so may the estimates change.

EXERCISE 17.1

1 The following table for a discrete variate gives the frequency with which x α-particles are emitted from a radioactive specimen in a given time.

x	0	1	2	3	4	5	6	7	8	9	10	11	12
f	55	204	384	524	533	405	276	138	46	26	11	4	2

(a) Calculate the mean number of α particles emitted. (b) State the median. (c) State the mode.

2 A foundry states that the mean mass of a certain type of casting is 10·0 kg. Using the given frequency table find (a) the mean, (b) the median, (c) the mode of this grouped frequency distribution. Note that the variate is *continuous*.

x = mass of casting in kilograms (mid-interval).

x	9·7	9·8	9·9	10·0	10·1	10·2	10·3
f	1	3	18	24	18	5	1

3 From trials of a certain detergent, the following table gives the number of plates washed by a weighed quantity of detergent before the foam on the water was reduced to a thin surface layer. Note that the variate is *discrete*.

x	25/27	28/30	31/33	34/36	37/39
f	2	7	35	5	1

Find (a) the mean, (b) the medial group and the modal group. (c) Estimate the median and mode as for a continuous variate and round the results.

4 Batches of components from current production are tested and the number of defective components x and the frequency f with which these occur are noted below. Note that the variate is *discrete*.

Number of defectives x	0	1	2	3	4	5
Frequency f		3	6	14	9	7	1

(a) Calculate the mean. (b) State the median and the mode.

5 The following table relates the diameter x mm (mid-interval) of mass produced rods with the frequency f with which these occur. Note that the variate is *continuous*.

x	8·3	8·4	8·5	8·6	8·7
f	2	8	23	15	2

Find (a) the mean, (b) the median. Draw the histogram to scale and determine the mode.

6 In 60 observations of a Wilson cloud chamber, the following frequencies of the number x of disintegrations in a given time were obtained. Note that the variate is *discrete*.

Number of disintegrations x	0	1	2	3	4	5
frequency f		7	12	8	9	8	6

(a) Find the mean. (b) State the median and the mode.

7 In an experiment carried out by a number of different groups to find the efficiency of a screw-jack, the following results were obtained. Note that the variate is *continuous*.

Efficiency x	0·22/0·24	0·24/0·26	0·26/0·28	0·28/0·30
Frequency f	3	15	10	4

Find (a) the mean, (b) the median. Draw the histogram to scale and determine the mode.

Measures of dispersion—range, mean deviation, semi-interquartile range and standard deviation

In addition to the previous averages discussed, a quantitative measure of the scatter or dispersion of the variate is required. For example, a man might be told that his average wage was likely to be £15 per week, but he would be vitally interested in the variation of his wage about this figure. In this case, if he were told that it was likely to vary between limits of £12 and £18 per week, he could consider how he could make arrangements to meet his expenditure. In this case the possible range would be £18—£12, i.e. £6 per week. The range is one measure of dispersion.

Range

The range is the difference between the highest and lowest values in a population. It is simple to calculate and to understand. For this reason it is used in statistical quality control in which control charts based on the range as a measure of variability are known as R-control charts. Since the range depends entirely on the extreme items, it is subject to fluctuation, and difficult to interpret mathematically.

Mean deviation

Consider the values 2, 4, 6, 12. Their mean is $(2 + 4 + 6 + 12)/4 = 24/4 = 6$. We can tabulate the values of these values of this variate x, say, and the deviation of these values from their mean of 6, i.e. the values of $x - 6$. We cannot average these deviations to get a measure of scatter because the average of these deviations is zero in all cases.

x	2	4	6	12	
Deviation from the mean $x - 6$	-4	-2	0	6	$\Sigma(x - 6) = -6 + 6 = 0$

Supposing that we tabulate $|x - 6|$, meaning the modulus of the $x - 6$ values, i.e. the signs are disregarded and the absolute values are taken, then we obtain

x	2	4	6	12	
$|x - 6|$	4	2	0	6	$\Sigma|x - 6| = 14$

The average of the absolute values of the deviations from the mean is known as the mean deviation.

In this case, mean deviation $= \dfrac{\Sigma|x - 6|}{4} = \dfrac{14}{4} = 3 \cdot 5$

In general, the mean deviation of the n values of a variate

$$= \frac{\Sigma|x - \bar{x}|}{n}, \quad \text{where} \quad \bar{x} = \frac{\Sigma x}{n}$$

For a frequency distribution of Σf items,

$$\text{mean deviation} = \frac{\Sigma f\,|\,x - \bar{x}\,|}{\Sigma f}, \quad \text{where} \quad \bar{x} = \frac{\Sigma f x}{\Sigma f}$$

The mean deviation has the disadvantage that it is not possible to obtain the mean deviation of the combination of several groups of observations from the mean deviations of the separate groups. It is not often useful for statistical inference. It has been largely superseded by the standard deviation, which has not these disadvantages.

Standard deviation

Consider the table previously considered. Another way to ensure that the dispersion measure does not vanish is to square the deviations, average the squared deviations, and then take the square root of the result. We thus obtain the root of the average of the deviations from the mean squared. This is a statistical average with the same dimensions or units as the variate.

$$\text{Standard deviation of the } n \text{ values of a variable} = \sqrt{\frac{\Sigma(x - \bar{x})^2}{n}}$$

$$\text{Standard deviation for a frequency distribution of } \Sigma f \text{ items} = \sqrt{\frac{\Sigma f(x - \bar{x})^2}{\Sigma f}}$$

It may be noted that the square of the standard deviation, i.e. the expression or value obtained before taking the final square root, is known as the *variance* and is much used in statistical theory.

We will now evaluate the standard deviation for the table previously considered.

x	2	4	6	12
$x - 6$	-4	-2	0	6
$(x - 6)^2$	16	4	0	36

$\bar{x} = 6$

$\Sigma(x - 6)^2 = 56$

$$\text{variance} = \frac{\Sigma(x - 6)^2}{N} = \frac{56}{4} = 14$$

$$\text{standard deviation} = \sqrt{\frac{(x - 6)^2}{N}} = \sqrt{14} = 3 \cdot 74$$

Meaning of standard deviation

Although more difficult to visualise than the other measures of dispersion, the standard deviation is of fundamental importance in statistics. It is analogous to the root-mean-squared value in alternating current theory, and to the radius of gyration in the theory of moments of inertia.

In a great many frequency distributions of the unimodal fairly symmetrical type, about 99 % of the values of the variate lie within three standard deviations above and below the mean value of the variate. In other words, the standard deviation is about one sixth of the range in many (although by no means all) frequency distributions. The greater the standard deviation, the greater the scatter or dispersion of the values concerned about their mean value.

Semi-interquartile range

The values of the variate corresponding to cumulative relative frequency values of 0·25 and 0·75 are known as the lower and upper quartiles. Let these be x_1 and x_2 respectively, then

$$\text{semi-interquartile range} = \frac{x_2 - x_1}{2}$$

where x_1 and x_2 may be calculated from the histogram. It has the advantages that it is easily calculated, and simple in meaning. Lacking simple algebraical properties, however, it is only suitable for the most elementary statistical work.

Calculation of standard deviation using a false mean M

$$\frac{\Sigma(x - \bar{x})^2}{N} = \frac{\Sigma(x^2 - 2\bar{x}x + \bar{x}^2)}{N}$$

$$= \frac{\Sigma x^2}{N} - 2\bar{x}\frac{\Sigma x}{N} + \frac{\bar{x}^2\Sigma 1}{N}$$

Now
$$\Sigma 1 = (1 + 1 + \ldots 1) \text{ to } N \text{ terms} = N,$$
$$\therefore \ \Sigma 1/N = 1$$

$$\therefore \ \frac{\Sigma(x - \bar{x})^2}{N} = \frac{\Sigma x^2}{N} - 2\bar{x} \cdot \bar{x} + \bar{x}^2$$

$$= \frac{\Sigma x^2}{N} - \bar{x}^2 \quad \ldots \quad \ldots \quad \text{(i)}$$

Also
$$\frac{\Sigma(x - M)^2}{N} = \frac{\Sigma x^2}{N} - 2M\frac{\Sigma x}{N} + \frac{\Sigma M^2}{N}$$

$$= \frac{\Sigma x^2}{N} - 2M\bar{x} + M^2\frac{\Sigma 1}{N}$$

$$\therefore \ \frac{\Sigma(x - M)^2}{N} = \frac{\Sigma x^2}{N} - 2M\bar{x} + M^2 \quad \ldots \quad \ldots \quad \text{(ii)}$$

Taking (i) from (ii),
$$\frac{\Sigma(x - M)^2}{N} - \frac{\Sigma(x - \bar{x})^2}{N} = \bar{x}^2 - 2M\bar{x} + M^2$$

$$= (\bar{x} - M)^2$$

$$\therefore \ \frac{\Sigma(x - M)^2}{N} - (\bar{x} - M)^2 = \frac{\Sigma(x - \bar{x})^2}{N}$$

i.e. *the square of the standard deviation based on a false mean M, decreased by the square of the difference between the true mean and the false mean, gives the square of the true standard deviation.* It will be noticed that the standard deviation based on the true mean is always less than a standard deviation based on a false mean. The use of a false mean simplifies the arithmetic in general, also most of the working can be completed before the true mean is known.

Example

17.3 The following data relates to the average hardness of annealed steel rods. Calculate the mean hardness and the standard deviation.

Choosing a false mean of 197, say,

x	194	195	196	196	197	198	200	201	
$(x - 197)$	-3	-2	-1	-1	0	1	3	4	$\Sigma(x - 197) = 8 - 7 = 1$
$(x - 197)^2$	9	4	1	1	0	1	9	16	$\Sigma(x - 197)^2 = 41$

$$\bar{x} = \frac{\Sigma(x - 197)}{N} + 197 = \frac{1}{8} + 197 = 197 \cdot 125, \text{ or } 197 \text{ correct to three significant figures}$$

$$\text{Standard deviation}^2 = \frac{\Sigma(\bar{x} - 197)^2}{N} - (\bar{x} - 197)^2 = \frac{41}{8} - (0 \cdot 125)^2$$

$$= 5 \cdot 11$$

$$\therefore \text{ Standard deviation } = 2 \cdot 26$$

EXERCISE 17.2

Calculate the mean and standard deviation of the following data in questions 1 to 6.

1 The life in hours of 20 electric bulbs:

932 1137 1103 1221   970 1048 947   958 1101 1096
963 1037 1039   998 1097 1065 983 1073 1041 1020

2 The lengths of span in metres over a stretch of overhead line:

320   311   350   339   400   375   385   380

3 The weight in kilograms of internal-type grinding machines with manual standard grinding cycle control, and various specifications of table travel and work diameter:

9250 9900 8400 8400 9250 8800 9500

4 The heights in millimetres of a sample of 20 compression springs:

101·0   100·2   100·9   100·5   100·6   100·2   100·7   100·7   101·1
100·2   100·7   100·8   100·3   100·2   100·1   100·7   100·5   100·0
100·8   100·8

5 The weights in grams of fifteen, 320 gram, packets delivered by an automatic packing machine:

$$321 \cdot 1 \quad 319 \cdot 7 \quad 321 \cdot 8 \quad 320 \cdot 3 \quad 322 \cdot 5$$
$$320 \cdot 7 \quad 320 \cdot 5 \quad 321 \cdot 4 \quad 321 \cdot 9 \quad 319 \cdot 8$$
$$320 \cdot 7 \quad 319 \cdot 9 \quad 321 \cdot 6 \quad 320 \cdot 3 \quad 321 \cdot 8$$

6 Coal consumption in kg/kilowatt hour at steam generating stations from 1943 to 1949:

$$0 \cdot 635 \quad 0 \cdot 642 \quad 0 \cdot 644 \quad 0 \cdot 650 \quad 0 \cdot 644 \quad 0 \cdot 642 \quad 0 \cdot 635$$

7 The following measurements were obtained by measures A and B. Calculate the mean and standard deviation of each set of measurements and state which is the better measuring performance.

(A) measurement in mm $\quad 59 \cdot 6 \quad 59 \cdot 7 \quad 60 \cdot 1 \quad 60 \cdot 2 \quad 59 \cdot 9 \quad 60 \cdot 5$

(B) measurement in mm $\quad 59 \cdot 7 \quad 59 \cdot 8 \quad 60 \cdot 2 \quad 60 \cdot 2 \quad 60 \cdot 1 \quad 60 \cdot 0$

Standard deviation of a frequency distribution using a false mean M

The expression $\dfrac{\Sigma(x - M)^2}{N} - (\bar{x} - M)^2 = \dfrac{\Sigma(x - \bar{x})^2}{N}$ previously obtained becomes

$$\frac{\Sigma f(x - M)^2}{\Sigma f} - (\bar{x} - M)^2 = \frac{\Sigma f(x - \bar{x})^2}{\Sigma f}$$

for a frequency distribution.

Example

17.4 The following table gives the frequency of stoppages of an automated process for a certain trial period. Calculate the mean number of stoppages, and the standard deviation.

Number of stoppages x	0	1	2	3	4	5
Frequency $\quad f$	13	18	20	19	6	4

Selecting a false mean $M = 2$, which has the highest frequency, subtract 2 from the x-row values to give the $(x - M)$-row values.

x	0	1	2	3	4	5	
f	13	18	20	19	6	4	$\Sigma f = 80$
$x - M$	-2	-1	0	1	2	3	
$f(x - M)$	-26	-18	0	19	12	12	$\Sigma f(x - M) = -1$
$f(x - M)^2$	52	18	0	19	24	36	$\Sigma f(x - M)^2 = 149$

$$\bar{x} = \frac{\Sigma f(x - M)}{\Sigma f} + M = -\frac{1}{80} + 2 = 1 \cdot 9875, \text{ say } 1 \cdot 99, \text{ stoppages}$$

Standard deviation squared or variance $= \dfrac{\Sigma f(x - \bar{x})^2}{\Sigma f}$

$$= \frac{\Sigma f(x - M)^2}{\Sigma f} - (\bar{x} - M)^2$$

$$= \frac{149}{80} - \left(\frac{1}{80}\right)^2$$

$$= 1{\cdot}8625 - 0{\cdot}001$$
$$= 1{\cdot}8624$$

$\therefore$ Standard deviation $= 1{\cdot}3646$, say $1{\cdot}36$, stoppages

Example
17.5 The following table gives the resistances in ohms of nominal 2 ohm resistors drawn from stock. Calculate the mean resistance and the standard deviation. State the range and the value of six standard deviations.

x	1·99	2·00	2·02	2·03	2·04	2·05	2·06	2·07
f	2	1	2	3	2	10	3	7

x	2·08	2·09	2·10	2·11	2·12	2·13	2·15	
f	8	6	4	2	4	5	1	

Selecting a false mean of 2·05, say, and noting $\Sigma f = 60$:

$x - 2{\cdot}05$	−0·06	−0·05	−0·03	−0·02	−0·01	0	0·01	0·02
$f(x - 2{\cdot}05)$	−0·12	−0·05	−0·06	−0·06	−0·02	0	0·03	0·14
$f(x - 2{\cdot}05)^2$	0·0072	0·0025	0·0018	0·0012	0·0002	0	0·003	0·0028

$x - 2{\cdot}05$	0·03	0·04	0·05	0·06	0·07	0·08	0·1	
$f(x - 2{\cdot}05)$	0·24	0·24	0·20	0·12	0·28	0·40	0·1	
$f(x - 2{\cdot}05)^2$	0·0072	0·0096	0·0100	0·0072	0·0196	0·0320	0·01	

$$\Sigma f(x - 2{\cdot}05) = 1{\cdot}44 \quad \bar{x} = \frac{1{\cdot}44}{60} + 2{\cdot}05 = 2{\cdot}074 \text{ ohm}$$

$$\Sigma f(x - 2{\cdot}05)^2 = 0{\cdot}1116. \quad \frac{\Sigma f(x - 2{\cdot}05)^2}{\Sigma f} = 0{\cdot}0018$$

$$\text{Variance} = 0{\cdot}0018 - (2{\cdot}074 - 2{\cdot}05)^2$$
$$= 0{\cdot}0013$$
$$\text{Standard deviation} = 0{\cdot}0360 \text{ ohm}$$
$$\text{The range is } 2{\cdot}15 - 1{\cdot}99 = 0{\cdot}16 \text{ ohm}$$
$$\text{Six times the standard deviation} = 6 \times 0{\cdot}0360$$
$$= 0{\cdot}216 \text{ ohm}$$

We might expect, on testing a larger number of resistors, that only very rarely would a resistance lie outside the range $2 \cdot 074 \pm \dfrac{0 \cdot 216}{2}$, i.e. $2 \cdot 074 \pm 0 \cdot 108$ ohm.

Standard deviation of a grouped frequency distribution
This is obtained, as in the previous examples, by replacing the class interval by the value of the variate at the middle of the class interval. This has little effect on the value of the mean, since the effects tend to compensate each other. The value of the standard deviation tends to be increased however. Sheppard's correction consists in deducting $C^2/12$ from the grouped data variance, where C is the class interval size. It is used for distributions of a continuous variable where the tails go gradually to zero in both directions, where the total frequency is of the order of 1000 or more, and the class interval is about one twentieth of the range.

EXERCISE 17.3

1 The following table gives the frequency with which x α-particles are emitted from a radioactive specimen in a given time.

x	0	1	2	3	4	5	6	7	8	9	10	11	12
f	55	204	384	524	533	405	276	138	46	29	11	4	2

Using a trial mean of 4, or otherwise, calculate the mean and standard deviation of the observed values of x.

2 A foundry states that the mean mass x of a certain type of casting is $10 \cdot 0$ kg. From the given frequency table, estimate the mean mass, and its standard deviation.

x	9·7	9·8	9·9	10·0	10·1	10·2	10·3
f	1	3	18	24	18	5	1

3 From trials of a certain detergent, the following table gives the number of plates x washed by a weighed quantity of detergent before the foam on the water was reduced to a thin surface layer.

x	25/27	28/30	31/33	34/36	37/39
f	2	7	35	5	1

Calculate the mean number of plates washed and its standard deviation.

4 Batches of components from current production are tested and the number of defective components x and the frequency f with which these occur are noted below. Calculate the mean and standard deviation of the observed values of x.

x	0	1	2	3	4	5
f	3	6	14	9	7	1

5 The following table relates the diameters x mm of mass-produced rods with the frequency f with which these occur. Calculate the mean diameter and its standard deviation.

x	8·3	8·4	8·5	8·6	8·7
f	2	8	23	15	2

6 From 50 observations of a Wilson Cloud Chamber, the following frequencies f of the number x of disintegrations in a given time were obtained.

x	0	1	2	3	4	5
f	7	12	8	9	8	6

Calculate the mean and its standard deviation.

7 In an experiment carried out by a number of different groups to find the efficiency of a screw-jack, the following results were obtained.

Efficiency x	0·22/0·24	0·24/0·26	0·26/0·28	0·28/0·30
Frequency f	3	15	10	4

Calculate the mean and standard deviation of the observed values of x.

8 Fifty samples of a material were tested in a hardness testing machine, yielding the following data.

Hardness value	200 to 220	220 to 240	240 to 260	260 to 280	280 to 300
Frequency	7	11	18	10	4

Determine the mean and standard deviation of these results.

ULCI specimen 01

Chapter Eighteen
The Use of Desk Calculating Machines

Setting register

The desk calculator described incorporates a setting register (S.R.) of 10 digits. The positions of the digits in S.R. are numbered from 1 to 10, reading from right to left. Corresponding to and vertically below each of these positions is a vertical slot. The positions in each of the vertical slots, reading from top to bottom, are numbered from 0 to 9. When a lever, which moves vertically in this slot, is moved to position 4, the digit 4 appears vertically above, in S.R.

The diagram (Fig. 18.1) shows the setting register containing the number 8174 with the setting levers in the corresponding positions. When the setting register is cancelled, by means of the cancelling lever, all setting

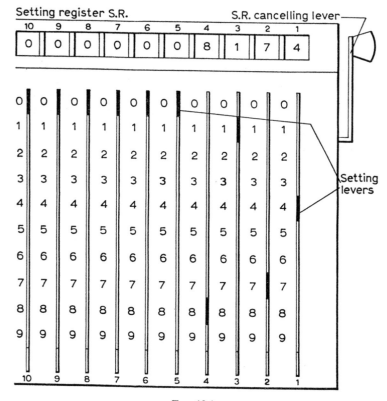

Fig. 18.1

levers are returned to the top 0 position, whilst each of the ten positions of the setting register show 0.

Carriage with counter register and product register or accumulator

Below the setting levers is mounted a movable carriage which can be traversed to any one of eight carriage positions, indicated by two downward pointing arrows. On the left of the carriage is an 8 digit counter register (C.R.), and on the right is a 13 digit product register (P.R.). The diagram (Fig. 18.2) shows the carriage in position 1. By pumping the carriage traversing lever away from the operator seven times in succession, the carriage is moved to positions 2, 3, 4, 5, 6, 7, 8 respectively. Then by pumping the carriage traversing lever towards the operator seven times in succession, the carriage is moved to positions 7, 6, 5, 4, 3, 2, 1 respectively. The C.R. and P.R. are cancelled together by the C.R., P.R. cancelling lever, but the P.R. contents may be retained by first actuating the P.R. locking lever.

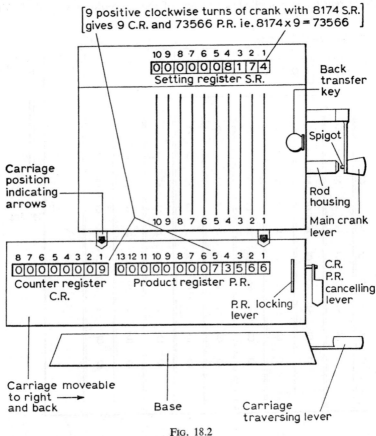

Fig. 18.2

When the crank is in the vertically downward position, a spring-loaded spigot engages in a slot in the rod housing, and moves the rod to the left, allowing the various levers to become free. If the crank is rotated, the rod springs to the right when the spigot is removed, and the various levers are locked.

When the crank is rotated once in the positive, i.e. clockwise, direction, the S.R. contents are entered in the P.R. whilst the C.R. contents are increased by one.

The diagram (Fig. 18.3) shows the state of the registers after each of nine successive positive turns with 8174 S.R. on carriage position 1. Windows shown empty contain 0.

Thus $8174 \times 9 = 73\,566$ is obtained by adding together 9 of 8174.

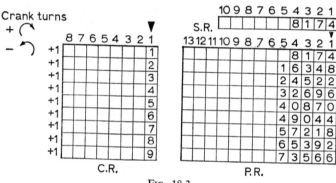

Fig. 18.3

Decimal-point markers

Decimal-point markers are mounted on slides above S.R. and below C.R. and P.R. They are used according to the rule

$$D_{\text{S.R.}} + D_{\text{C.R.}} = D_{\text{P.R.}}$$

where $D \equiv$ number of decimal places: i.e. number of decimal places on S.R. + number of decimal places on C.R. = number of decimal places on P.R.

Example
18.1 $0.008\,174 \times 0.009 =$
$0.000\,073\,566$
$(6 + 3 = 9)$

$0.8174 \times 0.9 =$
$0.735\,66$
$(4 + 1 = 5)$

$8.174 \times 0.0009 =$
$0.007\,356\,6$
$(3 + 4 = 7)$

Subtraction of two products

Suppose that we require 8174 × 9 − 6982 × 4. We obtain 8174 × 9 as before, cancel S.R. and C.R. but not P.R. Enter 6982 on S.R. Make 4 negative (anticlockwise) turns for subtraction. The diagram (Fig. 18.4) shows the states of the registers. Windows shown empty contain 0. i.e.   8174 × 9 − 6982 × 4 = 46 638.

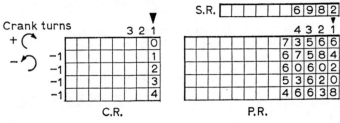

FIG. 18.4

Short cutting

Examples

18.2 To obtain 8174 × 9 as 8174 × 10 − 8174 × 1.

Check all registers cancelled. Enter 8174 in S.R. Move carriage to position 2. Make one positive turn. Move carriage to position 1. Make one negative turn. See diagram (Fig. 18.5).

Thus 8174 × 9 = 73 566. The result is obtained in 2 crank turns, saving 7 turns. Similarly, this method with 8174 × 8 saves 5 turns, with 8174 × 7 saves 3 turns, and with 8174 × 6 saves 1 turn.

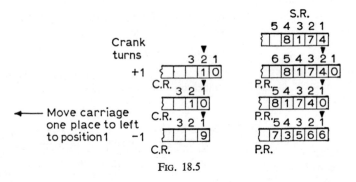

FIG. 18.5

18.3 To obtain 268 699 × 9 999 998.

Check all registers cancelled. Enter 268 699 in S.R. Move carriage to position 8. Make one positive turn. Move carriage to position 1. Make two negative turns. See diagram (Fig. 18.6).

The result 268 699 × 9 999 998 = 2 686 989 602 is obtained with three turns. It may be noted that 62 turns are required if nine turns and

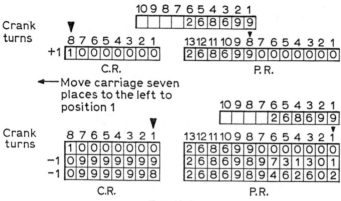

Fig. 18.6

one entry of eight turns are used in the various carriage positions. This is an extreme example of a useful principle.

Overflow

This will be illustrated by the examples

$$268\ 699 \times 19\ 999\ 998$$
$$268\ 699 \times 29\ 999\ 998$$
$$268\ 699 \times 39\ 999\ 998$$

The numbers used are in continuance of the previous problem, where, if the registers all retain their contents and the carriage is moved to position 8, one positive turn will give the required multiplicand in the first example (see Fig. 18.7).

$$268\ 699 \times 19\ 999\ 998 = 5\ 373\ 979\ 462\ 602$$
$$268\ 699 \times 29\ 999\ 998 = 8\ 060\ 969\ 462\ 602$$
$$268\ 699 \times 39\ 999\ 998 = \boxed{1}\ 0\ 747\ 959\ 462\ 602$$

↑   missing digit supplied by inspection

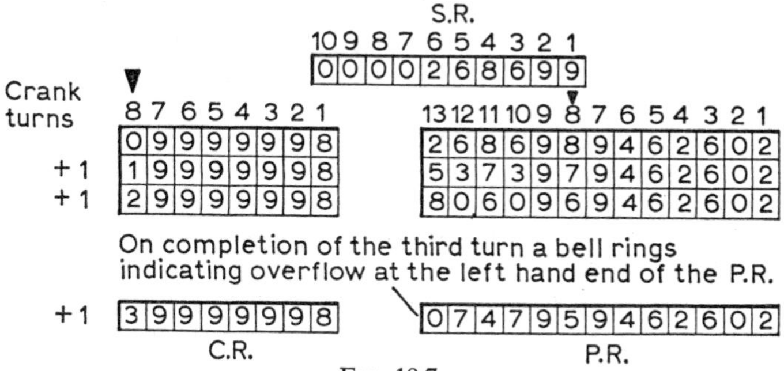

Crank turns

S.R.

On completion of the third turn a bell rings indicating overflow at the left hand end of the P.R.

Fig. 18.7

Back transfer
On depressing the back transfer key and then depressing the C.R. P.R.
cancelling lever, the contents of the P.R. are transferred to the S.R.,
facilitating continued multiplication.

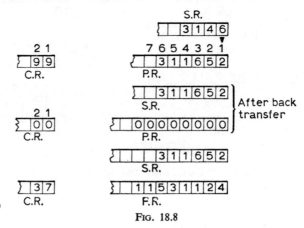

Fig. 18.8

Example
18.4 3148 × 99 × 37

Obtain 3148 × 99. Depress back transfer key and C.R. P.R. cancelling
lever in carriage position 1. The product 311 652 is transferred to the S.R.
Then multiply by 37 (see Fig. 18.8).
Result 3148 × 99 × 37 = 11 531 124

Use of complements
If 396 is entered in S.R. with carriage position 1 and C.R. and P.R.
cleared, and 1 negative turn is made, a bridge of nines appears in P.R. in
positions 13 to 4 (see Fig. 18.9).

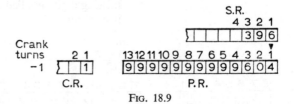

Fig. 18.9

The number in P.R. is $10^{13} - 396$, sometimes called the tens complement
of 396, since $(10^{13} - 396) + (396) = 10^{13}$, where 10^{13} is represented by
P.R. containing 13 zeros.
If back transfer is now applied, the number in S.R. becomes $10^{10} - 396$,
since digits in positions 13, 12 and 11 are not transferred from carriage
position 1 (see Fig. 18.10).

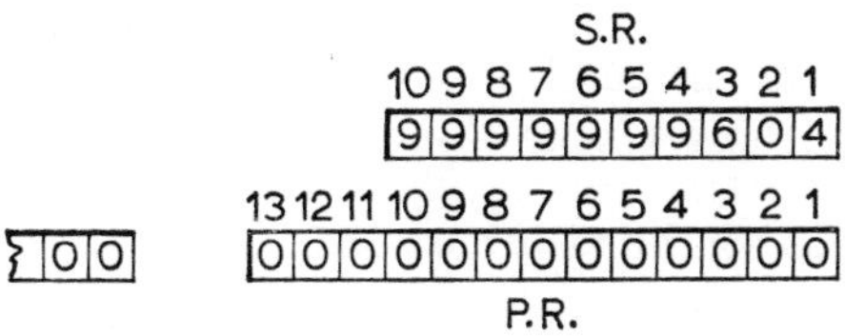

S.R.

10 9 8 7 6 5 4 3 2 1
| 9 | 9 | 9 | 9 | 9 | 9 | 9 | 6 | 0 | 4 |

13 12 11 10 9 8 7 6 5 4 3 2 1
| 0 | 0 | 0 | 0 | 0 | 0 | 0 | 0 | 0 | 0 | 0 | 0 | 0 | ⟩ | 0 | 0 |

P.R.

Fig. 18.10

The number in P.R. is 10^{13}, and in S.R. is $10^{10} - 396$. Hence, if one negative turn is now made, P.R. will contain $10^{13} - (10^{10} - 396)$, i.e. $10^{13} - 10^{10} + 396$, i.e. $10^{10}(1000 - 1) + 396$, i.e. $10^{10} \times 999 + 396$, i.e. 9 990 000 000 396 (see Fig. 18.11).

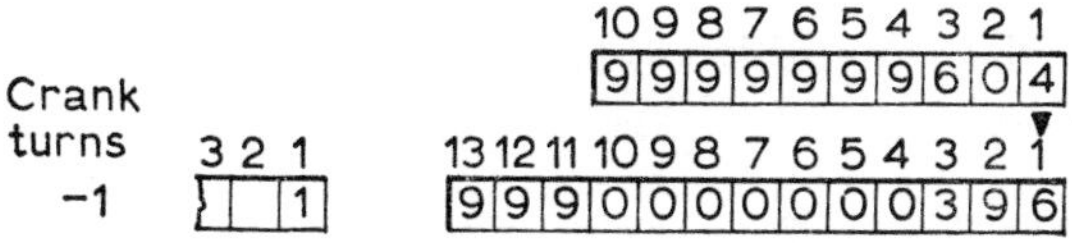

Crank
turns

10 9 8 7 6 5 4 3 2 1
| 9 | 9 | 9 | 9 | 9 | 9 | 9 | 6 | 0 | 4 |

3 2 1

13 12 11 10 9 8 7 6 5 4 3 2 1
−1   ⟩ | | 1 |

| 9 | 9 | 9 | 0 | 0 | 0 | 0 | 0 | 0 | 0 | 3 | 9 | 6 |

Fig. 18.11

Thus *if a negative number is represented in P.R. with the bridge of nines, the magnitude of the number may be obtained by back transfer followed by 1 negative turn. The number in S.R. in the right-hand position is recorded ignoring digits to the left of the string of zeros.*

Example
18.5 Evaluate $(392 \times 31 - 298 \times 49)28$.

392 in S.R. and 31 in C.R. gives 12 152 in P.R. On setting 298 in S.R. and making 49 *negative* turns C.R. we obtain

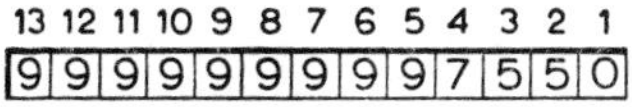

13 12 11 10 9 8 7 6 5 4 3 2 1
| 9 | 9 | 9 | 9 | 9 | 9 | 9 | 9 | 9 | 7 | 5 | 5 | 0 |

This is back transferred to S.R., and 28 *negative* turns C.R. gives

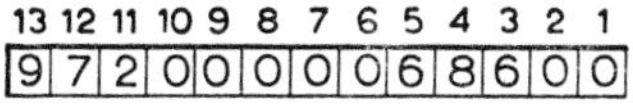

13 12 11 10 9 8 7 6 5 4 3 2 1
| 9 | 7 | 2 | 0 | 0 | 0 | 0 | 0 | 6 | 8 | 6 | 0 | 0 |

The result $= -68\ 600$. This result may be checked by evaluating $(298 \times 49 - 392 \times 31)28$.

298 in S.R. and 49 in C.R. gives 14 602 in P.R. Cancel S.R. and C.R. Set 392 in S.R., negative turns 31 C.R. gives 2450 in P.R. Back transfer 2450 in S.R. 28 turns C.R. gives 68 600, i.e. the previous result with a sign change.

Division
This operation is the inverse of multiplication. In multiplication the contents of S.R. and C.R. are known, and the content of P.R. is sought.

In division the contents of S.R. and P.R. are the divisor and dividend, and the content of C.R., the quotient, is sought (Fig. 18.12).

Example
18.6 To find 89·36/43·32 and 139·36/43·32.

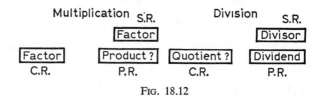

Fig. 18.12

Set the carriage in position 8, to ensure eight significant figures in the quotient (C.R.).

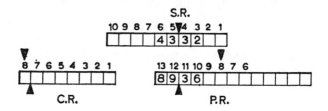

Since 89·36 is required in P.R. in positions 13, 12, 11, 10, set 43·32 in S.R. in alignment with this in positions 6, 5, 4, 3. P.R. is, of course, empty at this stage. Set the decimal-point markers in C.R. and P.R. (4 + 7 = 11).

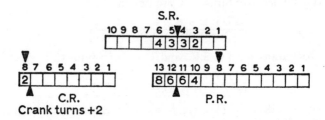

Move the carriage to position 7. N.B. P.R. must be increased.

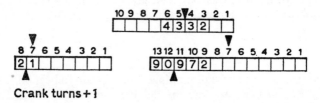

Move the carriage to position 6. N.B. P.R. must be reduced.

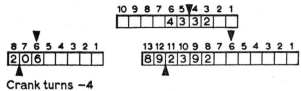

Crank turns −4

Move the carriage to position 5. N.B. P.R. must be increased.

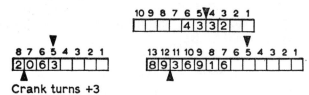

Crank turns +3

Move the carriage to position 4. N.B. P.R. must be reduced.

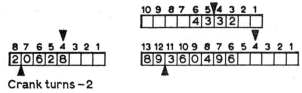

Crank turns −2

Move the carriage to position 3. N.B. P.R. must be reduced.

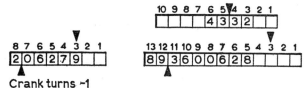

Crank turns −1

Move the carriage to position 2. N.B. P.R. must be reduced.

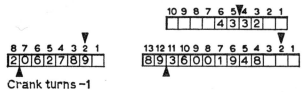

Crank turns −1

Move the carriage to position 1. N.B. P.R. must be reduced.

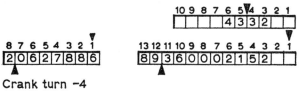

Crank turn −4

We now have $43{\cdot}32 \times 2{\cdot}062\ 788\ 6 = 89{\cdot}360\ 002\ 152\ 00$ and, by one more negative turn,

$$43{\cdot}32 \times 2{\cdot}062\ 788\ 5 = 89{\cdot}359\ 997\ 820\ 00$$

i.e. $$2{\cdot}062\ 788\ 5 < \frac{89{\cdot}36}{43{\cdot}32} < 2{\cdot}062\ 788\ 6$$

and $$\frac{89{\cdot}36}{43{\cdot}32} = 2{\cdot}062\ 789,\ \text{correct to seven significant figures.}$$

If $139{\cdot}36/43{\cdot}32$ is required, the initial setting and alignment are as before. On giving three positive turns the bell rings, indicating overflow at the left-hand end of the P.R.

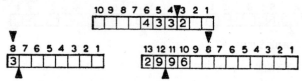

Bearing in mind that the most significant digit, 1, has been lost from the P.R., it will be required to obtain $39{\cdot}36$ in P.R., proceeding as previously. Corresponding results are then

i.e. $$3{\cdot}216\ 989\ 8 < \frac{139{\cdot}36}{43{\cdot}32} < 3{\cdot}216\ 989\ 9$$

and $139{\cdot}36/43{\cdot}32 = 3{\cdot}216\ 990$, correct to seven significant figures.

Division by the tearing down process

The method previously considered has been called the building-up method, since the dividend is built up in the P.R. The tearing-down method involves setting the dividend initially in P.R., with the divisor in S.R., and alternately applying negative and positive turns in subsequent carriage positions, with the object of emptying of P.R. as far as possible. In each carriage position the crank is turned until the bell rings, then the carriage is moved one position to the left and the crank turned in the opposite sense until the bell rings again. The process is quite automatic and can be carried out blindfolded. The bridge of nines alternately appears and disappears in P.R.

Example

18.7 To find $89{\cdot}36/43{\cdot}32$ by the tearing-down process.

Enter $89{\cdot}36$ in P.R. from S.R. with carriage position 8. Cancel C.R. Set $43{\cdot}32$ in S.R., as shown in Fig. 18.13.

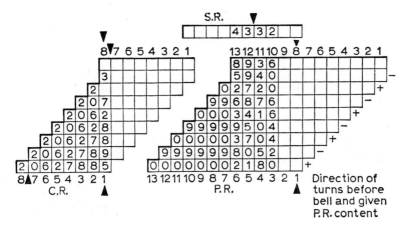

FIG. 18.13

If one negative turn is made in the final position

$$\boxed{2}\boxed{0}\boxed{6}\boxed{2}\boxed{7}\boxed{8}\boxed{8}\boxed{6}\qquad\boxed{9}\boxed{9}\boxed{9}\boxed{9}\boxed{9}\boxed{9}\boxed{9}\boxed{7}\boxed{8}\boxed{4}\boxed{8}\boxed{0}\boxed{0}$$

C.R.   P.R.

is obtained.

Hence $2{\cdot}062\ 788\ 5 < \dfrac{89{\cdot}36}{43{\cdot}32} < 2{\cdot}062\ 788\ 6$,   as before; the left-hand

limit corresponds to a positive remainder. Short cutting can be employed with this method if judgement is used rather than waiting for the bell to ring.

Square root of a number
The square root of a number can be found by a method based on the identities $1 = 1^2$, $1 + 3 = 2^2$, $1 + 3 + 5 = 3^2$, etc. The method is tedious and the iterative formula $x_{n+1} = \frac{1}{2}\left(x_n + \dfrac{N}{x_n}\right)$ is usually preferable to compute $\sqrt{N}$. We shall assume that a slide rule is available to give a good starting approximation, or four-figure tables with square roots and reciprocals.

Example
18.8 Find $\sqrt{(8642)}$ correct to five significant figures, using an initial slide rule approximation.

By slide rule $\sqrt{(8642)} \simeq 93$.
Then $0{\cdot}5\left(93 + \dfrac{8642}{93}\right) \simeq 0{\cdot}5(93 + 8642 \times 0{\cdot}010\ 75)$, using reciprocal
tables.

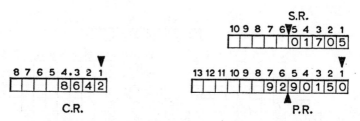

0·017 05 is set in S.R. and 8642 in C.R., giving 92·9015 in P.R. S.R. is cleared and 93 set in S.R. in position 76. 1 positive turn and back transfer gives 185·9015 in S.R. Then 0·5 in C.R. gives 92·950 75 in P.R.

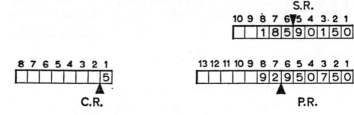

Then $0{\cdot}5\left(92{\cdot}95 + \dfrac{86{\cdot}42}{92{\cdot}95}\right) \simeq 0{\cdot}5(92{\cdot}95 + 8642 \times 0{\cdot}010\ 758\ 5)$

(evaluating as previously but obtaining the reciprocal by division)
$$= 0{\cdot}5(92{\cdot}95 + 92{\cdot}9749)$$
$$= 92{\cdot}9624$$

Check $92{\cdot}9624^2 \simeq 8642{\cdot}0078$
                    $92{\cdot}9623^2 \simeq 8641{\cdot}9892$

Hence $\sqrt{(8642)} = 92{\cdot}962$, correct to five significant figures.

Example
18.9 Find $\sqrt{(12\ 385)}$ correct to six significant figures, using four-figure tables of square roots.

$\sqrt{(1{\cdot}238)} = 1{\cdot}113$ by four-figure tables

hence $\sqrt{(12\ 385)}$

or $\sqrt{(1{\cdot}2385 \times 10^4)} = 100 \times \sqrt{(1{\cdot}2385)}$

$\simeq 111{\cdot}3$

$0{\cdot}5\left(111{\cdot}3 + \dfrac{12\ 385}{111{\cdot}3}\right) \simeq 0{\cdot}5(111{\cdot}3 + 12\ 385 \times 0{\cdot}008\ 985)$

$= 111{\cdot}290$

$0{\cdot}5\left(111{\cdot}290 + \dfrac{12\ 385}{111{\cdot}290}\right) = 111{\cdot}287\ 95$

Now $\qquad 111 \cdot 2879^2 \simeq 12\ 384 \cdot 9969$
$\qquad 111 \cdot 2880^2 \simeq 12\ 385 \cdot 0189$

Hence $\quad \sqrt{(12\ 385)} = 111 \cdot 288$, correct to six significant figures.

The use of reciprocal tables to avoid division fails when the reciprocal tables have insufficient figures. Access to a copy of *Barlow's Tables*, 1965 edition, published by E. and F. N. Spon is invaluable for numerical work.

Example
18.10 The flow diagram gives a Greek method of obtaining an approximation to $\sqrt{2}$. Show how a programme of instructions may be compiled for the desk calculator, based on this flow diagram.

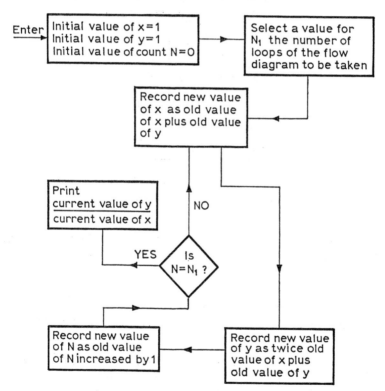

Fig. 18.14

Instructions: Enter x and y in S.R. in positions 6, 7, . . ., and 1, 2, . . . in carriage position 1. Give one positive turn. Interchange x and y in S.R. Give one positive turn. P.R. contains $x + y$ and $y + x$. Cancel y in S.R. Give one positive turn. P.R. contains $x + y$ and $y + 2x$. Back transfer.

S.R. now contains the new values of x and of y, which are recorded. The new value of N is recorded. If N is not equal to N_1, the agreed number of cycles, repeat the cycle until $N = N_1$, then print $\dfrac{\text{latest value of } y}{\text{latest value of } x}$.

If $N_1 = 10$, the following results are obtained.

x	y	N
1	1	0
2	3	1
5	7	2
12	17	3
29	41	4
70	99	5
169	239	6
408	577	7
985	1393	8
2378	3363	9
5741	8119	10

$$\frac{8119}{5741} = 1{\cdot}414\ 213\ 5 \text{ or } 1{\cdot}414\ 213\ 6$$
$$1{\cdot}414\ 213\ 6^2 = 2{\cdot}000\ 000\ 106\ 424\ 96$$
$$1{\cdot}414\ 213\ 5^2 = 1{\cdot}999\ 999\ 823\ 582\ 25$$

If all intermediate values of y/x are evaluated, it will be found that the process alternately underestimates and overestimates $\sqrt{2}$.

EXERCISE 18.1
(addition, subtraction and basic multiplication)

1 Using 19 in S.R., obtain a 19 times table. Using these results verify that

(a) $(19 \times 17) - (19 \times 12) + (19 \times 5) - (19 \times 3) - (19 \times 7) = 0$
(b) $(19 \times 9) - (19 \times 10) + (19 \times 8) - (19 \times 14) + (19 \times 7) = 0$
(c) $(19 \times 11) - (19 \times 15) + (19 \times 12) - (19 \times 14) + (19 \times 13)$
$\qquad\qquad\qquad\qquad\qquad\qquad - (19 \times 6) = 19$
(d) $(19 \times 12) - (19 \times 19) + (19 \times 14) - (19 \times 13) + (19 \times 18)$
$\qquad\qquad - (19 \times 15) + (19 \times 16) - (19 \times 17) = -76$

The following table shows a method of adding nine numbers in which the total 47 508 is obtained by adding the row sums and the column sums.

			Row sum
3 176	2 985	4321	10 482
9 216	5 847	3216	18 279
7 719	9 829	1199	18 747
Column sum 20 111	18 661	8736	47 508

In Examples 2 to 6, state the row sums, the column sums and the total.

2

5 189	6 321	65 136	3 182
6 762	5 631	9 292	7 819
3 462	7 169	3 518	4 312
7 789	6 792	1 357	2 468

3

2 891	−1 376	5 712	−3 143
−3 182	7 625	−8 193	2 279
6 168	−7 123	8 294	−7 162
−3 215	2 169	5 823	9 917

4

28·91	−13·76	57·12	−0·3143
−3 182	7 625	−81·93	0·2279
6 168	−7 123	82·94	−0·7162
−32·15	21·69	582·3	0·9917

5 (28·92 × 31·41)   (12·21 × 14·65)   (78·32 × 24·12)
     (61·51 × 29·83)   (53·12 × 29·87)   (34·65 × 23·65)
     (93·12 × 81·51)   (61·28 × 36·35)   (29·16 × 35·13)

6 (28·92 × 3·179)   (14·65 × 147·2)   (784·3 × 0·2313)
     (63·79 × 2·983)   (57·62 × 271·4)   (629·7 × 0·5361)
     (97·16 × 8·763)   (65·76 × 336·8)   (438·9 × 2·1352)

EXERCISE 18.2

A. Multiplication including decimal-point setting

1 2·43 × 26·5   **2** 358·2 × 0·9417
3 75·846 × 4·3215   **4** 0·003 17 × 0·0654
5 359·37 × 0·76142   **6** 0·000 123 5 × 0·000 765 2

B. Multiplication with back transfer

(check by multiplying in different order)

7 37 × 25 × 43   **8** 75 × 93 × 22
9 761 × 214 × 973   **10** 38 × 26 × 42 × 33
11 21 × 93 × 87 × 31 × 52   **12** Evaluate 37 × 53 × 21 × 15,
checking the result by evaluating the products in brackets first and multiplying the results (37 × 53)(21 × 15), (37 × 21)(53 × 15) and
(37 × 15)(53 × 21).

EXERCISE 18.3

Division and multiplication with division

A. Calculate the following quotients, correct to six significant figures.

1 $\dfrac{21}{23}$ **2** $\dfrac{123}{359}$ **3** $\dfrac{1876}{2439}$ **4** $\dfrac{19}{732}$ **5** $\dfrac{23}{5654}$ **6** $\dfrac{3\cdot7}{41\cdot4}$

7 $\dfrac{0\cdot07}{0\cdot001\,32}$ **8** $\dfrac{0\cdot0132}{21\cdot79}$ **9** $\dfrac{68\cdot51}{0\cdot0172}$ **10** $\dfrac{713}{0\cdot001\,23}$

B. Evaluate the following correct to six significant figures.

1 Evaluate $\dfrac{123 \times 236}{375}$ as $123 \times 236 \div 375$ and $123 \div 375 \times 236$

2 Evaluate $\dfrac{235 \times 269}{457 \times 125}$ as (a) $(235 \times 269) \div (457 \times 125)$
(b) $235 \div 457 \times 269 \div 125$
(c) $235 \div 125 \times 269 \div 457$
(d) $269 \div 457 \times 235 \div 125$

In examples 3, 4 and 5, use check methods as in example 2.

3 $\dfrac{73\cdot9 \times 0\cdot236}{0\cdot017}$ **4** $\dfrac{0\cdot013 \times 0\cdot002\,15}{0\cdot000\,376}$ **5** $\dfrac{318\cdot7 \times 2106}{53\cdot48}$

EXERCISE 18.4

Evaluation of the square root of a number N using the iterative formula
$$x_{n+1} = \tfrac{1}{2}\left(x_n + \frac{N}{x_n}\right)$$
Using a slide rule or four-figure tables to obtain a starting approximation evaluate the square root of each of the following numbers, correct to six significant figures. Check your result by squaring.

1 3   **2** 5   **3** 7   **4** 11   **5** 13   **6** 37   **7** 157   **8** 283
9 561   **10** 3127

EXERCISE 18.5

Use the iterative formula $x_{n+1} = \tfrac{1}{2}x_n\left(3 - \dfrac{x_n^2}{N}\right)$ to find the value of $\sqrt{N}$,
correct to four significant figures, when $N = 0\cdot1$ (ULCI specimen 01)

Here $N = 0\cdot1$: try $x = 0\cdot3$ and the following results are obtained.

x_n^2	$\dfrac{x_n^2}{N}$	$3 - \dfrac{x_n^2}{N}$	$x_n\left(3 - \dfrac{x_n^2}{N}\right)$	$\dfrac{x_n}{2}\left(3 - \dfrac{x_n^2}{N}\right)$
0·09	0·9	2·1	0·63	0·315
0·099 225	0·992 25	2·007 75	0·632 441	0·316 220
0·099 995	0·999 95	2·000 050	0·632 455	0·316 227
0·099 999	0·999 99	2·000 010	0·632 457	0·316 228
0·099 999	0·999 99	2·000 010	0·632 457	0·316 228

Only the first two iterations are necessary if a check is carried out.

$$0.316\ 22^2 = 0.099\ 995\ 088\ 4 \qquad 0.316\ 23^2 = 0.100\ 001\ 412\ 9$$

$$\therefore \quad \sqrt{0.1} = 0.3162, \text{ correct to four significant figures.}$$

Use this method to evaluate the square roots of the previous exercise, using the given starting approximations, correct to five significant figures.

1 $N = 3, \ x_1 = 2$ **2** $N = 5, \ x_1 = 2.2$ **3** $N = 7, \ x_1 = 2.6$
4 $N = 11, \ x_1 = 3.3$ **5** $N = 13, \ x_1 = 3.6$ **6** $N = 37, \ x_1 = 6.1$
7 $N = 157, \ x_1 = 12.5$ **8** $N = 283, \ x_1 = 16.8$ **9** $N = 561, \ x_1 = 23.7$
10 $N = 3127, \ x_1 = 55.9$

EXERCISE 18·6

1 The following flow diagram (Fig. 18.15) relates to the calculation of an approximation value of e. Evaluate the programme.

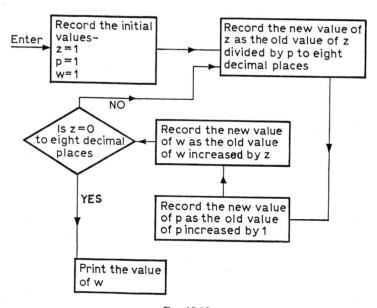

Fig. 18.15

2 Referring to the flow diagram on page 337, evaluate the programme when x and y are initially 10 and 12 respectively, and $N_1 = 6$.

3 Repeat the previous example when x and y are initially 33 125 and 34 250, and $N_1 = 6$.

4 The following flow diagram (Fig. 18.16) refers to the selection of $\sqrt{x}$, where x is a positive or negative number. Evaluate the programme when (*a*) $x = 5$, (*b*) $x = -2$, (*c*) $x = 23$, (*d*) $x = -17$.

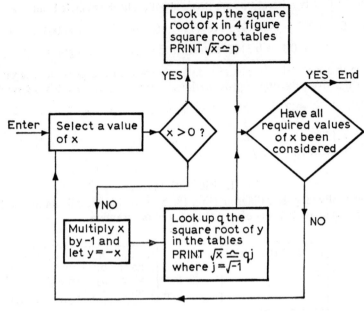

FIG. 18.16

Solution of two linear simultaneous equations in two unknowns by pivotal condensation with current sum checks

Consider the equations

$$\text{(i)} \quad 0{\cdot}210x - 0{\cdot}290y = 0{\cdot}699, \qquad \Sigma_{(i)} = 0{\cdot}619$$
$$\text{(ii)} \quad 0{\cdot}310x + 0{\cdot}790y = -0{\cdot}198, \qquad \Sigma_{(ii)} = 0{\cdot}902$$

The coefficients of x and y are inspected, and the largest is selected as the 'pivotal coefficient', i.e. $0{\cdot}790$. Equation (ii) is multiplied by

$-\dfrac{(0{\cdot}290)}{(0{\cdot}790)}$. By using the pivotal coefficient as denominator, the multiplier

used is less than one which tends to reduce rounding errors. $\Sigma_{(i)}$ and $\Sigma_{(ii)}$ denote the sums of the numbers in equations (i) and (ii) respectively.

Each new equation which is formed also has a new valve Σ. The operations which are carried out on equations are also carried out on the Σ values. Then, for each equation, the sum of its numbers, called the *cross sum*, should equal the current value of Σ. Thus, although $\Sigma_{(i)}$ and $\Sigma_{(ii)}$ are formed as a cross sum, subsequent values of Σ are formed by operations, and checked against the cross sum.

Since three significant figures are given in the data, the working is carried out to five significant figures, using two guarding figures.

$$-\frac{(-0\cdot290)}{(0\cdot790)} = 0\cdot367\ 09$$

Multiplying equation (ii) by 0·367 09,

$$0\cdot367\ 09 \times 0\cdot310x + 0\cdot367\ 09 \times 0\cdot790y = 0\cdot367\ 09(-0\cdot198)$$

$$\Sigma = 0\cdot367\ 09 \times 0\cdot902$$

i.e. $0\cdot113\ 80x + 0\cdot290\ 00y = -0\cdot072\ 68$ . . . . (iii)

$$\Sigma = 0\cdot331\ 12$$

Cross sum = $(0\cdot113\ 80 + 0\cdot290\ 00 - 0\cdot072\ 68) = 0\cdot331\ 12$

If at any stage the values of Σ and the cross sum differ appreciably, an inspection for error is made.

(i) $0\cdot210x - 0\cdot290y = 0\cdot699,$ $\Sigma_{(i)} = 0\cdot619$
(iii) $0\cdot113\ 80x + 0\cdot290\ 00y = -0\cdot072\ 68,$ $\Sigma = 0\cdot331\ 12$

Adding equations (i) and (iii),

(iv) $0\cdot323\ 80x + 0y = 0\cdot626\ 32,$ $\Sigma = 0\cdot619 + 0\cdot331\ 12$
                                                            $= 0\cdot950\ 12$

Cross sum $= 0\cdot323\ 80 + 0\cdot626\ 32 = 0\cdot950\ 12$

hence (v) $1 \cdot x = \dfrac{0\cdot626\ 32}{0\cdot323\ 80} = 1\cdot934\ 28,$ $\Sigma = \dfrac{0\cdot950\ 12}{0\cdot323\ 80} = 2\cdot934\ 28$

Cross sum $= 1 + 1\cdot934\ 28 = 2\cdot934\ 28$

Equation (v) is now multiplied by $-0\cdot310$ before adding to equation (ii).

$-0\cdot310x = -0\cdot310 \times 1\cdot934\ 28,$ $\Sigma = -0\cdot310 \times 2\cdot934\ 28$
i.e. (vi) $-0\cdot310x = -0\cdot599\ 63,$ $\Sigma = -0\cdot909\ 63$

Cross sum $= -0\cdot310 - 0\cdot599\ 63 = -0\cdot909\ 63$

(v) $-0\cdot310x + 0y = -0\cdot599\ 63,$ $\Sigma = -0\cdot909\ 63$
(ii) $0\cdot310x + 0\cdot790y = -0\cdot198,$ $\Sigma = 0\cdot902$

Adding equations (v) and (ii)

(vii) $0x + 0\cdot79y = -0\cdot797\ 63,$ $\Sigma = -0\cdot007\ 63$

Cross sum $= 0\cdot790 - 0\cdot797\ 63 = -0\cdot007\ 63$

hence (viii) $1 \cdot y = -\dfrac{0\cdot797\ 63}{0\cdot790} = -1\cdot009\ 66$

$$\Sigma = -\frac{0\cdot007\ 63}{0\cdot790} = -0\cdot009\ 66$$

Cross sum $= 1 - 1\cdot009\ 66 = -0\cdot009\ 66$

Having arrived at the solutions $x = 1·934\,28$, $y = -1·009\,66$, the residuals R_1 and R_2, which represent the differences between the two sides of equations (i) and (ii) respectively, may be calculated.

$$R_1 = 0·21 \times 1·934\,28 - 0·29(-1·009\,66) - 0·699 = 0·000\,000\,2$$
$$R_2 = 0·31 \times 1·934\,28 + 0·79(-1·009\,66) + 0·198 = -0·000\,004\,6$$

Finally, if the original coefficients have been obtained by measurement and are, in effect, rounded to three significant figures, the solutions are $x = 1·93$, $y = -1·01$, rounded to three significant figures. The working is often shown more concisely in a table.

Operation on given line to form next line	Equation	x	y	Constant	Σ	Cross sum
	(i)	0·210	−0·290	0·699	0·619	
Multiplier $-\left(\dfrac{0·29}{0·79}\right)$ $= 0·367\,09$	(ii)	0·310	0·790	−0·198	0·902	
(i) + (iii)	(iii)	0·113 80	0·290 00	−0·072 68	0·331 12	0·331 12
÷ 0·323 80	(iv)	0·323 80	0	0·626 32	0·950 12	0·950 12
× − 0·310	(v)	1	0	1·934 28	2·934 28	2·934 28
(ii) + (vi)	(vi)	−0·310	0	−0·599 63	−0·909 63	−0·909 63
÷ 0·79	(vii)	0	0·790	−0 797 63	−0·007 63	−0·007 63
	(viii)	0	1	−1·009 66	−0·009 66	−0·009 66

$R_1 = 0·21 \times 1·934\,28 - 0·29(-1·009\,66) - 0·699 = 0·000\,000\,2$
$R_2 = 0·31 \times 1·934\,28 + 0·79(1·009\,66) + 0·198 = 0·000\,004\,6$
$x = 1·93$, $y = -1·01$

Solution of three linear simultaneous equations in three unknowns by pivotal condensation with current sum checks

Example

18.11 Solve the following equations, working to five decimal places.

$$0·742x_1 - 0·384x_2 - 0·596x_3 = 0·497$$
$$0·396x_1 + 0·503x_2 + 0·640x_3 = 0·389$$
$$0·177x_1 + 0·640x_2 + 0·927x_3 = 0·649$$

Round the solutions off to a suitable number of significant figures.

ULCI specimen 01

The example on page 345, being lengthy, is of the project, rather than examination, type.

The solutions are rounded to three significant figures as it is impossible, in general, to obtain a result more accurate than the data used, i.e. $x_1 = 1·28$, $x_2 = 6·72$ and $x_3 = 5·09$. If the original equations had exact coefficients, the number of valid significant figures in the solutions depends upon the rounding errors introduced by the division and multiplication in the above process. It is not claimed that pivotal condensation is the best method of solution of linear equations in two and three unknowns, but that it forms a systematic procedure which can be used to solve linear equations in four or five unknowns, say, with the aid of a desk calculating machine.

	Equation	x_1	x_2	x_3	Constant	Σ	Cross sum
	(i)	0·742	−0·384	−0·596	0·497	0·259	
	(ii)	0·396	0·503	0·640	0·389	1·928	
	(iii)	0·177	0·640	0·927	0·649	2·393	
Multiply (iii) by −0·64/0·927 = −0·69040	(iv)	−0·122 20	−0·441 86	−0·640 00	−0·448 07	−1·652 13	−1·652 13
Multiply (iii) by 0·596/0·927 = 0·642 93	(v)	0·113 80	0·411 48	0·596 00	0·417 26	1·538 53	1·538 54
Add (i) and (v)	(vi)	0·855 80	0·027 48	0	0·914 26	1·797 53	1·797 54
Add (ii) and (iv)	(vii)	0·273 80	0·061 14	0	−0·059 07	0·275 87	0·275 87
Multiply (vi) by $\frac{0\cdot273\,80}{0\cdot855\,80}$ = −0·319 93	(viii)	−0·273 80	−0·008 79	0	−0·292 50	−0·575 08	−0·575 09
Add (vii) and (viii)	(ix)	0	0·052 35	0	−0·351 57	−0·299 21	−0·299 22
Divide (ix) by 0·05235	(x)	0	1	0	−6·715 76	−5·715 76	−5·715 76
Multiply (x) by −0·027 48	(xi)	0	−0·027 48	0	0·184 55	0·157 07	0·157 07
Add (vi) and (xi)	(xii)	0·855 80	0	0	1·098 81	1·954 60	1·954 61
Divide (xii) by 0·855 80	(xiii)	1	0	0	1·283 96	2·283 94	2·283 96
Multiply (xiii) by −0·177	(xiv)	−0·177	0	0	−0·227 26	−0·404 26	−0·404 26
Multiply (x) by −0·64	(xv)	0	−0·640	0	4·298 09	3·658 09	3·658 09
Add (iii) (xiv) and (xv)	(xvi)	0	0	0·927	4·719 83	5·646 83	5·646 83
Divide (xvi) by 0·927	(xvii)	0	0	1	5·091 51	6·091 51	6·091 51

$R_1 = 0·742(1·283\,96) − 0·384(−6·715\,76) − 0·596(5·091\,51) − 0·497 = 0·000\,01$
$R_2 = 0·396(1·283\,96) + 0·503(−6·715\,76) + 0·640(5·091\,51) − 0·389 = −0·000\,02$
$R_3 = 0·177(1·283\,96) + 0·640(−6·715\,76) + 0·927(5·091\,51) − 0·649 = 0·000\,00$

Solution of three linear simultaneous equations in three unknowns by Cramer's Rule

Consider the equations
$$a_1x + b_1y + c_1z = d_1$$
$$a_2x + b_2y + c_2z = d_2$$
$$a_3x + b_3y + c_3z = d_3$$

The solutions are $\quad x = \dfrac{\Delta_x}{\Delta}, \qquad y = \dfrac{\Delta_y}{\Delta}, \qquad z = \dfrac{\Delta_z}{\Delta}$

where Δ, Δx, Δy and Δz are obtained as follows.

$$\Delta = (a_1b_2c_3 + b_1c_2a_3 + c_1a_2b_3)$$
$$- (a_3b_2c_1 + b_3c_2a_1 + c_3a_2b_1)$$

Replacing the x column, a_1, a_2, a_3, by d_1, d_2, d_3,

$$\Delta x = (d_1b_2c_3 + b_1c_2d_3 + c_1d_2b_3)$$
$$- (d_3b_2c_1 + b_3c_2d_1 + c_3d_2b_1)$$

Replacing the y column, b_1, b_2, b_3, by d_1, d_2, d_3,

$$\Delta y = (a_1d_2c_3 + d_1c_2a_3 + c_1a_2d_3)$$
$$- (a_3d_2c_1 + d_3c_2a_1 + c_3a_2d_1)$$

Replacing the z column, c_1, c_2, c_3, by d_1, d_2, d_3,

$$\Delta z = (a_1b_2d_3 + b_1d_2a_3 + d_1a_2b_3)$$
$$- (a_3b_2d_1 + b_3d_2a_1 + d_3a_2b_1)$$

Thus exact solutions can be obtained if the original coefficients are exact.

Referring to the equations previously solved by pivotal condensation, denoting the variables by x, y, z instead of x_1, x_2 and x_3,

$$0 \cdot 742x - 0 \cdot 384y - 0 \cdot 596z = 0 \cdot 497$$
$$0 \cdot 396x + 0 \cdot 503y + 0 \cdot 640z = 0 \cdot 389$$
$$0 \cdot 177x + 0 \cdot 640y + 0 \cdot 927z = 0 \cdot 649$$

$$\Delta = (0 \cdot 345\,980\,502 - 0 \cdot 043\,499\,520 - 0 \cdot 151\,050\,240)$$
$$- (0 \cdot 530\,624\,76 - 0 \cdot 303\,923\,200 - 0 \cdot 140\,963\,328)$$
$$= 0 \cdot 415\,333\,46$$

$$\Delta_x = (0 \cdot 231\,741\,657 - 0 \cdot 159\,498\,240 - 0 \cdot 148\,380\,160)$$
$$- (-0 \cdot 194\,562\,412 + 0 \cdot 203\,571\,200$$
$$- 0 \cdot 138\,471\,552) = 0 \cdot 053\,326\,021$$

Similarly

$\Delta_y = -0.278\ 912\ 254$

$\Delta_z = 0.211\ 456\ 611$

$$\therefore \quad x = \frac{0.053\ 326\ 021}{0.041\ 533\ 326} = 1.283\ 932\ 70$$

$$y = \frac{-0.278\ 912\ 254}{0.041\ 533\ 326} = -6.715\ 381\ 27$$

$$z = \frac{0.211\ 456\ 611}{0.041\ 533\ 326} = 5.091\ 249\ 11$$

for which $\quad R_1 = 15.2 \times 10^{-10}, R_2 = 7.9 \times 10^{-10}, R_3 = 0.7 \times 10^{-10}.$

If the original coefficients were exact, then the solutions for x, y and z before the final division are exact. If the coefficients were rounded, then the solutions are $x = 1.28$, $y = -6.72$ and $z = 5.09$, as before. Note that the solutions obtained by pivotal condensation, $x = 1.283\ 96$, $y = -6.715\ 76$, $z = 5.091\ 51$, are not correct to four significant figures considered as solutions to the equations with exact coefficients, having a discrepancy of 1 in the least significant figure of y, since $y = -6.716$ rounded to four significant figures, whereas $y = -6.715$ is correct.

EXERCISE 18.7

1 Solve the equations $\quad 2.81i_1 - 0.521i_2 = 3.75$
$\qquad\qquad\qquad\qquad\quad 3.72i_1 + 0.31i_2 = 11.25$

Give the solutions correct to four significant figures, assuming that the coefficients are exact. What would the solutions be if the coefficients were rounded? Calculate the residuals in each case.

2 Solve the equations $\quad 5.62x - 8.23y = 86.4$
$\qquad\qquad\qquad\qquad\quad 9.13x + 6.71y = 37.9$

if the coefficients are rounded, and state the residuals.

3 Solve the equations $\quad 15u + 29v = 56$
$\qquad\qquad\qquad\qquad\quad 13u - 15v = 39$

correct to five significant figures. The coefficients are exact. Estimate the residuals.

4 Solve the equations $\quad 3.15i_1 - 3.28i_2 = 6.91$
$\qquad\qquad\qquad\qquad -2.72i_1 + 7.16i_2 = 10.37$

where all coefficients are rounded, and estimate the residuals.

5 Solve the equations $\quad 15x - 17y = 62$
$\qquad\qquad\qquad\qquad\quad 14x + 38y = -11$

correct to five significant figures, and estimate the residuals.

6 Solve the equations $22{\cdot}3p - 0{\cdot}76q = 31{\cdot}8$
$-0{\cdot}29p + 41{\cdot}6q = -32{\cdot}3$

where the coefficients are rounded, and estimate the residuals.

EXERCISE 18.8

1 Solve the equations $15i_1 - 87i_2 + 92i_3 = 139$
$-37i_1 - 21i_2 + 27i_3 = -29$
$19i_1 - 20i_2 + 21i_3 = 63$

Give the solutions correct to four significant figures, and calculate the residuals.

2 Solve the equations $27x - 15y + 12z = 135$
$-6x + 19y - 87z = 371$
$39x - 61y + 93z = 673$

Give the solutions correct to two decimal places, and estimate the residuals.

3 Solve the equations (with rounded coefficients)

$$2{\cdot}13p - 3{\cdot}15q + 5{\cdot}12r = 4{\cdot}03$$
$$9{\cdot}31p + 7{\cdot}18q - 9{\cdot}57r = 37{\cdot}5$$
$$1{\cdot}05p + 1{\cdot}72q - 2{\cdot}60r = 4{\cdot}83$$

Calculate the residuals.

4 Solve the following equations (with rounded coefficients) and calculate the residuals.

$$0{\cdot}21I_1 + 0{\cdot}22I_2 + 0{\cdot}23I_3 = 4{\cdot}17$$
$$0{\cdot}33I_1 - 0{\cdot}34I_2 + 0{\cdot}35I_3 = -3{\cdot}12$$
$$0{\cdot}57I_1 + 0{\cdot}42I_2 - 0{\cdot}99I_3 = 5{\cdot}13$$

5 Solve the following equations, which have rounded coefficients, and calculate the residuals.

$$3{\cdot}9u - 6{\cdot}1v + 9{\cdot}3f = 67{\cdot}3$$
$$2{\cdot}7u - 1{\cdot}5v + 1{\cdot}2f = 13{\cdot}5$$
$$-0{\cdot}6u + 1{\cdot}9v - 8{\cdot}7f = -47{\cdot}5$$

6 Solve the following equations, which have exact coefficients.

$$121\theta + 37\omega + 16\alpha = 1124$$
$$97\theta - 28\omega + 161\alpha = 733$$
$$-57\theta + 67\omega - 398\alpha = 265$$

Give the solutions rounded to five significant figures, and calculate the residuals.

Chapter Nineteen
Rounding Errors

Exact and approximate numbers

$\frac{3}{7}$ and $\sqrt{3}$ are exact numbers in which no approximation or uncertainty is involved. 0·4285 is an approximation to $\frac{3}{7}$, as is 0·4286. Since $7 \times 0·4285 = 2·9995$, and $7 \times 0·4286 = 3·0002$, we can say that the exact number $\frac{3}{7}$ lies between the bounding approximations 0·4285 and 0·4286,

i.e. $0·4285 < \frac{3}{7} < 0·4286$.

Since $1·7320^2 = 2·999\,824$ and $1·7321^2 = 3·000\,170\,41$, we can say that the exact number $\sqrt{3}$ lies between the bounding approximations 1·7320 and 1·7321,

i.e. $1·7320 < \sqrt{3} < 1·7321$.

Significant figures

Any one of the digits 0, 1, 2, 3, 4, 5, 6, 7, 8, 9 may be a significant figure, as shown in the following table.

Number	Significant figures	Number of significant figures
8	8	1
79	7, 9	2
381	3, 8, 1	3
7049	7, 0, 4, 9	4
0·001 73	1, 7, 3	3

The *exact* number 17 500 has five significant figures 1, 7, 5, 0, 0, but if the figures 0, 0 are used to fill the places of unknown or discarded numbers, then the *approximate* value 17 500 has three significant figures.

The importance of significant figures arises from the fact that the accuracy of a measurement or a computation is expressed by the number of significant figures obtained, rather than the number of decimal places. The relative error bounds are divided by ten for each additional significant figure correctly obtained in the number.

Rounding of numbers

We have previously shown that $0·4285 < \frac{3}{7} < 0·4286$. Now $\frac{3}{7} = 0·428\,5714$ ... If we wish to use an approximation to $\frac{3}{7}$ rounded to four significant figures, we have to decide between the values 0·4285 and 0·4286. This is done so as to cause the least possible error. 4, 2, 8, 5 are retained and 7, 1, 4 . . . are discarded. In this case the leading or most significant figure discarded is 7, and, since this is greater than 5, the fourth digit is increased by one, i.e. 4, 2, 8, 6 are the significant figures, $\frac{3}{7} \simeq 0\cdot4286$ rounded to four significant figures. Had the leading or most significant figure discarded been less than 5, then the fourth digit would have been unchanged, i.e. $0\cdot428\ 537\ 5$ becomes $0\cdot4285$ rounded to four significant figures.

Example
19.1 Round the following numbers to three significant figures: 325 962, 21·743, 0·012 463, 12 549

$$325\ 962 \simeq 326\ 000$$
$$21\cdot743 \simeq 21\cdot7$$
$$0\cdot012\ 463 \simeq 0\cdot0125$$
$$12\ 549 \simeq 12\ 500$$

Special case
If the most significant figure discarded is a 5, carry out the rounding so as to keep the least significant figure retained an *even* digit. It may be noted that this rule contradicts the rule which says that the least significant figure retained should always be increased by one. The latter rule tends to accumulate rounding errors as is shown in the following example in the right hand column.

Example
19.2 Find the sum of the numbers given in the left-hand column. Find also the sum of the numbers if they are rounded to two significant figures before addition, using the 'special case' rule, as in the centre column, and the general rule, as in the right-hand column.

7152	7200	7200
8257	8200	8300
6855	6800	6900
5359	5400	5400
1752	1800	1800
2651	2600	2700
2153	2200	2200
1852	1800	1900
1453	1400	1500
1759	1800	1800
39 243	39 200	39 700

$$\text{Relative error for column 2} = \frac{39\ 243 - 39\ 200}{39\ 243} = \frac{43}{39\ 243}$$

$$\text{Relative error for column 3} = \frac{39\ 700 - 39\ 243}{39\ 243} = \frac{457}{39\ 243}$$

The rounding method used in column 3 involves more than ten times the relative error involved in column 2, where the rounding errors tend to cancel each other.

Rounding errors in a sum or difference

If 12·18 is the rounded value of a number, it must have been obtained from a number between 12·18 − 0·005 and 12·18 + 0·005. If numbers which may be positive or negative are added, the possible error in the total is in magnitude equal to the sum of the separate error magnitudes. Thus, if the rounded numbers 12·18 and 9·23 are added, the sum is given by (12·18 ± 0·005) + (9·23 ± 0·005). The errors could tend to cancel, but allowance must be made for the worst cases when they do not cancel. The sum is thus between (12·18 + 9·23) ±(0·005 + 0·005)

i.e. 21·41 ± 0·01   or   21·40 and 21·42

and is 21·4, rounded to three significant figures.

Example

19.3 $P = 0·214 + 0·318 − 0·174 − 0·125 + 0·781 + 0·923 − 0·301 − 0·21$ where all the numbers are rounded. Evaluate P to as many significant figures as possible.

Here there are eight numbers involved, hence

$$P = (0·214 + 0·318 − 0·174 − 0·125 + 0·781 + 0·923 − 0·301 − 0·212)$$
$$± (8 × 0·0005) = 1·424 ± 0·004$$

P is between 1·420 and 1·428, and the value 1·4, correct to two significant figures, is all that can be guaranteed.

Rounding errors in a product

Let x and y represent the rounded value of two numbers.

Δx and Δy represent the rounding error of each number, which may be positive or negative, so that the true value of each number is $x + \Delta x$ and $y + \Delta y$.

The true value of the product $= (x + \Delta x)(y + \Delta y)$
$$= xy + y\Delta x + x\Delta y + \Delta x\Delta y$$

Approximate value of the product $= xy$

$\therefore$ rounding error of the product by subtraction $= y\Delta x + x\Delta y + \Delta x\Delta y$

Relative error of the product $= \dfrac{\text{rounding error of the product}}{\text{approximate value of the product}}$ (i)

$$= \frac{y\Delta x + x\Delta y + \Delta x\Delta y}{xy}$$

$$= \frac{\Delta x}{x} + \frac{\Delta y}{y}$$

$$= \text{relative error in } x + \text{relative error in } y$$

neglecting $\left(\dfrac{\Delta x}{x}\right)\left(\dfrac{\Delta y}{y}\right)$, which is of the second order of small quantities.

It may be noted that no great error is introduced in line (i) by dividing by the approximate rather than the true value of the product, since the relative error is only required approximately.

Examples

19.4 Calculate the product $32·8 \times 13·2$, and give the product to as many significant figures as are justified if both numbers are rounded.

$$32·8 \times 13·2 = 432·96$$

$$\text{Relative error} = \frac{0·05}{32·8} + \frac{0·05}{13·2} < \frac{0·05}{30} + \frac{0·05}{10}$$
$$= 0·0067 \text{ approx.}$$
$$\text{Approximate error} = 0·0067 \times 433 = 2·9$$

The product is between $432·96 - 2·9$ and $432·96 + 2·9$, i.e. between 430 and 436.

Hence the product is 400 rounded to one significant figure. The most significant figure is the only one definitely known. If rounded to two significant figures, as 430, there is a possible error of 1 unit in the second significant figure. It may be noted that the relative error is worked out approximately by rounding down the divisors $32·8$ and $13·2$ to 30 and 10 respectively, so as to increase the relative error.

As a check on the above the upper and lower limits of the product may be computed exactly:

$$32·85 \times 13·25 = 435·2625$$
$$32·75 \times 13·15 = 430·6625$$

19.5 Calculate the product $1258 \times 72·18 \times 1·375$, and give the product to as many significant figures as are justified if all the numbers are rounded.

$$1258 \times 72·18 \times 1·375 = 90\,802·44 \times 1·375 = 124\,853·355$$

$$\text{Relative error} = \frac{0·5}{1258} + \frac{0·005}{72·18} + \frac{0·0005}{1·375} < 0·0005 + 0·00001 + 0·0005$$
$$= 0·0011$$

Treating the divisors as 1000, 50 and 1 respectively,

$$\text{Approximate error} = 125\,000 \times 0·0011 = 137·5, \quad \text{say } 140.$$

The product is between $124\,853 \pm 140$, i.e. between 124 993 and 124 713, i.e. 125 000, rounded to three significant figures.

Computing the exact upper and lower limits, as a check,

$$1258·5 \times 72·185 \times 1·3755 = 124\,957·05 \dots$$
$$1257·5 \times 72·175 \times 1·3745 = 124\,749·97 \dots$$

Rounding errors in a quotient

Let $\dfrac{x}{y}$ be the approximate quotient and $\dfrac{x + \Delta x}{y + \Delta y}$ the exact quotient, then

$$\text{rounding error of the quotient} = \frac{x + \Delta x}{y + \Delta y} - \frac{x}{y}$$

$$= (x + \Delta x)(y + \Delta y)^{-1} - x/y$$

$$= (x + \Delta x)\left(y^{-1} - \frac{1}{1} \cdot y^{-2} \cdot \Delta y \ldots\right) - x/y$$

$$= (x + \Delta x)\left(\frac{1}{y} - \frac{\Delta y}{y^2} \ldots\right) - x/y$$

$$= \frac{x}{y} - \frac{x\Delta y}{y^2} + \frac{\Delta x}{y} - \frac{x}{y}$$

$$= -\frac{x\Delta y}{y^2} + \frac{\Delta x}{y}$$

neglecting the second and higher orders of small quantities.
Hence the relative error of the quotient

$$= \frac{\text{rounding error of the quotient}}{\text{approximate value of the quotient}}$$

$$= \frac{\dfrac{-x\Delta y}{y^2} + \dfrac{\Delta x}{y}}{\dfrac{x}{y}} = -\frac{\Delta y}{y} + \frac{\Delta x}{x}$$

Since Δy and Δx may themselves be positive or negative, we may ignore the minus sign, and say that
relative error of the quotient = relative error of x + relative error of y

Example

19.6 Calculate the quotient $\dfrac{131 \cdot 5}{2 \cdot 1}$ and give the result to as many significant figures as are justified if both the numbers are rounded.

$$\frac{131 \cdot 5}{2 \cdot 1} = 62 \cdot 619 \ldots$$

$$\text{Relative error of quotient} = \frac{0 \cdot 05}{131 \cdot 5} + \frac{0 \cdot 05}{2 \cdot 1} < \frac{0 \cdot 05}{100} + \frac{0 \cdot 05}{2} = 0 \cdot 0255$$

$$\text{Approximate error} = 63 \times 0 \cdot 0255 = 1 \cdot 61$$

The quotient is between 61·00 and 64·23, i.e. 60, rounded to one significant figure.

Computing the upper and lower limits exactly, as a check,

$$\frac{131\cdot55}{2\cdot05} = 64\cdot17\ldots$$

$$\frac{131\cdot45}{2\cdot15} = 61\cdot13\ldots$$

N.B. It may appear that it is as quick to compute the exact upper and lower limits as to estimate them from the relative error. When the computations are extended to more numbers, however, this ceases to be so.

Rounding errors in a combined product and quotient

The previous result may be extended, so that the relative error in a combined product or quotient is equal to the sum of the relative errors of all the numbers present in both the numerator and denominator.

Example

19.7 Calculate $\dfrac{21\cdot5 \times 30\cdot2}{3\cdot72}$, and give the result to as many significant

figures as are justified if all the numbers are rounded.

$$\frac{21\cdot5 \times 30\cdot2}{3\cdot72} = \frac{649\cdot3}{3\cdot72} = 174\cdot543\ldots$$

Relative error $= \dfrac{0\cdot05}{21\cdot5} + \dfrac{0\cdot05}{30\cdot2} + \dfrac{0\cdot005}{3\cdot72} < \dfrac{0\cdot05}{20} + \dfrac{0\cdot05}{30} + \dfrac{0\cdot005}{3}$

$$= 0\cdot0025 + 0\cdot006 + 0\cdot002 = 0\cdot0105$$

Approximate error $= 175 \times 0\cdot0105 = 1\cdot9$

The result is between $174\cdot5 \pm 1\cdot9$, i.e. 172·6 and 176·4, and can be written as 200, rounded to one significant figure, or as 170, rounded to two significant figures with a possible error of 1 unit in the second significant figure. If it is written as 174, rounded to three significant figures, there is a possible error of two units in the third significant figure according to this estimate.

Computing the upper and lower limits, as a check,

$$\frac{21\cdot55 \times 30\cdot25}{3\cdot715} = \frac{651\cdot8875}{3\cdot715} = 175\cdot47\ldots$$

$$\frac{21\cdot45 \times 30\cdot15}{3\cdot725} = 173\cdot61$$

We find that 174 represents the result to three significant figures with a possible error of 1 unit in the third significant figure in fact.

The previous examples illustrate that *it is not in general possible to determine the result of a computation to more significant figures than the number of significant figures present in the least accurately known number involved in the computation.*

EXERCISE 19.1

1 Evaluate the following (*a*) if all the numbers are exact, (*b*) if all the numbers are rounded, stating the limits between which the result may lie, and giving the result rounded to as many significant figures as possible.

   (i) $21\cdot2 - 37\cdot9 + 15\cdot6 - 21\cdot3 + 81\cdot7 + 92\cdot3 - 15\cdot6 - 21\cdot9 + 86\cdot2 - 10\cdot6$

   (ii) $0\cdot316 - 0\cdot525 + 0\cdot736 - 0\cdot917 + 0\cdot869 + 0\cdot795 + 0\cdot213 - 0\cdot318 - 0\cdot172 + 0\cdot911$

   (iii) $21\cdot2 - 0\cdot525 + 15\cdot6 - 0\cdot917 + 92\cdot3 + 0\cdot795 - 15\cdot6 - 0\cdot318 + 86\cdot2 - 0\cdot911$

   (iv) $213\cdot5 - 27\cdot92 + 9\cdot625 - 0\cdot2135 + 8\cdot937 - 27\cdot31 + 371\cdot6 - 1150 + 317\ 52$

In examples 2 to 8 (using desk calculators in (*a*) and (*b*)(v)) (*a*) calculate the product if the numbers are exact, (*b*) if the numbers are rounded, estimate (i) the approximate relative error in the product (ii) the approximate error in the product, (iii) the approximate limits between which the product lies, (iv) state the product to as many significant figures as are justified, (v) check by evaluating exact limits for the product.

2 $61\cdot97 \times 27\cdot87$ **3** $0\cdot537 \times 21\cdot6$ **4** $0\cdot023 \times 31\cdot41$

5 $0\cdot07 \times 1802$ **6** $2\cdot5 \times 3\cdot7 \times 8\cdot6$ **7** $9\cdot76 \times 4\cdot83 \times 3\cdot62$

8 $23 \times 37 \times 59 \times 21$

In examples 9 to 12 (*a*) calculate the quotient rounded to five significant figures if the numbers are exact, (*b*) if the numbers are rounded, estimate (i) the approximate relative error of the quotient, (ii) the approximate error of the quotient, (iii) approximate limits between which the quotient lies, (iv) state the quotient rounded to as many significant figures as are justified, (v) check by evaluating the exact limits for the quotient rounded to five significant figures so that the lower limit is decreased and the upper limit increased.

9 $\dfrac{21\cdot3}{93\cdot5}$ **10** $\dfrac{3\cdot742}{5\cdot163}$ **11** $\dfrac{2\cdot163}{0\cdot097}$ **12** $\dfrac{9\cdot6}{2\cdot36}$

In examples 13 to 16, evaluate the combined product and quotient if the numbers are   (a) exact,   (b) rounded.   In case (b) a possible error of 1 unit in the least significant figure of the result is acceptable. Give the result in (a) rounded to six significant figures.

13 $\dfrac{21 \cdot 5 \times 32 \cdot 2}{47 \cdot 1}$ **14** $\dfrac{29 \cdot 32 \times 0 \cdot 015}{0 \cdot 23}$ **15** $\dfrac{1721 \times 389}{572}$

16 $\dfrac{2131 \times 0 \cdot 5}{0 \cdot 7}$

Rounding errors in circular functions when the angle in radians is rounded

Using the formula $\Delta y \simeq \dfrac{dy}{dx} . \Delta x,$

when $y = \sin x,$ $\dfrac{dy}{dx} = \cos x$

$\therefore$ $\Delta(\sin x) \simeq (\cos x)\Delta x,$

where Δx is the rounding error in x radians and $\Delta(\sin x)$ is the rounding error in $\sin x$.

Example
19.8 Find (a) the rounding error in $\sin(0 \cdot 12)$ if the angle $0 \cdot 12$ radians is rounded, (b) state the value of $\sin(0 \cdot 12)$.

(a) $\Delta(\sin[0 \cdot 12]) \simeq \cos(0 \cdot 12) \times 0 \cdot 005$
$= 0 \cdot 9928 \times 0 \cdot 005$
$= 0 \cdot 0050$

(b) $\sin(0 \cdot 12) = 0 \cdot 1197$ (from four-figure tables) $\pm 0 \cdot 005$
$= 0 \cdot 1147$ to $0 \cdot 1247$

$\therefore$ $\sin(0 \cdot 12) = 0 \cdot 1$ (or $0 \cdot 12$, with a possible error of 1 unit in the second significant figure).

It may be noted that these results may be checked using the values of $\sin(0 \cdot 115)$, $\sin(0 \cdot 12)$ and $\sin(0 \cdot 125)$, but interpolation is needed with Castle's four-figure tables.

$$\sin(0 \cdot 125) \text{ and } \sin(0 \cdot 115) = \sin(0 \cdot 12) \pm \frac{\sin(0 \cdot 14) - \sin(0 \cdot 12)}{4}$$

$$= 0 \cdot 1197 \pm \frac{0 \cdot 1395 - 0 \cdot 1137}{4}$$

$$= 0 \cdot 1197 \pm 0 \cdot 0050, \text{ as before}$$

A similar procedure is used for the functions $\cos x$ and $\tan x$.

19.9 Find (a) the rounding error in cos(0·54) if the angle 0·54 radians is rounded, (b) state the value of cos(0·54).

(a) Using $\Delta y \simeq \dfrac{dy}{dx}. \Delta x,$

when $y = \cos x,$ $\dfrac{dy}{dx} = -\sin x$

then $\Delta(\cos x) \simeq (-\sin x)\Delta x$

$$\therefore \quad \Delta(\cos[0·54]) \simeq -\sin(0·54) \times 0·005$$
$$= -0·5141 \times 0·005$$
$$= -0·0026$$

(the minus sign indicates that $\cos x$ decreases as x increases in this neighbourhood).

(b) $\cos(0·54) = 0·8577 \mp 0·0026$
$$= 0·8551 \text{ to } 0·8603$$
$$\cos(0·54) = 0·86$$

Using the tables,

$$\cos(0·545) \text{ and } \cos(0·535) = \cos(0·54) \pm \frac{\cos(0·56) - \cos(0·54)}{4}$$

$$= 0·8577 \pm \frac{0·8473 - 0·8577}{4}$$

$$= 0·8577 \mp 0·0026, \text{ as before.}$$

19.10 Find (a) the rounding error in tan(1·02) if the angle 1·02 radians is rounded, (b) state the value of tan(1·02).

(a) Using $\Delta y \simeq \dfrac{dy}{dx}.\Delta x,$

when $y = \tan x,$ $\dfrac{dy}{dx} = \sec^2 x = \dfrac{1}{\cos^2 x}$

$\therefore \quad \Delta(\tan x) \simeq \dfrac{1}{\cos^2 x}. \Delta x$

$\therefore \quad \Delta(\tan[1·02]) \simeq \dfrac{0·005}{\cos^2(1·02)} = \dfrac{0·005}{(0·5234)^2} = \dfrac{0·005}{0·2739} = 0·0183$

(b) $\tan(1·02) = 1·628 \pm 0·018$
$$= 1·646 \text{ to } 1·610$$
$$\therefore \quad \tan(1·02) = 1·6$$

Using the tables,

$$\tan(1 \cdot 025) \text{ and } \tan(1 \cdot 015) = \tan(1 \cdot 02) \pm \frac{\tan(1 \cdot 04) - \tan(1 \cdot 02)}{4}$$

$$= 1 \cdot 628 \pm \frac{1 \cdot 704 - 1 \cdot 628}{4}$$

$$= 1 \cdot 628 \pm 0 \cdot 019$$
$$= 1 \cdot 647 \text{ to } 1 \cdot 609$$

Rounding errors in the exponential functions e^x and e^{-x} when the independent variable x is rounded

Using the formula $\Delta y \simeq \dfrac{\mathrm{d}y}{\mathrm{d}x} \cdot \Delta x$,

when $y = e^x$, $\dfrac{\mathrm{d}y}{\mathrm{d}x} = e^x$

$\therefore$ $\Delta(e^x) \simeq e^x \Delta x$

Also, when $y = e^{-x}$, $\dfrac{\mathrm{d}y}{\mathrm{d}x} = -e^{-x}$

$\therefore$ $\Delta(e^{-x}) = -e^{-x}\Delta x$

Example
19.11 If $x = 0 \cdot 19$, and is rounded, find for $e^{0 \cdot 19}$ and $e^{-0 \cdot 19}$ (*a*) the rounding errors, (*b*) their values.

(*a*)
$$\Delta(e^{0 \cdot 19}) \simeq e^{0 \cdot 19}\Delta x$$
$$= 1 \cdot 2092 \times 0 \cdot 005$$
$$= 0 \cdot 0060$$
$$\Delta(e^{-0 \cdot 19}) \simeq -e^{-0 \cdot 19}\Delta x$$
$$= -0 \cdot 8270 \times 0 \cdot 005$$
$$= -0 \cdot 0041$$

(*b*) $e^{0 \cdot 19} = 1 \cdot 2092 \pm 0 \cdot 0060 = 1 \cdot 2032$ to $1 \cdot 2152$, i.e. $1 \cdot 2$ (or $1 \cdot 21$, with a possible error of 1 unit in the third significant figure).

$e^{-0 \cdot 19} = 0 \cdot 8270 \pm 0 \cdot 0041 = 0 \cdot 8311$ to $0 \cdot 8229$, i.e. $0 \cdot 8$ (or $0 \cdot 83$, with a possible error of 1 unit in the second significant figure).

19.12 (*a*) Find formulae for the rounding error in y due to a rounding error in x when (i) $y = x^2$, (ii) $y = x^3$, (iii) $y = \dfrac{1}{x}$, (iv) $y = x^n$, where n is exact. (*b*) Find the relative error in each case.

(a) (i) $\Delta y = 2x\Delta x$, (ii) $\Delta y = 3x^2\Delta x$, (iii) $\Delta y = -\dfrac{1}{x^2}\Delta x$, (iv) $\Delta y = nx^{n-1}\Delta x$.

(b) (i) relative error $= \dfrac{\Delta y}{y} = \dfrac{2x\Delta x}{x^2} = \dfrac{2\Delta x}{x}$

(ii) relative error $= \dfrac{\Delta y}{y} = \dfrac{3x^2\Delta x}{x^3} = \dfrac{3\Delta x}{x}$

(iii) relative error $= \dfrac{\Delta y}{y} = \dfrac{-1/x^2\Delta x}{1/x} = \dfrac{-\Delta x}{x}$

(iv) relative error $= \dfrac{\Delta y}{y} = \dfrac{nx^{n-1}\Delta x}{x^n} = \dfrac{n\Delta x}{x}$

EXERCISE 19.2

In examples 1 to 6 find (a) the rounding error, (b) the value to the appropriate number of significant figures if the given numbers are rounded.

1 (i) $\sin(0\cdot46)$ (ii) $\cos(0\cdot48)$ (iii) $\tan(0\cdot14)$
2 (i) $\sin(1\cdot32)$ (ii) $\cos(1\cdot44)$ (iii) $\tan(1\cdot04)$
3 (i) $\sin(0\cdot07)$ (ii) $\cos(1\cdot54)$ (iii) $\tan(0\cdot56)$
4 (i) $e^{3\cdot5}$ (ii) $e^{-3\cdot9}$
5 (i) $e^{5\cdot3}$ (ii) $e^{-5\cdot9}$
6 (i) $e^{2\cdot1}$ (ii) $e^{-2\cdot4}$
7 Find the values of

   (a) $(1\cdot213)^2$ (b) $(1\cdot213)^3$ (c) $(1\cdot213)^4$

   if (i) $1\cdot213$ is exact, (ii) $1\cdot213$ is rounded.

8 Find the values of $\dfrac{1}{1\cdot2775}$, $\dfrac{1}{1\cdot2780}$, $\dfrac{1}{1\cdot2785}$, rounded to six significant figures, by division. Estimate the relative error and error in the reciprocal of the rounded number $1\cdot278$, and state its value.

9 (a) Calculate $(3\cdot579)^2$ and $(3\cdot579)^4$ if $3\cdot579$ is an exact number.
   (b) Calculate $(3\cdot579)^2$ if $3\cdot579$ is a rounded number, and use your result to calculate $(3\cdot579)^4$.
   (c) Calculate $(3\cdot5795)^2$ and $(3\cdot5795)^4$, $(3\cdot5785)^2$ and $(3\cdot5795)^4$.

10 (a) Calculate $2\cdot37^3$, $2\cdot36^3$ and $2\cdot37^3 - 2\cdot36^3$.
   (b) Calculate $2\cdot37^3 - 2\cdot36^3$, using $(2\cdot37 - 2\cdot36)(2\cdot37^2 + 2\cdot37 \times 2\cdot36 + 2\cdot36^2)$.
   (c) If $2\cdot37$ and $2\cdot36$ are rounded numbers, calculate the range of $2\cdot37^3 - 2\cdot36^3$, and of $\dfrac{1}{2\cdot37^3 - 2\cdot36^3}$, rounded to four significant figures.

11 (a) If P_1, V_1, P_2, V_2, and n are rounded numbers, calculate the values of P_1V_1, P_2, V_2, $n-1$, and $\dfrac{P_1V_1 - P_2V_2}{n-1}$, where $P_1 = 31$, $V_1 = 2\cdot6$ $P_2 = 15$, $V_2 = 1\cdot7$, and $n = 1\cdot4$.

(b) Calculate $\dfrac{31\cdot5 \times 2\cdot65 - 14\cdot5 \times 1\cdot65}{1\cdot35 - 1}$

and $\dfrac{30\cdot5 \times 2\cdot55 - 15\cdot5 \times 1\cdot75}{1\cdot45 - 1}$

12 Estimate the value of $(x^3 - y^3)/(y^2 - x^2)$, where x and y are the rounded numbers $2\cdot16$ and $1\cdot93$.

13 Evaluate (a) $(16\cdot28 + 7\cdot03)/10\cdot67$, (b) $\sin(0\cdot76)$. Estimate the error in each if all the numbers have been rounded correctly.

Answers

Exercise 2.2
12 (a) $y = x(x - 1)$, $a = \pm 1$ (b) $\frac{3}{4}, \frac{9}{16}, 1$
13 (a) $x = 1, y = 1$ (b) $x = 3, y = 1$ (c) $x = a, y = 2a$

Exercise 2.3
2 (a) $a = -1, b = 1, c = 1$ (b) $x = 5\frac{1}{2}, y = -\frac{5}{6}$
3 $A = 2\cdot5, B = 1\cdot5, C = 0\cdot5$ **4** $a = 3, b = 8, c = -15$
5 $a = 1\cdot62, b = 0\cdot35, c = 0\cdot02$
6 (a) $A = 12, B = 4$ (b) $A = 2, B = 4, C = 5$

Exercise 2.4
10 $x = 1, y = -1$ **11** $x = 2, y = -3$ **12** $x = 4, y = 3$
13 $x = (2b + c)/(b^2 + 1)$, $y = (2 - bc)/(b^2 + 1)$ **14** $i_1 = 0\cdot5, i_2 = 0\cdot25$
15 $i_1 = (RE - re)/(R^2 - r^2)$, $i_1 = (Re - rE)/(R^2 - r^2)$, except when $R = r$ and the matrix is singular

Exercise 3.1
1 $4, 125, \sqrt{8}$ **2** $\frac{1}{4}, \frac{1}{2^5}, 100, 16$ **3** $4, 3\frac{3}{8}, 4$ **4** $\frac{1}{6}, 8, \frac{1}{32}$
5 $5\cdot196, 15\cdot588, 2\cdot828$ **6** $31\cdot62, 0\cdot01, 0\cdot1, 0\cdot3162$ **7** $1\cdot19, 2\cdot38$
8 $0\cdot125, 0\cdot0442$ **9** $a^{5/2}$ or $\sqrt[2]{a^5}$ **10** $1/\sqrt[6]{a}$
11 $4\cdot64$ **12** $\pm 9a^2b, 2a^2b$
21 (a) $x = 9$ (b) $A = 1, B = 1, C = 2$ (c) $x = -2, x = -2$
22 (a) (i) $a^\circ = a^{1-1} = a/a = 1$ (ii) $a^{-b} = a^\circ/a^b = 1/a^b$
(b) $m^{2x}, 1 \pm \sqrt{6}$

Exercise 3.2
1 $q = \log_p r$, $V = \log_u w$, $3 = \log_8 512$, $3 = \log_{100} 1\,000\,000$
$x = \log_e y$, $\mu\theta = \log_e (T_1/T_2)$, $-Rt/L = \log_e (iR/V)$
2 $2^4 = 16$, $3^5 = 243$, $4^5 = 1024$, $a^N = x$, $b^M = y$, $q^R = p$
3 $a^{\log_a N} = N$, $10^{\log_{10} 1000} = 1000$, $2^{\log_2 32} = 32$

Exercise 3.5
1 $0\cdot4831$	**2** $0\cdot0028$	**3** $0\cdot079\,36$	**4** $0\cdot8265$
5 3109	**6** $0\cdot3423$	**7** $0\cdot9330$	**8** $0\cdot7071$
9 $2\cdot174$	**10** $0\cdot2026$	**11** $2\cdot048$	**12** $0\cdot9906$
13 $54\cdot67$	**14** $0\cdot2394$	**15** $0\cdot8247$	**16** $5\cdot081$
17 $4\cdot357$	**18** $5\cdot122$	**19** $0\cdot9913$	**20** $10\cdot43$

21 $80\cdot09$ **22** (a) $318\cdot6$ (b) $2\cdot463$ (c) $1\cdot406$
23 (a) (i) 3 (ii) -3 (b) $6\cdot76$
24 (a) x (b) $\log(\sqrt[3]{(A^2B)}/C)$ (c) $17\cdot1$ **25** $6\cdot152$
26 (a) (i) a^4b^4 (ii) $2\sqrt{b}/a^2$ (b) (i) $\sqrt{3} + \sqrt{2}, 3\cdot146$ (ii) $4\sqrt{2} + 5, 10\cdot656$
27 (a) $0\cdot0559$ (b) $0\cdot2206$ (c) $5\cdot15$ km
28 (a) $3\cdot499$ (b) $(\sqrt[3]{b^2}/a^3)(b + 1/\sqrt[3]{a^2})$
29 (a) $1\cdot5$ (b) $-1\cdot04$ **30** (a) $47\cdot65$ (b) 160 m
31 (a) (i) $64\cdot19$ (ii) $0\cdot5763$ (b) $2\cdot512$
32 (a) $0\cdot9646$ (b) $c = k^2(f - p)/pl^2$ **33** (a) $5\cdot76$ (b) 3719

34 (a) $10^{-0.42}$, $10^{0.14}$ (b) 1.177 (c) 0.7886
35 $T_1 = 296$ N, $T_2 = 136$ N, T_1/T_2 is squared   **36** $x = \log_{10} 2$ or 0
37 $x = 1$ or -1 **38** (a) $x = 1$ (double root)   (b) $x = 0$ or 1
39 (a) (i) $0.389\,08$ (ii) $0.176\,09$ (iii) $0.698\,97$ (b) 1

Exercise 4.1

2 $A = (L/R) \log_e E$
5 (a) $x = 2$ or $3\frac{1}{8}$ (b) $\log_e i = \log_e a + Kt$ (c) $t = D^a \cdot 10^b$
6 (a) $(100/3) \log_e 2 = 23.10$ (b) $P/P_0 = e^{3nt1/100} = (e^{3t1/100})^n = 2^n$
   (c) $x = 1.13$ or 8.52

Exercise 4.2

1 (a) 3 (b) 3 (c) 6 (d) 3 (e) 2.5 (f) 0.5 **2** (a) 2.431 (b) 2.460
3 (a) 1.5261 (b) 2.0149 (c) 2.2618 (d) 3.688 (e) 4.024
   (f) $\bar{3}.2195$ (or -2.7805)
5 (a) 4.015 (b) 6.424 (c) 9.07 (d) 0.4299 **6** 0.9163
7 (a) 54.7 (b) (i) x (ii) x^2 (iii) $a/\log_e a$
   (c) (i) 0.403 (ii) 15.28 (iii) 233.4
8 (a) 4 (b) 1.861 (c) 4.1431 **9** (a) 0.6627 (b) 1.549 **10** 6.77 bar
11 (a) (i) 2.29 (ii) $\bar{1}.206$ (b) 0.2965 (c) 3.573
12 14.9, 6.64 **13** 1.7148, additional time $= 28\frac{1}{2}$ min   **14** 598m
16 0.827 **17** 76 **18** (a) -5 (b) 0.8489 **19** (a) 2.72 (b) 0.250
20 (a) 0.2721 (b) 988 (c) 0.6352 **21** (a) 0.1030 (b) $2, -4$

Exercise 5.2

3 (a) 4 (b) 3.96 **4** (a) 20 (b) 20.9

Exercise 5.3

1 142, 142.2 **2** 1.968, 1.967 **3** $183\,000$, $182\,800$
4 1.519, 1.518 **5** 2.985, 2.986 **6** 0.457, 0.4564

Exercise 5.4

1 1.231 **2** 1.377 **3** 346 **4** 1635 **5** 735
6 19.6 **7** 0.0209 **8** 1.192 **9** 0.555 **10** 1472
11 1.705 **12** $50\,800$

Exercise 6.1

1 $x = 6, y = 5$; $x = -5, y = -6$
2 $x = 5, y = 4$; $x = -5, y = -11$
3 $x = -5, y = 4$; $x = 5, y = -4$
4 $x = 3, y = -2$; $x = 2, y = -1$
5 $x = -\frac{12}{5}, y = \frac{17}{5}$; $x = 3, y = -2$
6 $x = \frac{4}{3}, y = \frac{34}{9}$; $x = 1, y = 4$ **7** $x = 6, y = 2$; $x = -4, y = -3$
8 $x = 4, y = 1$; $x = -1, y = -4$
9 $x = 8, y = 5$; $x = -8, y = -5$; $x = 5, y = 8$; $x = -5, y = -8$
10 $x = -\frac{10}{9}, y = -\frac{33}{2}$; $x = 3, y = 2$
11 $x = 3, y = -\frac{1}{3}$; $x = 5, y = -1$
12 $x = \frac{1}{2}, y = \frac{3}{4}$; $x = 1\frac{1}{2}, y = -\frac{1}{4}$
13 (a) -2.06, 0.728 (b) $-3, 2$ (c) $5\frac{1}{2}, -\frac{5}{6}$
14 (a) -0.162, -1.439 (b) -0.543, -2.458
15 (a) $y^2/(x + 2y)$ (b) x^2/y (c) $x = -1, -4$; $y = -4, -1$

16 (a) 8·6, 3·4 (approx.) (b) $p = 11$, $x = \frac{4}{3}, -\frac{3}{2}$
17 (a) $1\frac{6}{13}$ (b) $A = \{(E - 100d)b/100c\}^2$ (c) $x = \frac{1}{3}, y = -\frac{1}{4}$
18 1·65. $-3·65$
19 (a) $-\frac{1}{3}, 4$ (b) $d = \pm C. A. B/\sqrt{(2w^2r^2 - A^2B^2)}$ (c) $x = 3, y = -4$

Exercise 6.2
1 $-2, -1, 0·67$ **2** $79° 24'$ **3** (a) 1 (b) $5, -3$
4 (a) $-2·73, 0·73, 2$ (b) $0·75, 1·97$
5 $0·34, -3·05, 0·52, 2·55$ **6** $-2·62, 0·55, 2·15$
7 $-1·66, -0·55, 2·22; -2·4, -0·52, 2·67$ **8** $(0·24, 1·98); (-1·44, -1·38)$
9 $-0·732, 1·0, 2·732$ **10** (a) $3, -1·75$ **11** 1·83

Exercise 6.3
1 (a) (i) -168 (ii) 0 (b) $(x - 1)(x + 1)(x - 2)$
2 $(x - a)(x - b)(x - c)$ **3** $(t + p)(t + r)(t - q)$
4 $(y - 1)(y + 1)(y - 2)(y + 3)$
6 $(x + y - 2)(x^2 + y^2 - xy + 2x + 2y + 4)$ **7** $(a - b)(b - c)(c - a)$
8 (a) $(x + j)(x - j)$ (b) $(x + \frac{1}{2} + \sqrt{3}j/2)(x + \frac{1}{2} - \sqrt{3}j/2)$
9 $-7, (x + 5)(x - 2)(x + 1)$
10 $A = -7, B = 6, (x - 2)(x - 1)(x + 1)(x + 3)$
11 $11, (x + 1)(x + 2)(x + 3)$

Exercise 7.1
1 (a) $-4·5, 1·67$, min. $= 57$ (b) $-3·9, 1·06$
2 $-4·5, 1·67$ **3** 8·7 when $x = 2·4$ **4** Min. 48·5, when $M = 2·32$ **5** 0·32
6 $0, 3, -4$; max. 41·5, min. $-25·2$ **7** 2·8   **8** 3·532   **9** $-1·4, 0·65, 2·75$
10 79·2 gives max. of 162·5   **11** £$(2c^2 + 13\,500/c)$, 15 knots
12 $-0·57, 0·75, 2·3$ **13** $-2·7, 0·55, 2·15$ **14** $-1·4, 0·8, 3·6$
15 (a) $-1·7, 2·5, 5·2$, (b) $-1·4, 1·9, 5·5$ **16** $-3·72, -2·68, 1·40$

Exercise 7.2
1 $0·135, 0·86$ **2** $0·17, 0·88$ **3** (a) 0·53 (b) 0·67 (c) 0·42
4 (a) 0·72 (b) 0·21   **5** $0·11, 0·82$ **6** $0·15, 0·75$ **7** 0·87

Exercise 8.1
1 $a = 2·3, b = -10$ **2** $L = 2540 - 2·91\theta$, 2250 kJ
3 Missing values of (a) $x \ldots 0·6$ (b) $y \ldots -8, -3, 4, 22$
 (The law is $y = 5x - 3$)
4 $E = 0·45W$ **5** $m = 5·3, c = 10·5$
6 Missing values of (a) $x \ldots 0, 1·2$ (b) $y \ldots 8, 1·6$ $(a = 8, b = 2)$
7 $a = 0·42, b = 7·5$ **8** $a = 40, b = 0·03$
9 $a = 0·3, b = 0·01$ **10** $P = 1·08v^3 + 420$
11 $a = 0·0424, b = 0·15$ **12** $a = 500, n = 0·28$
13 $a = 5·7, n = 0·36$ **14** $a = 16, \mu = 0·3$

Exercise 8.2
1 $-2·7, 2·3, y_{max} = 5·6$ **2** $30°, 90°, 150°, 270°$
3 $a = 1·4, b = 1·6$ **4** $a = 5·7, n = 0·36$
5 $a = 6·07, b = 0·925$ **6** $y_{max} = 2\frac{1}{3}$
7 $a = 90, b = 19·2$ **8** $a = 41·4, b = 11·4$
9 $0·41, 3·14$ **10** $A = 10, \mu = 0·138$

11 $a = 0.077$, $b = 0.033$ **12** $C = 46$, $k = 0.086$
13 $v = 20$ knots   **14** 0, 6; $y_{min} = -4\frac{1}{2}$ when $x = 3$
15 $a = 2.3$, $b = 8.2$ **16** 0.37, 4.44
17 2.31 radians, 4.14 length units
18 $y_{min} = 4\sqrt{10}$, for sides of square $\sqrt{10}$
19 (a) (i) 0.8375   (ii) 1.193s   (b) (i) 0.9969   (ii) 0.005 255
20 $m = 0.92$, $b = -23.4$ **22** 1.4
23 $a = 100$, $b = 0.5$ **24** $n = 1.8$, $a = 0.069$
25 $a = 420$, $b = -4.5$ **26** $a = 4.6 \times 10^{-7}$, $n = 4.1$
27 $a = 20$, $b = -0.006$ (approx)

Exercise 9.1

1 -15 **2** $-25/256$ **3** 3.48   **4** 0.000 109 3
5 $5n - 3$ **6** 20   **7** 3   **8** 0.6277
9 0.1   **10** 60, 48, 38.4   **11** 11th   **12** 19.92
13 $-6, 0, 6, \ldots, 18, 24$ **14** 216.8
15 $(3x + y)/4$, $(x + y)/2$, $(x + 3y)/4$ **16** -4.918 **17** -5
18 $10\frac{2}{3}$ **19** 192   **20** $\frac{5}{9}, \frac{4}{11}, 3\frac{5}{16}$ **21** 11
22 20 m   **23** £13.75   **24** £2700, 11 yr **25** £103.20, £2.08
26 (a) The first clerk, £480 more   (b) The second clerk, £540 more
27 19 rows and 10 over   **28** $19\frac{1}{2}$ km/h   **29** 973
30 120, 1860   **31** £75 000   **32** 208.4 m

Exercise 9.2

1 (a) 14 or 15   (b) 1.412, 1.5   **2** (a) 10   (b) $5\frac{1}{3}$, $42\frac{2}{3}$
3 (a) 12 or 16   (b) 0.9, 0.6561   **4** (a) 12   (b) (i) 4.58 m   (ii) 6 m
5 (a) 45, 780   (b) 1.779, 63.24   **6** (a) -195 (b) £721
7 (a) 16 or 25, 5 or -4 (b) 0.003 311, 2
8 £10 000, £4098   **9** (a) 37   (b) 29, -2, -29
10 (a) 7   (b) 342   **11** (a) 2.044, 23.44   (b) 1.8803
12 $7\frac{1}{2}$ h, $42\frac{2}{3}$ m   **13** $\frac{2}{3}$, 6, 18   **14** 175 mm and 520 mm
15 Nothing, £387.20

Exercise 10.1

3 (a) 1.004 21   (b) $x^5 - 20x^3 + 160x - 640/x + 1280/x^3 - 1024/x^5$
    (c) $10\ 206/x$
4 (a) 1.769   (b) (i) $3^8 x^8 - 8 \times 3^7 x^6 + 28 \times 3^6 x^4 - 56 \times 3^5 x^2 + 70 \times 3^4$
    (ii) 0.984 10   (iii) 63/8
5 $11\frac{1}{4}$
6 (a) $2(x^4 + 30x^2 + 25)$ (b) -20 (c) (i) 0.786 mm² (ii) 4.91 mm³
7 (a) 0.02   (b) (i) 1.2155   (ii) 0.994 01
8 $a^6 + 6a^5 x + 15a^4 x^2 + 20a^3 x^2 + 15a^2 x^4 + 6ax^5 + x^6$
9 0.960 60

Exercise 10.2

1 $x^5 - 15x^4 + 90x^3 - 270x^2 + 405x - 243$
2 $81x^4 + 216x^3 y + 216x^2 y^2 + 96xy^3 + 16y^4$
3 $x^6 - 12x^4 + 60x^2 - 160 + 240/x^2 - 192/x^4 + 64/x^6$
4 $1 - 7xy + 21x^2 y^2 - 35x^3 y^3 + 35x^4 y^4 - 21x^5 y^5 + 7x^6 y^6 - x^7 y^7$
5 $a^8 - 8a^6 c + 24a^4 c^2 - 32a^2 c^3 + 16c^4$
6 $8x^3 - 36x^2 y + 54xy^2 - 27y^3$

7 $1 - 10/x + 45/x^2 - 120/x^3 + 210/x^4 - 252/x^5 + 210/x^6 - 120/x^7 +$
$$45/x^8 - 10/x^9 + 1/x^{10}$$
8 $1/256 + a/16 + 7a^2/16 + 7a^3/4 + 35a^4/8 + 7a^5 + 7a^6 + 4a^7 + a^8$
9 $1/x^2 - 2\triangle x/x^3 + 3(\triangle x)^2/x^4 \ldots$ **10** $1/m^4 - 4n/m^5 + 10n^2/m^6 \ldots$
11 $1 \cdot 012$ **12** $1 \cdot 003$ **13** $1 \cdot 000\,036$ **14** $945a^3b^4/16$
15 $-15\,120x^4y^3$ **17** (a) $1 \cdot 219$ (b) 26, $24\frac{1}{2}$, 23, etc.
18 (a) $x^{11} + 22x^{10} + 220x^9 + 1320x^8 \ldots$ (b) 4592 m^2
19 (a) 13 (b) 4096 **20** $1 + 16n + 96n^2 + 256n^3 + 256n^4$
21 (a) $1 + 14x + 84x^2 + 280x^3 + 560x^4$
(b) $1 - x - x^2/2 - x^3/2 - 5x^4/8$
22 (a) 8, common ratio $\frac{1}{2}$, $15\frac{15}{16}$, 16
23 (a) $1 + 6x + 15x^2 + 20x^3 + 15x^4 + 6x^5 + x^6$,
$1 - 6x + 15x^2 - 20x^3 + 15x^4 - 6x^5 + x^6$
(b) $1 + x - x^2 \ldots$; $1 + x + \frac{3}{2}x^2 \ldots$
25 (a) $a^8 + 8a^7x + 28a^6x^2 + 56a^5x^3 + 70a^4x^4 + 56a^3x^5 + 28a^2x^6 + 8ax^7 + x^8$
(b) $x = 1\frac{1}{7}$

Exercise 11.2

1 (a) $0 \cdot 9781$, $-0 \cdot 2079$, $-4 \cdot 7046$ (b) $-0 \cdot 8873$, $-0 \cdot 4612$, $1 \cdot 9237$
(c) $-0 \cdot 730$, $0 \cdot 7112$, $-0 \cdot 9884$ (d) $0, 1, 0$ (e) $1, 0, \infty$
(f) $-0 \cdot 0610$, $0 \cdot 9981$, $-0 \cdot 0612$ (g) $-0 \cdot 7771$, $0 \cdot 6293$, $-1 \cdot 2349$
(h) $0 \cdot 866$, $0 \cdot 5000$, $1 \cdot 7321$ (i) $0 \cdot 9703$, $-0 \cdot 2419$, $-4 \cdot 0108$
(j) $-0 \cdot 3420$, $-0 \cdot 9397$, $0 \cdot 3640$ (k) $0 \cdot 5068$, $-0 \cdot 8621$, $-0 \cdot 5879$
(l) $-0 \cdot 3619$, $-0 \cdot 9322$, $0 \cdot 3882$ (m) $-0 \cdot 2901$, $0 \cdot 9570$, $-0 \cdot 3032$
2 (a) $2 \cdot 281$ (b) $-1 \cdot 744$ (c) $-1 \cdot 235$ (d) $-1 \cdot 058$ (e) 2
(f) $-3 \cdot 271$
3 $-0 \cdot 9165$, $0 \cdot 4365$ **4** $94° 8'$, $265° 52'$ **5** $28° 41'$, $151° 19'$
6 $\sec \theta = \mp \frac{13}{12}$, $\cot \theta = \pm \frac{12}{5}$ **7** $0 \cdot 9798$, $0 \cdot 2041$ **8** $-1 \cdot 732$ **9** $\frac{5}{13}$
10 $1/\sqrt{(1 - m^2)}$, $\sqrt{(1 - m^2)}/m$ **11** $-4 \cdot 428$ **12** 0 **13** $0 \cdot 966$
14 $0 \cdot 2588$ **15** $0 \cdot 9511$ **25** $26°$, $110°$ **26** $28°$, $120°$
27 (a) (i) $50° 6'$ or $309° 54'$ (ii) $84°$ or $264°$
(b) (i) $50° 6' + n_1 360°$ or $309° 54' + n_2 360°$
(ii) $84° + n_1 360°$ or $264° + n_2 360°$
28 $10° 6'$, $109° 54'$, $130° 6'$ **29** $11° 53'$, $43° 8'$, $-46° 53'$, $-78° 8'$
30 $30°$, $150°$, $194° 29'$, $345° 31'$
31 $84° 16'$, $113° 35'$, $246° 25'$, $275° 44'$ **32** $0°$, $120°$, $240°$, $360°$
33 (a) $89° 12'$, $269° 12'$ (b) $44° 36'$, $134° 36'$, $224° 36'$, $314° 36'$

Exercise 12.1

1 $0 \cdot 94$, $0 \cdot 34$ **2** $-0 \cdot 93$
3 $-0 \cdot 8545$ **4** $0 \cdot 8945$, -2
5 $3 \cdot 0777$, $0 \cdot 5407$ **6** $0 \cdot 3919$
7 $0 \cdot 477$ **8** $0 \cdot 96$, $0 \cdot 28$, $3 \cdot 428$
9 $0 \cdot 9917$, $-0 \cdot 1288$ **10** $0 \cdot 6151$, $0 \cdot 3253$
11 $4 \cdot 5 \sin (6t + 0 \cdot 927)$ **12** $39 \cdot 04 \sin (2\pi nt + 0 \cdot 6946)$
13 $5 \cdot 83 \sin (4t - 0 \cdot 54)$ **32** $0°$, $180°$, $80° 25'$, $279° 35'$
33 $26° 34'$, $45°$, $206° 34'$, $225°$ **34** $0°$, $90°$, $180°$
35 $60°$, $120°$, $240°$, $300°$ **36** $43° 52'$, $136° 8'$
37 $60°$, $300°$
38 $26° 35'$, $153° 25'$, $206° 35'$, $333° 25'$
39 $45°$, $135°$, $225°$, $315°$ **40** $60°$, $120°$, $240°$, $300°$

41 $10°\,33'$, $259°\,27'$

42 $30°$, $150°$, $210°$, $330°$

43 $78°\,28'$, $281°\,32'$

44 $45°$, $135°$, $225°$, $315°$

45 $0°$, $120°$, $180°$, $240°$

46 $0°$, $60°$, $180°$, $300°$

47 $0°$, $120°$, $240°$

48 $180°$

49 (a) $\frac{1}{2}(\sin 16y + \sin 4y)$ (b) $\sin 8\theta + \sin 4\theta$
(c) $\frac{7}{2}(\sin 14y + \sin 6y)$

50 (a) $\frac{1}{2}(\sin 16y - \sin 4y)$ (b) $\sin 8z - \sin 2z$
(c) $\frac{5}{2}(\sin 14\theta - \sin 4\theta)$

51 (a) $-\frac{1}{2}(\cos 12\phi - \cos 4\phi)$ (b) $-(\cos 8\theta - \cos 6\theta)$
(c) $-\frac{3}{2}(\cos 18\phi - \cos 4\phi)$

52 (a) $\frac{1}{2}(\cos 14x + \cos 2x)$ (b) $\cos 4\theta + \cos 2\theta$
(c) $4(\cos 8z + \cos 2z)$

53 (a) $2 \sin 3\theta \cos \theta$ (b) $2 \sin 4\phi \cos \phi$
(c) $2 \sin 7y \cos y$

54 (a) $2 \cos 3\theta \sin \theta$ (b) $2 \cos 4\phi \sin \phi$
(c) $2 \cos 7y \sin y$

55 (a) $2 \cos 3\theta \cos \theta$ (b) $2 \cos 4\phi \cos \phi$
(c) $2 \cos 7y \cos y$

56 (a) $-2 \sin 3\theta \sin \theta$ (b) $-2 \sin 4\phi \sin \phi$
(c) $-2 \sin 7y \sin y$

Exercise 12.2

2 $26°\,45'$, $3\cdot345$m

3 (b) $0°$, $60°$, $180°$, $300°$ (c) 7270 mm², $104°$

4 (a) $35°\,18'$, $144°\,42'$, $215°\,18'$, $324°\,42'$ (b) $326°\,18'$ (c) $62°\,18'$

5 (a) $5\cdot96$ m   (b) 272 m²   (c) $6\cdot9$ m (approx)   (d) 11 m (approx)

6 $60°$, $131°\,48'$, $228°\,12'$, $300°$; $61\cdot4$ m, $87°\,35'$, $55°\,25'$, 2578 mm²

7 (a) (i) $y = 3x + 5$ (ii) $-\frac{2}{3}x + \frac{1}{3}$ (b) $-3\frac{2}{3}$ (c) $-\frac{14}{11}$, $\frac{13}{11}$
(d) $1\cdot736$, $138°\,7'$

8 $\angle B = 61°\,19'$, $\angle C = 65°\,41'$, $b = 11\cdot53$ m; or
$\angle B = 12°\,41'$, $\angle C = 114°\,19'$, $b = 2\cdot89$ m

9 (a) $1°\,27'$, $37°\,25'$ (b) $192°\,58'$, $347°\,2'$

10 (b) $6\cdot44$ m   **11** $4\cdot27$, $74\frac{1}{2}°$

13 (a) $6\cdot75$ m   (b) $10\cdot5$ m (approx)   (c) $16\cdot62$ m   (d) $607\cdot5$ m²

15 $37 \cos(\theta + 18°\,56')$ **16** (a) $2\cdot49$ m³   (b) $48°\,13'$

17 (a) (i) $0°$, $180°$, $360°$ (ii) $63°\,26'$, $135°$, $243°\,26'$, $315°$ (c) 176 mm

18 (a) $\frac{-3}{5}$, $\frac{119}{169}$, $-41\sqrt{10}/130$ (b) $30°$, $150°$, $228°\,35'$, $311°\,25'$

19 $28°\,11'$, $121°\,49'$, $105\cdot9$ mm, 4500 mm²

20 (b) $0°$, $70°\,32'$, $180°$, $289°\,28'$, $360°$
(c) $0\cdot500$, $-0\cdot9511$, $-0\cdot5000$, $8\cdot144$, $-0\cdot2867$

21 (a) $5°\,59'$ (b) $2\cdot388$ m   (c) $3\cdot591$ m   **22** $6\cdot072$ m, $12\cdot57$ m/s

23 $DB = 2\cdot935$ m, $CD = 2\cdot46$ m, $DE = 1\cdot89$ m

24 (a) 625 mm   (b) $12°\,50'$

25 (a) $12\cdot43$ m   (b) $33\cdot04$ N at $36°\,40'$ E of N

26 (a) $\frac{1}{2}$, $\sqrt{3}/2$ (b) $\frac{24}{25}$, $\frac{7}{25}$, $\frac{24}{7}$

27 (a) (i) $210°$, $330°$ (ii) $0°$, $45°$, $135°$, $180°$, $225°$, $215°$, $360°$ (b) $0\cdot1481$

28 (a) $c = 3\cdot84$ m, $\angle A = 87°\,36'$, $\angle B = 42°\,14'$ (b) $2\cdot5$ m

29 (a) $30°$, $90°$, $150°$; $20°$, $100°$, 140 (b) $0°$, $14°\,2'$, $180°$, $194°\,2'$, $360°$

30 (a) (i) $\frac{323}{325}$ (ii) $\frac{-323}{36}$ (b) $56°\,44'$, $236°\,44'$

31 (b) $-1\cdot1326$, $1\cdot1918$ **32** $9\cdot93$ knots   **33** (c) $R = 1\cdot12$, $\theta = 63°\,26'$

34 $104°\,24'$, $330°\,20'$ **35** (a) $\frac{120}{169}$, $\frac{-7}{25}$, $\frac{33}{65}$ (b) $30°$, $150°$, $228°\,35'$, $311°\,25'$

36 14° 48′, 23·11 mm, 12·73 mm   **37** (a) 0·548   (b) 3460 mm², 7070 mm²
38 1·7 m   **39** (a) 4·65, 304°   (b) 5·8, 156° 22′
40 (b) 196° 18′, 343° 42′;
   (c) (i) $\frac{1}{50}$ s   (ii) 25   (iii) 50 Hz, $i = 12·5$
41 2·81 m   **42** 291 mm, 626 400 mm²   **43** 150°, 210°
44 (a) $\frac{3}{4}$ (b) 46° 27′
45 (a) $\frac{63}{65}$ (ii) $\frac{56}{65}$ (b) 19° 28′, 270°
46 (a) 4·7 m, 40° 38′   (b) 7·34 m²
47 (a) 8·09 km   (b) 40·1 newtons, 17° 50′
48 (a) $\frac{-120}{169}, \frac{119}{169}, \frac{-120}{119}$ (b) 25, 16° 16′; 69° 24′, 143° 8′
49 (a) $-78° 14′, 45° 42′$ (b) 0·4084, $\pi/2$, 4   **50** (a) 120°
51 $a = 127·8$ mm, $\angle B = 33° 35′$, $\angle C = 41° 25′$, 3090 mm²
52 (a) 0·574, 0·911; $2\pi/5$, 3   (b) $\pm 81° 14′$; (c) $\frac{840}{1369}, \frac{1508}{1517}$
53 (a) 320 mm (approx)   (b) 81 000 mm² (approx)
54 (b) 2, 4   (c) 54° 43′
55 (a) 91° 16′   (b) $A = 168, \alpha = 0·464$
   (c) $a/\sqrt{(a^2 + b^2)}, b/\sqrt{(a^2 + b^2)}, 2ab/(a^2 + b^2), (b^2 - a^2)/(a^2 + b^2)$
   (d) 0°, 146° 26′, 180°, 213° 34′, 360°
56 (a) 110° 27′, 326° 53′
   (b) $a = 134·6$ mm, $\angle B = 49° 13′$, $\angle C = 76° 47′$, 8256 mm²
57 (b) 1·278 m   **58** (a) 642 mm   (b) 125° 41′, $\arctan \frac{1}{4}$
59 $c = 40·3$ mm, $\angle A = 38° 20′$, $\angle B = 111° 30′$, 80·7 mm
60 (a) (i) $\frac{56}{65}$ (ii) $\frac{33}{56}$ (b) $a = 7·0, b = 7·2$ (c) 10, 6·97, 90   (d) $\pm 77°$

Exercise 13.1
1 (a) (i) l (ii) t (iii) r (b) (i) A (ii) s (iii) V
2 (a) $3a^2 + 2a + 5$ (b) $3b^2 + 20b + 38$ (c) 21   (d) 5
3 (a) $5c + 7$ (b) 7   (c) 2   **4** (a) 2   (b) 0   (c) 0
6 1, −3   **8** $a = 2\frac{1}{2}, b = -\frac{1}{2}, 1\frac{1}{2}$
9 $(a/b) = 1, c = -2$ **10** $a = 2, b = 1, c = 1$
11 (b) $x^2 - 5x + 6$; 2, 3
12 (a) $2x^2 + 4xh + 2h^2 - 1$ (b) $4x + 2h$ (c) $2h^2 - 1$
   (d) $3 + 2h$ (e) −1, 2

Exercise 13.2
1 (a) indeterminate   (b) $\frac{1}{4}$
2 (a) a, a (b) $2ax + ah + b, 2ax + b$ (c) $3x^2 + 3xh + h^2, 3x^2$

Exercise 13.3
3 (a) $\mathrm{d}y/\mathrm{d}x = 2ax$ (b) $\mathrm{d}s/\mathrm{d}t = -5t^4$ (c) $\mathrm{d}v/\mathrm{d}u = 2u - 1$

Exercise 13.4
1 $1 + x/1! + x^2/2! + x^3/3! + x^4/4!$
2 $-1 + t - t^2/2 + t^3/3 - t^4/4$
5 (a) $9u^2 + 4u$ (b) $y = (x + 1)^2, \mathrm{d}y/\mathrm{d}x = 2(x + 1)$
   (c) $x^5/5 - x^4/4 + x^3/3$

Exercise 13.5
1 (a) (i) $10x^4$ (ii) $3/2\sqrt{x}$ (iii) $-5/2x^{3/2}$
   (b) 51, 36; $x = 1/\sqrt{2}, y = 6 - \sqrt{2}$; $x = -1/\sqrt{2}, y = 6 + \sqrt{2}$
2 (a) $2at$ (b) (i) 8 rad/s   (ii) 78 rad

3 (a) $-1/x^2$ (b) $3x^2 - 6x$, $6x - 6$, 6, 0 **4** (b) 12
5 (a) $6x^2 - 3$ (b) $2t + 3$ (c) $2x + 1/\sqrt{x}$ (d) $1 \cdot 2u^{0 \cdot 2} - 1 \cdot 25/u^{3 \cdot 5}$
    (e) $-6/x^3 + 1/2\sqrt{x} + 0 \cdot 252x^{0 \cdot 26}$ (f) $8x^{-1/3}/3 - 6x^{-2}$
6 Each differentiation has the effect of removing the term of highest degree.
Finally $d^6z/dt^6 = 0$

Exercise 13.6

1 (a) 2 (b) $10t - 2$ (c) $12t^2 + 2t$ (d) $-3/x^4$
2 (a) $20x^3 + 3$ (b) $18t^2 - 6t + 1$ (c) $3/2\sqrt{x}$ (d) r
    (e) $4x/3 - 5/6 + 2/x^2 - 8/x^3$ (f) $1/2\sqrt{x} - 1/2\sqrt{x^3}$
    (g) $2 \cdot 76x^{0 \cdot 2} - 1/\sqrt{x} - 7/x^2$ (h) $10 - 6t$ (i) $0 \cdot 405x^{0 \cdot 35}$
3 (a) $6u + 1$ (b) $1 \cdot 2x^{0 \cdot 2} + 0 \cdot 2/x^{1 \cdot 2} + 3x^{0 \cdot 5}$ (c) $-560/v^{2 \cdot 4}$
    (d) $0 \cdot 1125v^{0 \cdot 5}$ (e) $-0 \cdot 001/h^2$
4 $54t^2 + 6t + 14$ **5** $(12 - 3x)/x^3$ **6** $24t - 18$
7 $22 \cdot 5$ mm²/s   **8** 347×10^3 mm³/s
9 50 mm   **10** $+6, -8$ **11** $-2, +1$ **12** $7, 0 \cdot 25$
13 (a) $\frac{2}{3}$ or -1 (b) 0 or $-\frac{1}{3}$ **14** $16 \cdot 97$
15 (a) $3 \cdot 5$ (b) $2 \cdot 5$ (c) $1 \cdot 75$ **16** 23
17 (a) $4x - 3$ (b) $10x/3 - \frac{3}{4} - 3/x^2 + 4/3x^3$
    (c) (i) -1 (ii) $\frac{3}{4}$ (iii) 2

Exercise 13.7

1 (a) $100 - 8t$ m/s (b) 68 m/s (c) acceleration $= -8$ m/s²
2 (a) 41 m/s (b) 6 m/s² (c) $t = 4$ seconds (d) $t = 8 \cdot 2$ seconds
3 55 m/s, 12 m/s²
4 (a) 57 m/s (b) $8 \cdot 89$ seconds (c) 14 m/s² (d) $4\frac{1}{3}$ seconds
5 (a) 35 m/s (b) 3 m/s² (c) 1 or 5 seconds (d) $6 \cdot 42$ seconds
    (e) 3 seconds
6 (a) $t = 1$ or 2 seconds (b) $t = 1\frac{1}{2}$ seconds (c) $-1\frac{1}{2}$ m/s (d) ± 6 m/s²
7 (a) $2x^{-1/3} - \frac{1}{2}x^{-3/2} + 6x^{0 \cdot 2}$ (b) $\dot{\theta} = 28 - 6t$, $\ddot{\theta} = -6$, $t = 4\frac{2}{3}$ seconds,
    $\theta = 65\frac{1}{3}$ rad
8 (a) (i) $2x + 3$ (ii) $-1/x^2$ (b) $\dot{P} = 12s^2 - 2$ (i) $s = \pm\frac{1}{6}$ (ii) $s = \pm\sqrt{3}$
    (c) $t = 5$ seconds
9 (a) (i) $3x^3 - 10x^2 - 8x$ (ii) $20x^9 + 48/5x^5 - 35x^{3/2}/2$
    (iii) $3 + 4/x^2 - 10/x^3$
    (b) (i) $t = \frac{1}{2}$ or 2 seconds (ii) $9\frac{3}{4}$ m/s (iii) $8\frac{7}{16}$ m
10 (a) $\dot{s} = 3 - 2t + 3t^2$, $\ddot{s} = -2 + 6t$, $t = \frac{1}{3}$ seconds
    (b) (i) $12x^2 - 7$ (ii) $2x - 2/x^3$

Exercise 13.8

1 (a) $x(2\cos x - x\sin x)$ (b) $3(\sin x + x\cos x)$
    (c) $1 \cdot 2\theta^{0 \cdot 2}\cos\theta - \theta^{1 \cdot 2}\sin\theta$
2 $x(2 - 5x^3)$ **3** $\sin x + x\cos x$ **5** $(x\cos x - 5\sin x)/x^6$

Exercise 13.9

1 (a) $-12(2 - 3x)^3$ (b) $-3\cos(2 - 3x)$ (c) $3\sin(2 - 3x)$
2 $5\cos(5x + 3)$, $-25\sin(5x + 3)$
3 (a) $2\cos 2x$ (b) $2\cos^2 x - 2\sin^2 x \equiv 2\cos 2x$
4 (a) $6\theta(1 + \theta^2)^2$ (b) $2\theta\cos(1 + \theta^2)$ (c) $-2\theta\sin(1 + \theta^2)$
5 $-3\sin 3x - 5\sin 5x + 2\cos 2x + 4\cos 4x$

6 (*b*) (i) $-6/(1 + 2x)^4$ (ii) $\cos x \cos ax - a \sin x \sin ax$
7 (i) $6x - 8/x^3$ (ii) $15(3x + 2)^4$ (iii) $3x(2 \sin x + x \cos x)$

Exercise 13.10
1 (*a*) $0 \cdot 031$, $0 \cdot 031 \, 024 \, 008 \, 001$ (*b*) $0 \cdot 005$, $0 \cdot 005 \, 006 \, 004 \, 001$
2 $0 \cdot 002$, $0 \cdot 002 \, 003$ **3** (*a*) $\delta y \simeq -\sin x \delta x$ (*b*) $-0 \cdot 009 \, 490$
   (*c*) $0 \cdot 3058$
4 (*a*) $3x^2 \delta x + 3x \delta x^2 + \delta x^3$ (*b*) $3x^2 \delta x$ (*c*) $3x \delta x^2 + \delta x^3$
   (*d*) $x(3x + \delta x)/(3x^2 + 3x \delta x + \delta x^2)$ or $1 - 3/(3 + 3 \delta x/x + (\delta x/x)^2)$
   (*e*) (i) $0 \cdot 0999\%$ (ii) 100%

Exercise 14.1
(arbitrary constant of integration understood in this and following exercises)
1 (*a*) $x^8/8$ (*b*) $x^6/6$ (*c*) $x^4/4$ (*d*) $-1/2x^2$ (*e*) $-1/4x^4$ (*f*) $-1/6x^6$
2 (*a*) $z^{4 \cdot 5}/4 \cdot 5$ (*b*) $2u^{3/2}/3$ (*c*) $3u^{2/3}/2$ (*d*) $-1/2 \cdot 5y^{2 \cdot 5}$
3 (*a*) $4x^9/9$ (*b*) $-3x^7/7$ (*c*) $5/3x^3$ (*d*) $-4/7x^7$
4 (*a*) $z^{6 \cdot 5}/32 \cdot 5$ (*b*) $8u^{5/2}/5$ (*c*) $-9u^{1/3}$ (*d*) $5/y^{2 \cdot 2}$
5 (*a*) $10p^{1/5}$ (*b*) $12p^{13/12}/13$ (*c*) $v^4/4 - v^2/2$ (*d*) $80^{7/8}/7 - 80^{5/8}/5$
6 (*a*) $2x + 3x^2/2 + 2x^3/3 + 3x^4/4$ (*b*) $x^3/3 - 2x - 1/x$
8 (*a*) $(5x + 1)^4/20$ (*b*) $-(3 - 2\theta)^5/10$ (*c*) $5(z/5 + 2)^6/6$

Exercise 14.2
1 (*a*) $-\cos x$ (*b*) $-\frac{1}{2} \cos 2x$ (*c*) $-\frac{1}{3} \cos 3x$ (*d*) $-(\cos kx)/k$
2 (*a*) $\sin \theta$ (*b*) $\frac{1}{2} \sin 2\theta$ (*c*) $\frac{1}{3} \sin 3\theta$ (*d*) $(\sin k\theta)/k$
3 (*a*) $-\frac{1}{3} \cos(3\theta - \pi)$ (*b*) $-(\cos(k\theta + \pi/2))/k$ (*c*) $+(\cos(\pi - A\theta))/A$
4 (*a*) $\frac{1}{3} \sin(3\theta - \pi)$ (*b*) $(\sin(k\theta + \pi/2))/k$ (*c*) $-(\sin(\pi - A\theta))/A$

Exercise 14.3
1 (*a*) (i) x^4 (ii) $10x^{3/2}/3$ (iii) $-7/x$ (iv) $(2x + 1)^3/6$
   (*b*) $x^4 + x^2 + 7x + 8$
2 (*a*) (i) $x^4/4 + 2x^3/3 + 5x$ (ii) $2(1 + x/2)^3/3$
   (*b*) $s = t^3/3 + 3t^2/2 - 7t + 20\frac{1}{3}, 6\frac{5}{6}$
3 (*a*) $4x^{1/3} + 6/x^4 + 3/2\sqrt{x}$ (*b*) $x^3/9 - 4x^{1/2} + x^{6/5}/6$
   (*c*) $y = 5x^3 - x^2 + 3x - 4$ (*d*) $43\frac{1}{2}$m
4 (*b*) $5x^2/2 - 2/3 + 8/3x^3 - 4/3x^4$
   (*c*) (i) $y = 2x^3 - 3x + 3$ (ii) $y = -3/x + x + 3$
5 (*a*) $\frac{1}{2}$ m   (*b*) (i) $-(2 - 3x)^6/8$ (ii) $-1/2(x + 2)^2$ (iii) $\frac{1}{3} \sin(3x - 2)$
   (iv) $\frac{1}{4} \cos(3 - 4x)$
6 (*a*) (i) $5x^4/4$ (ii) $4\frac{2}{3}$ (iii) $-\frac{3}{4}$ (*b*) $y = x^3 - 5x^2/2 - x + 6$

Exercise 14.4
1 (*a*) -1 (*b*) $2\pi r^3/3$ (*c*) $3\frac{3}{8}$ **2** (*a*) $\frac{2}{3}$ (*b*) 2   (*c*) $0 \cdot 7321$
3 (*a*) impossible to give a finite value.   (*b*) 6   (*c*) $1/(1 - m)$, if m less than 1
4 (*a*) $6\frac{1}{5}$ (*b*) 78   **5** (*a*) $BH^3/4$ (*b*) $BH^3/12$ (*c*) $BH^3/36$
6 (*a*) $-5\frac{1}{3}$ (*b*) $\frac{1}{3}$ (*c*) $\frac{1}{2}(\pi/12 + \frac{1}{6})$ **7** (*e*) $\pi/2p$
8 (*b*) $\frac{5}{24}$ **9** (*b*) $\frac{5}{24}$

Exercise 14.5
1 (*a*) (i) $t^4 + t^3 - 4t^2 - 6t$ (ii) $2x^{3/2} - 1/2x$ (*b*) $2\frac{2}{3}$
2 (*a*) $58 \cdot 85$ (*b*) 16   (*c*) $6 \cdot 04$ (*d*) $20 \cdot 83$
3 (*a*) (i) $x^3/3 + 2x^{3/2}/3 - 4x^{1/2}$ (ii) $-3/2t^2 - t/2$ (iii) $1\frac{1}{3}$ (*b*) $4\frac{1}{2}$

5 $(2, -2)$, $(-5, 19)$, $57\frac{1}{6}$
6 (b) (i) $2\frac{2}{3}$ (ii) $5\frac{1}{3}$ (c) $2\frac{2}{3}$; $\frac{1}{3}(0 + 4 \times 1 + 4) = 2\frac{2}{3}$;
$\frac{1}{3}(0 + 4 + 3 + 4) = 5\frac{1}{3}$; $\frac{1}{3}(0 + 4 \times 2 + 0) = 2\frac{2}{3}$ **7** (b) $3\frac{2}{3}$

Exercise 15.1
1 min. at $x = \frac{1}{3}$ **2** min. at $x = -\frac{3}{2}$ **3** max. at $x = 2\frac{1}{2}$
4 min. at $x = -\frac{1}{4}$ **5** max. at $x = 12$ **6** min. at $x = -1\frac{1}{2}$
7 max. 25 when $x = -1$, min. $- 71$ when $x = 3$
8 max. 0 when $x = 1$, min. -32 when $x = 5$
9 min. $-18\frac{10}{27}$ when $x = -\frac{4}{3}$. max. 63 when $x = 3$
10 max. $23\frac{1}{4}$ when $x = \frac{3}{2}$, min. 23 when $x = 2$
11 max. 2 when $x = 3$, min. $- 2$ when $x = 1$
12 25, 25   **13** $\frac{1}{2}$ **14** 1   **15** $h/d = 1$
16 depth $= 0{\cdot}816$ dia, breadth $= 0{\cdot}577$ dia,   **17** $9{\cdot}21$ knots

Exercise 15.2
1 (a) (i) $(12x^4 - 3\sqrt{x})/4x^3$ (ii) $7x^{15/8}/24$ (iii) $(8x + 1)(2x + 1)^2$
   (b) $(2, 5)$, $(-2, -3)$, $4\sqrt{5}$
2 $(1, -4)$, gradients $+4, -4$ **3** max. $(1, 5)$, min. $(3, 1)$
4 (a) (i) 7   (ii) $5/2\sqrt{x}$ (iii) $-3/x^4$ (iv) $10x - 2$
   (b) 30 m/s up, 10 m/s down, 5 s, 125 m   (c) $y_{min} = 2$
5 (a) (i) $-a/x$ (ii) $(2x\sqrt{x})/3 - \sqrt{x}$ (iii) $x^3/3 + ax^2/2 + bx$
   (b) $y = x^3/3 - 2x + 2$
6 (a) $3x^2$ (b) (i) $6x^2 + 2x$ (ii) $(x^2 - 1)/x^2$ (iii) $2x^4(7x^2 + 9x - 5)$
   (c) max. $(-\frac{1}{3}, \frac{1}{27})$, min. $(0, 0)$
7 (a) (i) $(\frac{1}{2}, -6\frac{1}{4})$ (ii) min.   (iii) 3, -2 (iv) -7
   (b) (i) $a^2 x^{a-1}$ (ii) $-2a/bx^3$
8 (a) $6x$ (b) $6x - 13$ (c) $a = -4$, $b = 6$
9 (a) $1/(x + 1)^2$ (b) $\frac{1}{9}$ **10** $(5 - r)r$, 3900 mm^2 **11** $12x^2 - 3$, $0{\cdot}4524$
12 (a) $3x^{1/2} - \frac{1}{2}x^{-1/2} - 2x^{-3/2}$ (b) $4{\cdot}155$, $1{\cdot}845$
13 (a) $8x$ (b) (i) $a(n + 1)x^n$ (ii) $-x^{-4/3}/3$ (iii) $50x^4 - 2x^{-3} - \frac{3}{4}x^{-4}$
   (c) $2 - 3V^{-2}$, $6V^{-3}$
14 1333 m^3 per m   **15** (a) $3x^2$ (b) gradient $= 5$; $(\frac{1}{2}, -2\frac{1}{4})$
16 (a) 22 m/s   (b) $40{\cdot}3$ m   (c) 6 m/s^2 **17** $y = 3x - 8$
18 $0{\cdot}6$ m/s, $-0{\cdot}37$ m/s^2, $0{\cdot}33$ m/s^2, $0{\cdot}77$ m/s^2
19 2 m/s^2, -2 m/s^2 **20** (a) $1/(5 - x)^2$, $\frac{1}{4}$
21 (a) (i) $5x^4 - 12x^3 + 12x^2 - 14x$ (ii) $\frac{1}{2}x^{-1/2} - \frac{1}{2}x^{-3/2}$
   (b) 200 mm $\times$ 200 mm $\times$ 100 mm
22 (a) $6x^2$ (b) $t = 1$ (max.), $t = 3$ (min.)
23 (a) $6x$ (b) $x = 2$ (max.), $x = 3$ (min.)
24 (a) -2 (b) (i) 52 m/s, 76 m/s^2 (ii) -8 m/s, 4 m/s^2, $t = 1$ s
   (c) $y = x^3/3 - x^2/2 - 2x + 7\frac{1}{6}$
25 (a) (i) $0{\cdot}52x^{-0.6}$ (ii) $5\sqrt{x^3}/2$
   (b) (i) 15 m/s   (ii) 20 m/s   (iii) 45 m
26 $16x$ **27** (a) $12x^2 - 6/x^3 + 1/\sqrt{x}$ (b) (i) 69 m/s   (ii) $5\frac{1}{3}$ s
28 (a) $2x + 3$ (b) 5 m/s up, 5 m/s down, $2\frac{1}{2}$ s, $31{\cdot}25$ m
29 (a) (i) $12x^3 - 6x^2 - 10x$ (ii) $2/\sqrt{x}$ (b) $-2x^{-3}$
30 (a) 4, $V = 11{\cdot}2$ (b) max. $= 17$, min. -10
31 (a) (i) $8ax$ (ii) $-2/3x^3$ (iii) $-1/2x^{3/2} + 1/x^{1/2}$
   (b) (i) 0 m/s, 43 m/s   (ii) -2 m/s^2 (iii) 4 s, $\frac{2}{9}$ s
32 $a = 4$, $b = 8$, $c = -1$, $d = -2$, $x = +\frac{1}{2}$, $-\frac{1}{2}$, -2

33 (a) $21x^2 + 6x - 10/x^3$
  (b) (i) 6 rad/s (ii) -4 rad/s^2 (iii) 5 s (iv) 20 rad/s   (c) $\frac{2}{27}$ m^3
34 1, 1·5   **35** $\omega = -9$, $\alpha = -6$, $t = 1, 5$
36 (a) (i) $(2x^3 - 3)/10$ (ii) $4(2x - 1)$ (iii) $\frac{1}{2}$ (iv) $5/\sqrt{x}$
  (b) $88 - 32t$, -32, $S_{max} = 221$
37 (a) $6x^2 - 1$ (b) $12x^2 + 6x - 5$, 2·92, $-2·75$, $38° 53'$

Exercise 16.1

1 (a) (i) $\frac{1}{8}$ (ii) $\frac{3}{8}$ (iii) $\frac{1}{4}$ (b) (i) $\frac{1}{12}$ (ii) $\frac{5}{12}$ (iii) $\frac{5}{18}$
2 (a) $\frac{1}{2}$ (b) $\frac{1}{26}$ (c) $\frac{3}{52}$ (d) $\frac{7}{13}$
3 (a) (i) $\frac{25}{36}$ (ii) $\frac{15}{22}$ (b) (i) $\frac{5}{18}$ (ii) $\frac{10}{33}$
  (c) (i) $\frac{1}{36}$, $\frac{25}{36} + \frac{5}{18} + \frac{1}{36} = 1$ (ii) $\frac{1}{66}$, $\frac{15}{22} + \frac{10}{33} + \frac{1}{66} = 1$
4

Score	2	3	4	5	6	7
Theoretical probability	$\frac{1}{36}$ 0·0278	$\frac{2}{36}$ 0·0556	$\frac{3}{36}$ 0·0833	$\frac{4}{36}$ 0·1111	$\frac{5}{36}$ 0·1389	$\frac{6}{36}$ 0·1667
Experimental probability	0·0208	0·0625	0·0903	0·1250	0·1320	0·1528

Score	8	9	10	11	12
Theoretical probability	$\frac{5}{36}$ 0·1389	$\frac{4}{36}$ 0·1111	$\frac{3}{36}$ 0·0833	$\frac{2}{36}$ 0·0556	$\frac{1}{36}$ 0·0278
Experimental probability	0·1389	0·1042	0·0923	0·0486	0·0278

6 $\frac{1}{8}, \frac{3}{8}, \frac{3}{8}, \frac{1}{8}$

Exercise 16.2

2 six
6 4129·7, 3983·8, 2001; $x = 4129·7/2001$, $y = 3983·8/2001$
7 (a) (i) 6·855 655 (ii) 8·426 150 (iii) 9·848 858
  (b) (i) 0·021 276 60 (ii) 0·014 084 51 (iii) 0·010 309 28
8 (a) $x = 3·019$, $y = 1·239$ (b) $x = 4·138$, $y = -2·173$

Exercise 17.1

1 (a) 3·87 (b) 4 (c) 4
2 (a) 10·0 kg (b) $9·95 + 0·1(35 - 22)/24 = 10·0$ kg (c) 10·0 kg
3 (a) 31·76 (b) 31/33, 31/33 (c) 32, 32   **4** (a) 2·35 (b) 2, 2
5 (a) 8·5 mm (b) $8·45 + 0·1(25 - 10)/23 = 8·5$ mm, mode $= 8·5$ mm
6 (a) 2·34 (b) 2, 1
7 (a) 0·26 (b) $0·24 + 0·02(16 - 3)/15 = 0·26$, mode 0·25

Exercise 17.2

1 1041, 71·2 h   **2** 357·5, 30·29 m   **3** 9071, 522·4 kg
4 100·6, 0·314 mm   **5** 320·9, 0·839 g   **6** 0·642, 0·006 42 kg/kWh
7 (A) 60·0, 0·305 mm (B) 60·0, 0·191 mm. B is better

Exercise 17.3

1 $\Sigma f(x - 4) = -329$, $\Sigma f = 2608$, $\bar{x} = -329/2608 + 4 = 3·8739$,
  $\Sigma f(x - 4)^2 = 9633$, $\Sigma f(x - 4)^2/\Sigma f = 3·6936$, $(4 - \bar{x})^2 = 0·0159$,
  Variance $= 3·6777$, standard deviation $= 1·9177$

2 $\Sigma f(x - 10) = 0.4$, $\Sigma f = 70$, $\bar{x} = 0.4/70 + 10 = 10.006$ kg,   i.e.   10.0 kg,
$\Sigma f(x - 10)^2 = 0.86$, $\Sigma f(x - 10)^2/\Sigma f = 0.0122$, $(\bar{x} - 10)^2 = 0.0000$,
variance $= 0.0122$, standard deviation $= 0.1104$, i.e. 0.11 kg

3 $\Sigma f(x - 32) = -12$, $\Sigma f = 50$, $\bar{x} = 32 - 12/50 = 31.76$ plates, $\Sigma f(x - 32)^2 =$
216, $\Sigma f(x - 32)^2/\Sigma f = 4.32$, $(32 - 31.76)^2 = 0.0576$, variance $= 4.2624$,
standard deviation $= 2.0645$

4 $\Sigma f(x - 2) = 14$, $\Sigma f = 40$, $\bar{x} = 2 + 14/40 = 2.35$, $\Sigma f(x - 2)^2 = 64$,
$\Sigma f(x - 2)^2/\Sigma f = 1.6$, $(2.35 - 2)^2 = 0.1225$, variance $= 1.4775$, standard
deviation $= 1.2155$

5 $\Sigma f(x - 8.5) = 0.7$, $\Sigma f = 50$, $\bar{x} = 8.5 + 0.7/50 = 8.514$, say 8.5 mm,
$\Sigma f(x - 8.5)^2 = 0.39$, $\Sigma f(x - 8.5)^2/\Sigma f = 0.0078$ $(\bar{x} - 8.5)^2 = 0.000\,19$,
variance $= 0.007\,61$, standard deviation $= 0.087\,235$, say 0.087 mm

6 $\Sigma f(x - 1) = 67$, $\Sigma f = 50$, $\bar{x} = 67/50 + 1 = 2.34$, $\Sigma f(x - 1)^2 = 219$,
$\Sigma f(x - 1)^2/\Sigma f = 4.38$, $(2.34 - 1)^2 = 1.7956$, variance $= 2.5844$, standard
deviation $= 1.6076$

7 $\Sigma f(x - 0.25) = 0.3$, $\Sigma f = 32$, $\bar{x} = 0.25 + 0.3/32 = 0.259$, $\Sigma f(x - 0.25)^2$
$= 0.0116$, $\Sigma f(x - 0.25)^2/\Sigma f = 0.000\,362\,5$, $(\bar{x} - 0.25)^2 = 0.000\,087\,8$, variance $= 0.000\,274\,7$, standard deviation $= 0.0166$, say 0.017

8 $\Sigma f(x - 250) = -140$, $\Sigma f = 50$, $\bar{x} = -140/50 + 250 = 247.2$, $\Sigma f(x - 250)^2$
$= 26\,000$, $\Sigma f(x - 250)^2/50 = 520$, $(250 - 247.2)^2 = 7.84$, variance $= 512.16$,
standard deviation $= 22.63$

Note also that the units of the variate can be coded on dividing by 20.

x	210	230	250	270	290	becomes
x	10.5	11.5	12.5	13.5	14.5	
f	7	11	18	10	4	

Take 12.5 as false mean: $\Sigma f(x - 12.5) = -7$, $\Sigma f = 50$, $\bar{x} = -7/50 + 12.5 =$
12.36, $\Sigma f(x - 12.5)^2 = 65.0$, $\Sigma f(x - 12.5)^2/\Sigma f = 1.30$, $(12.5 - 12.36)^2 = 0.0196$,
variance $= 1.2804$, standard deviation $= 1.1315$. Since we have worked in units
of 20, mean $= 12.36 \times 20 = 247.2$, standard deviation $= 1.1315 = 20 = 22.63$,
as before.

Exercise 18.1

2 Row sums $= 79\,828 + 29\,504 + 18\,461 + 18\,406 = 146\,199 =$ Col. sums
$= 23\,202 + 25\,913 + 79\,303 + 17\,781$

3 Row sums $= 4084 - 1471 + 177 + 14\,694 = 17\,484 =$ Col. sums $= 2662$
$+ 1295 + 11\,636 + 1891$

4 Row sums $= 71.9557 + 4361.2979 - 872.7762 + 572.8317 = 4133.3091 =$
Col. sums $= 2982.76 + 509.93 + 640.43 + 0.1891$

5

908.3772	178.8765	1889.0784	2976.3321
1834.8433	1586.6944	819.4725	4241.0102
7590.2112	2227.5280	1024.3908	10 842.1300
10 333.4317	3793.0900	3732.9417	18 059.4723

6

91.936 68	2156.480 00	181.408 59	2429.825 27
190.285 57	15 638.068 00	337.582 17	16 165.935 74
851.413 08	22 147.968 00	937.139 28	23 936.520 36
1133.635 33	39 942.516 00	1456.130 04	42 532.281 37

Exercise 18.2

1 64·395	**2** 337·316 94	**3** 327·768 489
4 0·000 207 318	**5** 273·631 505 4	**6** 0·000 000 094 502 2
7 39 775	**8** 153 450	**9** 158 456 942
10 1 369 368	**11** 273 896 532	

12 $1961 \times 315 = 777 \times 795 = 555 \times 1113 = 617 \times 715$

Exercise 18.3

A1 0·913 043	**2** 0·342 618	**3** 0·769 168
4 0·025 956 3	**5** 0·004 067 92	**6** 0·089 327 0
7 53·0303	**8** 0·000 605 782	**9** 3983·14
10 579 675·0		

B1 $29\ 028/375 = 0·328 \times 236 = 77·4080$

2 (*a*) $63\ 215/57\ 125 = 1·106\ 61$
  (*b*) $0·514\ 223\ 19 \times 269 = 138·326\ 04$; divide by 125, result 1·106 61
  (*c*) $1·880\ 00 \times 269 = 505·7200$; divide by 457, result 1·106 61
  (*d*) $0·588\ 621\ 44 \times 235 = 138·326\ 04$; divide by 125, result 1·106 61

3 $4347·0588 \times 0·236 = 1025·91$
  $13·882\ 353 \times 73·9 = 1025·91$

4 $0·000\ 027\ 95/0·000\ 376 = 0·074\ 335\ 1$
  $5·718\ 085 \times 0·013 = 0·074\ 335\ 1$

5 $671\ 182·2/53·48 = 12\ 550·2$
  $39·379\ 207 \times 318·7 = 12\ 550·2$

Exercise 18.4

1 1·732 05,	$1·732\ 05^2 = 2·999\ 997\ 202\ 5$	
	$1·732\ 06^2 = 3·000\ 031\ 843\ 6$	
2 2·236 06,	$2·236\ 06^2 = 4·999\ 964\ 323\ 6$	
	$2·236\ 07^2 = 5·000\ 009\ 044\ 9$	
3 2·645 75,	$2·645\ 75^2 = 6·999\ 993\ 062\ 5$	
	$2·645\ 76^2 = 7·000\ 045\ 977\ 6$	
4 3·316 62,	$3·316\ 62^2 = 10·999\ 968\ 224\ 4$	
	$3·316\ 63^2 = 11·000\ 034\ 556\ 9$	
5 3·605 55,	$3·605\ 55^2 = 12·999\ 990\ 802\ 5$	
	$3·605\ 56^2 = 13·000\ 062\ 913\ 6$	
6 6·082 76,	$6·082\ 76^2 = 36·999\ 969\ 217\ 6$	
	$6·082\ 77^2 = 37·000\ 090\ 872\ 9$	
7 12·5300,	$12·5299^2 = 156·998\ 394\ 01$	
	$12·5300^2 = 157·000\ 900\ 00$	
8 16·8226,	$16·8226^2 = 282·999\ 870\ 76$	
	$16·8227^2 = 283·003\ 235\ 29$	
9 23·6854,	$23·6854^2 = 560·998\ 173\ 16$	
	$23·6855^2 = 561·002\ 910\ 25$	
10 55·9196	$55·9195^2 = 3126·990\ 480\ 25$	
	$55·9196^2 = 3127·001\ 664\ 16$	

Exercise 18.5

	x_n^2	$\dfrac{x_n^2}{N}$	$3 - \dfrac{x_n^2}{N}$	$x_n\left(3 - \dfrac{x_n^2}{N}\right)$	$\dfrac{x_n}{2}\left(3 - \dfrac{x_n^2}{N}\right)$	$\sqrt{N}$ correct to five sig. figs
1	2·89	0·963 333	2·036 667	3·462 333	1·731 166	
	2·996 935	0·998 978	2·001 022	3·464 101	1·732 050	$1·7321 \simeq$
	2·999 997	0·999 999	2·000 001	3·464 101	1·732 050	$\sqrt{3}$
2	4·84	0·968	2·032	4·4704	2·2352	
	4·996 119	0·999 223	2·000 777	4·472 136	2·236 068	$2·2361 \simeq$
	5·000 000	1·000 000	2·000 000	4·472 136	2·236 068	$\sqrt{5}$
3	6·76	0·965 714	2·034 286	5·289 143	2·644 571	
	6·993 755	0·999 107	2·000 893	5·291 503	2·645 751	$2·6458 \simeq$
	6·999 998	0·999 999	2·000 001	5·291 504	2·645 752	$\sqrt{7}$
4	10·89	0·99	2·01	6·633	3·3165	
	10·999 172	0·999 924	2·000 076	6·633 252	3·316 626	$3·3166 \simeq$
	11·000 008	1·000 000	2·000 000	6·633 252	3·316 626	$\sqrt{11}$
5	12·96	0·996 923	2·003 077	7·211 077	3·605 538	
	12·999 904	0·999 992	2·000 008	7·211 104	3·605 552	$3·6055 \simeq$
	13·000 005	1·000 000	2·000 000	7·211 104	3·605 552	$\sqrt{13}$
6	37·21	1·005 675	1·994 325	12·165 382	6·082 691	
	36·999 129	0·999 976	2·000 024	12·165 527	6·082 763	$6·0828 \simeq$
	37·000 005	1·000 000	2·000 000	12·165 526	6·082 763	$\sqrt{37}$
7	156·25	0·995 222	2·004 778	25·059 725	12·529 862	
	156·997 441	0·999 983	2·000 017	25·059 937	12·529 968	$12·530 \simeq$
	157·000 098	1·000 000	2·000 000	25·059 936	12·529 968	$\sqrt{157}$
8	282·24	0·997 314	2·002 686	33·645 124	16·822 562	
	282·998 592	0·999 995	2·000 005	33·645 208	16·822 604	$16·823 \simeq$
	283·000 000	1·000 000	2·000 000	33·645 208	16·822 604	$\sqrt{283}$
9	561·69	1·001 229	1·998 771	47·370 872	23·685 436	
	560·999 878	0·999 999	2·000 001	47·370 895	23·685 447	$23·685 \simeq$
	561·000 399	1·000 000	2·000 000	47·370 894	23·685 447	$\sqrt{561}$
10	3124·810 000	0·999 299	2·000 701	111·839 185	55·919 592	
	3127·000 769	1·000 000	2·000 000	111·839 184	55·919 592	$55·920 \simeq$
						$\sqrt{3127}$

Exercise 18.6

1 z	1	1	0·5	0·166 666 66	0·041 666 66	0·008 333 33	0·001 388 88
p	1	2	3	4	5	6	7
w	1	2	2·5	2·666 666 66	2·708 333 32	2·716 666 65	2·718 055 53

0·000 198 41	0·000 248 0	0·000 002 75
8	9	10
2·718 253 94	2·718 278 74	2·718 281 49

0·000 000 27	0·000 000 02	0·000 000 00
11	12	13
2·718 281 76	2·718 281 78	2·718 281 78

$e = 2·718\,282$, correct to six decimal places

2

x	10	22	54	130	314	758	1830	$2588/1830 = 1·414\,208$, correct
y	12	32	76	184	444	1072	2588	to seven sig. figs
N	0	1	2	3	4	5	6	

3

x	33 125	67 375	167 875	403 125	974 125	2 351 375	5 676 875
y	34 250	100 500	235 250	571 000	1 377 250	3 325 500	8 028 250
N	0	1	2	3	4	5	6

$8\,028\,250/5\,676\,875 = 1·414\,202$, correct to seven sig. figs

4 (a) $\sqrt{5} \simeq 2·236$ (b) $\sqrt{-2} \simeq 1·414j$ (c) $\sqrt{23} \simeq 4·796$
   (d) $\sqrt{-17} = 4·123j$

Exercise 18.7
1 $i_1 = 2·500$, $i_2 = 6·296$; $R_1 = 0·001\,08$, $R_2 = 0·001\,76$
   $i_1 = 2·5$, $i_2 = 6·3$; $R_1 = -0·001$, $R_2 = 0·003$
2 $x = 7·90$, $y = -5·10$; $R_1 = -0·029$, $R_2 = 0·006$
3 $u = 3·2741$, $v = 0·237\,54$; $R_1 = 0·000\,16$, $R_2 = 0·000\,20$
4 $i_1 = 6·12$, $i_2 = 3·77$; $R_1 = 0·0024$, $R_2 = -0·0232$
5 $x = 2·6844$, $y = -1·2785$; $R_1 = 0·0005$, $R_2 = -0·0014$
6 $p = 1·4$, $q = -0·77$; $R_1 = 0·0052$, $R_2 = -0·038$

Exercise 18.8
1 $i_1 = 2·307$, $i_2 = 3·206$, $i_3 = 4·211$; $R_1 = 0·045$, $R_2 = 0·002$, $R_3 = 0·014$
2 $x = -8·56$, $y = -33·14$, $z = -10·91$; $R_1 = 0·06$, $R_2 = -0·13$, $R_3 = 0·07$
3 $p = 3·00$, $q = 3·98$, $r = 1·98$; $R_1 = -0·0394$, $R_1 = 0·0578$, $R_3 = 0·0176$
4 $I_1 = 2·6$, $I_2 = 14$, $I_3 = 2·3$; $R_1 = -0·015$, $R_2 = 0·023$, $R = -0·045$
5 $u = 1·2$, $v = -3·1$, $f = 4·7$; $R_1 = 0·00$, $R_2 = 0·03$, $R_3 = 0·00$
6 $\theta = 11·300$, $\omega = -5·2079$, $\alpha = -3·1608$; $R_1 = 0·0349$, $R_2 = 0·0324$,
   $R_3 = -0·0309$

Exercise 19.1
1 (i) (a) 189·7   (b) $189·7 \pm (10 \times 0·05)$; i.e. 184·7 to 194·7, 200 guaranteed,
   190 with a possible error of 1 unit in the second significant figure.
   (ii) (a) 1·908   (b) $1·908 \pm (10 \times 0·0005)$; i.e. 1·904 to 1·913, 1·9 guaranteed,
   1·91 with a possible error of 1 unit in the third significant figure
   (iii) (a) 197·824   (b) $197·824 \pm (5 \times 0·05 + 5·0 \times 0·0005)$; i.e. 197·5715 to
   198·0765
   (iv) (a) 31 150·2185 (b) $31\,150·2185 \pm (5 + 0·5 + 2·0 \times 0·05 + 2 \times 0·005$
   $+\, 2·0 \times 0·0005 + 0·000\,05)$; i.e. 31 144·607 45 to 31 155·829 55, 31 000
   guaranteed, 31 200 with a possible error of 1 unit in the third significant figure
2 (a) 1726·4842   (b) (i) $0·005/60 + 0·005/20 \simeq 0·000\,34$ (ii) $1800 \times 0·000\,34$
   $\simeq 0·61$ (iii) 1725·8 to 1727·1   (iv) 1730   (v) 1726·035 075 to
   1726·933 375
3 (a) 11·5992   (b) (i) $0·0005/0·5 + 0·05/20 = 0·0035$ (ii) $12 \times 0·0035 =$
   0·042   (iii) 11·5572 to 11·6412   (iv) 11·6   (v) 11·561 575 to 11·636 875
4 (a) 0·722 43   (b) (i) $0·0005/0·02 + 0·005/30 = 0·0252$ (ii) $0·8 \times 0·0252 =$
   0·02   (iii) 0·070 243 to 0·074 243   (iv) 0·07   (v) 0·706 612 5 to 0·738 252 5

5 (a) 126·14   (b) (i) 0·005/0·05 + 0·5/1800 = 0·1003   (ii) 0·1003 × 126 = 12·7, say (iii) 113·44 to 138·84 (iv) 100, 130 with a possible error of 1 unit in the second significant figure (v) 117·0975 to 135·1875

6 (a) 79·55   (b) (i) 0·05/2 + 0·05/3 + 0·05/8 ≏ 0·05   (ii) 80 × 0·05 = 4 (iii) 75·55 to 83·55 (iv) 80 (v) 76·458 375 to 82·715 625

7 (a) 170·649 696   (b) (i) 0·005/9 + 0·005/4 + 0·005/3 ≏ 0·003   (ii) 17 × 0·003 = 0·51 (iii) 170·03 to 171·16 (iv) 170 rounded to two significant figures, 171 with a possible error of 1 unit in the third significant figure (v) 170·150 368 125 to 171·149 934 375

8 (a) 1 054 389   (b) (i) 0·5/20 + 0·5/30 + 0·5/50 + 0·5/20 ≏ 0·077 (ii) 1 060 000 × 0·77 ≏ 82 000 (iii) 972 390 to 1 136 390 (iv) 1 100 000 (v) 984 884·0625 to 1 127 339·0625

9 (a) 0·227 81   (b) (i) 0·05/20 + 0·05/90 ≏ 0·0031   (ii) 0·230 × 0·0031 ≏ 0·0007 (iii) 0·227 11 to 0·228 51 (iv) 0·23, 0·228 with a possible error of 1 unit in the third significant figure (v) 0·227 15 to 0·228 47

10 (a) 0·724 77   (b) (i) 0·0005/3 + 0·0005/5 ≏ 0·000 27   (ii) 0·000 27 × 0·73 ≏ 0·0002 (iii) 0·724 57 to 0·724 97 (iv) 0·725 (v) 0·724 60 to 0·724 94

11 (a) 22·299   (b) (i) 0·0005/0·09 + 0·0005/2 ≏ 0·0058   (ii) 0·0058 × 22·3 ≏ 0·13 (iii) 22·169 to 22·429 (iv) 22,22·3 with a possible error of 1 unit in the third significant figure (v) 22·179 to 22·419

12 (a) 4·0678   (b) (i) 0·05/9 + 0·005/2 ≏ 0·008   (ii) 0·008 × 4 = 0·032 (iii) 4·0357 to 4·0998 (iv) 4,4·1 with a possible error of 1 unit in the second significant figure (v) 4·0380 to 4·0977

13 (a) 14·7442 (b) 14·7

14 (a) 1·912 17 (b) 1·9

15 (a) 1170·40 (b) 1170

16 (a) 1522·14 (b) 2000

Exercise 19.2
(values with a possible error of 1 unit in the least significant figure are included)

1 (i) ±0·0045; 0·4, 0·44 (ii) ∓0·0023; 0·9, 0·89 (iii) ±0·0055; 0·1, 0·14

2 (i) ±0·0012; 0·97, 0·969 (ii) ∓0·0050; 0·1, 0·13 (iii) ±0·0195; 1·7

3 (i) ±0·0050; 0·1, 0·07 (ii) ∓0·0050; 0·0, 0·03 (iii) ±0·0070; 0·6, 0·63

4 (i) ±1·655, 30 (ii) ∓0·0010; 0·02, 0·020

5 (i) ±10·02, 200 (ii) ∓0·000 137, 0·003

6 (i) ±0·4083, 8 (ii) ∓0·0045; 0·1, 0·09

7 (i) (a) 1·471 369 (b) 1·784 769 384 (c) 2·164 925 262 792 (ii) (a) 1·47, error ≏ ±0·0012 (b) 1·8, 1·78, error ≏ ±0·0022 (c) 2·2, 2·16, error ≏ 0·0035

8 0·782 779, 0·782 473, 0·782 167. Relative error < −0·000 05/1·2 Error ≏ −0·000 05, value 0·78 242 ∓ 0·000 05; 0·78, 0·782

9 (a) 12·809 241, 164·076 654 996 081 (b) 12·8, 12·81; 12·8² = 163·84, 12·81² = 164·10; 164 (c) 12·812 820 25, 164·168 362 76, 12·805 662 25, 163·984 985 66

10 (a) 13·312 053, 13·144 256, 0·167 797 (b) 0·01 × 16·7797 = 0·167 797 (c) 0 to 0·3356, ∞ to 2·980

11 (a) 80·6, 3·3, 25·5, 1·7, 0·4, 0·05, 138, 31 (b) 59·55/0·35 = 170·1, 50·65/0·45 = 112·6. The result cannot be guaranteed to one significant figure.

12 −3

13 (a) 2·1826, error ≏ ±0·002 (b) 0·6889, error ≏ ±0·004